CONTENTS

1 WHOLE NUMBERS

2 ADDITION

3 SUBTRACTION

CONTENTS

ABOUT CATCH-UP MATHS

The Catch-Up Maths series enables students to start from scratch when they are struggling with their year level maths. Each book takes maths back to the foundation and ensures that all basic concepts are consolidated before moving forward. Lots of revision and opportunities to practise and build confidence are provided before moving on to new topics.

Each new strand and sub-strand of the primary maths curriculum is introduced clearly with simple explanations, examples and trial questions (with answers), before children move to the Practice section. To ensure concepts are understood fully, videos of the author working through and explaining new concepts are included in every chapter.

A QR code on the topic page provides access to the subtitled video.

This book has 10 chapters divided between the Year 4 Number & Algebra and Statistics & Probability strands of the Australian Maths Curriculum. The chapters are:

1 Whole Numbers
2 Addition
3 Subtraction
4 Multiplication
5 Division
6 Fractions
7 Decimals
8 Patterns & Algebra
9 Chance
10 Data

★ Review section that can be used as assessment and to check students' progress is included at the end of each chapter.

★ Answers are at the back of the book.

How to use this book

Children can work through the pages from front to back, or choose individual topics to reinforce areas where they are struggling.

The topics are introduced with:

- clear instructions, using simple language
- completed examples and incomplete examples for students to tackle before moving on to the **Your Turn** section
- a video linked by QR code that shows the incomplete example and has the author giving extra instruction about the page.

Each Your Turn section contains a SELF CHECK for students to reflect and give self-assessment on their understanding.

HOW TO USE THE QR CODES IN CATCH-UP MATHS

A unique aspect of the **Catch-Up Maths** series is the **instructional video** created by the author for every new topic.

The videos give a step-by-step explanation of the examples and work through them to provide a helpful lesson for the student. The videos are simply accessed via the QR code on the same page and can be watched on a phone, tablet, computer or interactive whiteboard.

Access the video using a QR code reader app

Each video shows the page from the book. The author talks through the concepts and examples, and demonstrates what students need to do. Her words are shown as easy-to-read captions so that the audio can be turned down in noisy classrooms, and for students with hearing difficulties. The solutions to the examples are presented before students are expected to tackle the 'Your Turn' section. This careful instruction ensures that students can confidently move on to the following Practice questions. Those assisting students should encourage them to check their 'Your Turn' answers before moving on.

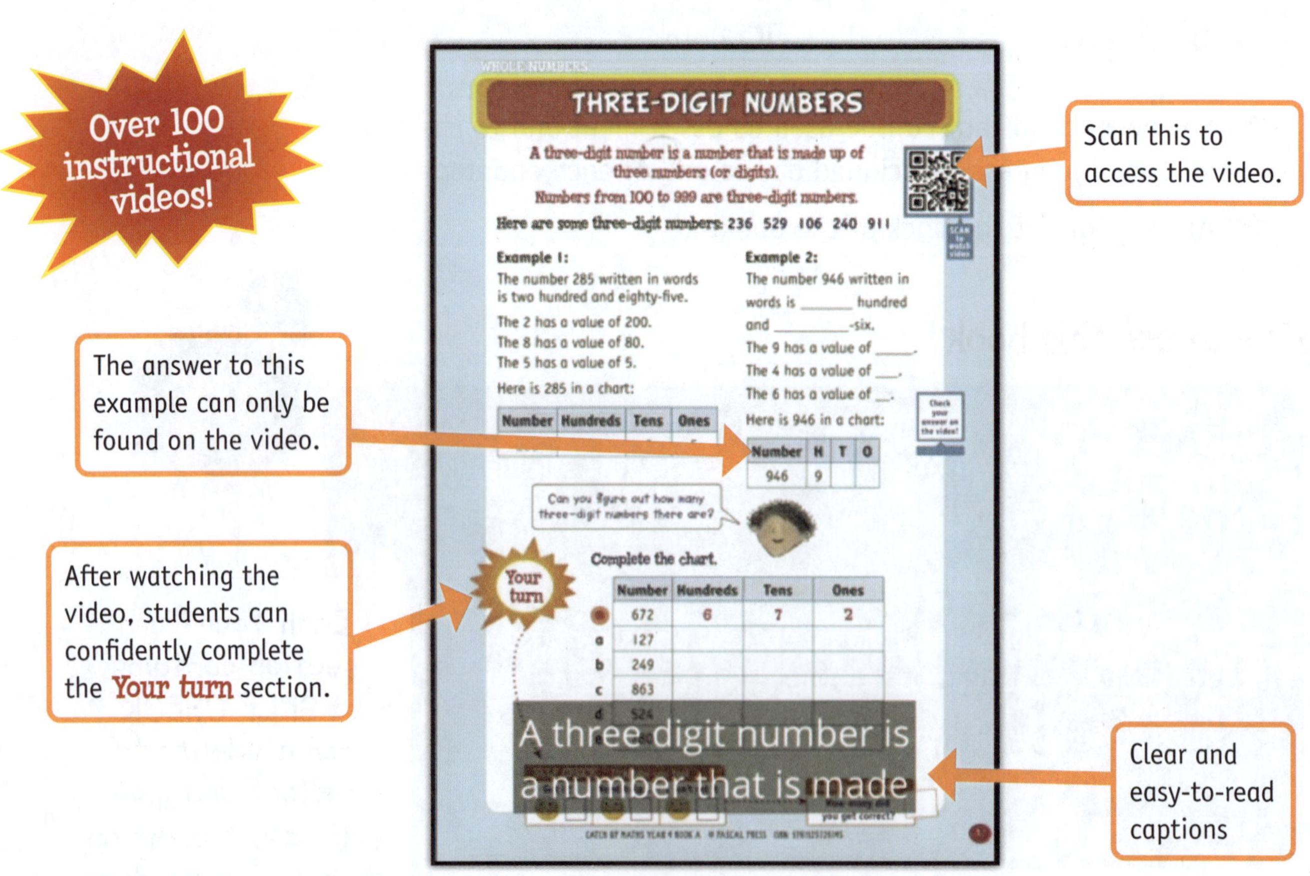

CATCH UP MATHS YEAR 4 BOOK A © PASCAL PRESS ISBN: 9781925726145

AUSTRALIAN CURRICULUM CORRELATIONS

ACARA CODE	Content Description	Pages
1. Whole Numbers		**PAGES 1–65**
ACMNA027 Year 2	Recognise, model, represent and order numbers to at least 1000	1, 2, 11, 12, 13, 14
ACMNA028 Year 2	Group, partition and rearrange collections up to 1000 in hundreds, tens and ones to facilitate more efficient counting	3, 4, 5, 6, 7, 8, 9, 10
ACMNA052 Year 3	Recognise, model, represent and order numbers to at least 10 000	17, 18, 27, 28, 29, 30, 31, 32, 33, 34
ACMNA053 Year 3	Apply place value to partition, rearrange and regroup numbers to at least 10 000 to assist calculations and solve problems	19, 20, 21, 22, 23, 24, 25, 26, 35, 36
ACMNA060 Year 3	Describe, continue, and create number patterns resulting from performing addition or subtraction	15, 16
ACMNA071 Year 4	Investigate and use the properties of odd and even numbers	55, 56
ACMNA072 Year 4	Recognise, represent and order numbers to at least tens of thousands	37, 38, 47, 48, 49, 50, 51, 52, 53, 54, 57, 58
ACMNA073 Year 4	Apply place value to partition, rearrange and regroup numbers to at least tens of thousands to assist calculations and solve problems	39, 40, 41, 42, 43, 44, 45, 46
2. Addition		**PAGES 66–103**
ACMNA029 Year 2	Explore the connection between addition and subtraction	68, 69, 70, 71, 77
ACMNA030 Year 2	Solve simple addition and subtraction problems using a range of efficient mental and written strategies	68, 69, 70, 71, 72, 73
ACMNA054 Year 3	Recognise and explain the connection between addition and subtraction	68, 69, 70, 71
ACMNA055 Year 3	Recall addition facts for single-digit numbers and related subtraction facts to develop increasingly efficient mental strategies for computation	66, 67, 72, 73, 74, 75, 76, 77, 78, 79, 80, 81, 82, 83, 84, 85, 86, 87, 88, 89, 90, 91
ACMNA059 Year 3	Represent money values in multiple ways and count the change required for simple transactions to the nearest five cents	92, 93
ACMNA073 Year 4	Apply place value to partition, rearrange and regroup numbers to at least tens of thousands to assist calculations and solve problems	74, 75, 76, 77, 78, 79, 80, 81, 82, 83, 84, 85, 86, 87, 88, 89, 90, 91
ACMNA080 Year 4	Solve problems involving purchases and the calculation of change to the nearest five cents with and without digital technologies	94, 95, 96
ACMNA099 Year 5	Use estimation and rounding to check the reasonableness of answers to calculations	74, 75

ACARA CODE	Content Description	Pages
3. Subtraction		**PAGES 104–143**
ACMNA029 Year 2	Explore the connection between addition and subtraction	124, 125
ACMNA030 Year 2	Solve simple addition and subtraction problems using a range of efficient mental and written strategies	116, 117, 118, 119
ACMNA053 Year 3	Apply place value to partition, rearrange and regroup numbers to at least 10 000 to assist calculations and solve problems	124, 125, 126, 127, 128, 129
ACMNA054 Year 3	Recognise and explain the connection between addition and subtraction	126, 127
ACMNA055 Year 3	Recall addition facts for single-digit numbers and related subtraction facts to develop increasingly efficient mental strategies for computation	104, 105, 106, 107, 108, 109, 110, 111, 112, 113, 114, 115, 120, 121, 122, 123
ACMNA073 Year 4	Apply place value to partition, rearrange and regroup numbers to at least tens of thousands to assist calculations and solve problems	104, 105, 106, 107, 108, 109, 110, 111, 112, 113, 114, 115, 130, 131, 132, 133, 134, 135, 136
ACMNA099 Year 5	Use estimation and rounding to check the reasonableness of answers to calculations	134, 135, 136
4. Multiplication		**PAGES 144–171**
ACMNA026 Year 2	Investigate number sequences, initially those increasing and decreasing by twos, threes, fives and tens from any starting point, then moving to other sequences	152, 153
ACMNA031 Year 2	Recognise and represent multiplication as repeated addition, groups and arrays	144, 145, 146, 147
ACMNA056 Year 3	Recall multiplication facts of two, three, five and ten and related division facts	148, 149, 150, 151, 152, 153, 154, 155, 156, 157, 158, 159, 160, 161
ACMNA074 Year 4	Investigate number sequences involving multiples of 3, 4, 6, 7, 8, and 9	152, 153
ACMNA075 Year 4	Recall multiplication facts up to 10 × 10 and related division facts	154, 155, 156, 157, 158, 159, 160, 161, 162, 163, 164, 165, 166
ACMNA076 Year 4	Develop efficient mental and written strategies, and use appropriate digital technologies for multiplication and for division where there is no remainder	150, 151

Australian Curriculum Correlations continued

ACARA CODE	Content Description	Pages
5. Division		**PAGES 172–189**
ACMNA032 Year 2	Recognise and represent division as grouping into equal sets and solve simple problems using these representations	172, 173, 174, 175, 176, 177
ACMNA056 Year 3	Recall multiplication facts of two, three, five and ten and related division facts	178, 179
ACMNA074 Year 4	Investigate number sequences involving multiples of 3, 4, 6, 7, 8, and 9	176, 177
ACMNA075 Year 4	Recall multiplication facts up to 10 × 10 and related division facts	178, 179, 180, 181
ACMNA076 Year 4	Develop efficient mental and written strategies, and use appropriate digital technologies for multiplication and for division where there is no remainder	178, 179, 180, 181, 182, 183
ACMNA101 Year 5	Solve problems involving division by a one digit number, including those that result in a remainder	182, 183
6. Fractions		**PAGES 190–215**
ACMNA033 Year 2	Recognise and interpret common uses of halves, quarters and eighths of shapes and collections	192, 193,194, 195, 196, 197, 198, 199
ACMNA058 Year 3	Model and represent unit fractions including $\frac{1}{2}$, $\frac{1}{4}$, $\frac{1}{3}$, $\frac{1}{5}$ and their multiples to a complete whole	190, 191, 192, 193, 194, 195, 196, 197, 204, 205
ACMNA077 Year 4	Investigate equivalent fractions used in contexts	202, 203, 206, 207, 210, 211
ACMNA078 Year 4	Count by quarters, halves and thirds, including with mixed numerals. Locate and represent these fractions on a number line	208, 209
7. Decimals		**PAGES 216–236**
ACMNA079 Year 4	Recognise that the place value system can be extended to tenths and hundredths. Make connections between fractions and decimal notation	216, 217, 218, 219, 220, 221, 222, 223, 224, 225, 226, 227, 228, 229
8. Patterns & Algebra		**PAGES 237–249**
ACMNA035 Year 2	Describe patterns with numbers and identify missing elements	237, 238, 239, 240
ACMNA036 Year 2	Solve problems by using number sentences for addition or subtraction	243, 244
ACMNA057 Year 3	Represent and solve problems involving multiplication using efficient mental and written strategies and appropriate digital technologies	243, 244
ACMNA060 Year 3	Describe, continue and create number patterns resulting from performing addition or subtraction	241, 242, 245, 246
ACMNA081 Year 4	Explore and describe number patterns resulting from performing multiplication	241, 242

ACARA CODE	Content Description	Pages
9. Chance		**PAGES 250–271**
ACMSP024 Year 1	Identify outcomes or familiar events involving Chance and describe them using everyday language such as 'will happen', 'won't happen' and 'might happen'.	250, 251
ACMSP047 Year 2	Identify practical activities and everyday events that involve chance. Describe outcomes as 'likely' or 'unlikely' and identify some events as 'certain' or 'impossible'	252, 253, 254, 255
ACMSP067 Year 3	Conduct chance experiments, identify and describe possible outcomes and recognise variation in results	260, 261, 262, 263, 264, 265, 266, 267
ACMSP093 Year 4	Identify everyday events where one cannot happen if the other happens	256, 257
ACMSP094 Year 4	Identify events where the chance of one will not be affected by the occurrence of the other	258, 259
10. Data		**PAGES 272–291**
ACMSP048 Year 2	Identify a question of interest based on one categorical variable. Gather data relevant to the question	272, 273, 274
ACMSP049 Year 2	Collect, check and classify data	272, 273, 274
ACMSP050 Year 2	Create displays of data using lists, table and picture graphs and interpret them	275, 276, 277, 278, 279
ACMSP069 Year 3	Collect data, organise into categories and create displays using lists, tables, picture graphs and simple column graphs, with and without the use of digital technologies	280, 281
ACMSP070 Year 3	Interpret and compare data displays	280, 281
ACMSP096 Year 4	Construct suitable data displays, with and without the use of digital technologies, from given or collected data. Include tables, column graphs and picture graphs where one picture can represent many data values	280, 281, 282, 283, 284, 285, 286, 287

THREE-DIGIT NUMBERS

A three-digit number is a number that is made up of three numbers (or digits).

Numbers from 100 to 999 are three-digit numbers.

Here are some three-digit numbers: **236 529 106 240 911**

Example 1:

The number 285 written in words is two hundred and eighty-five.

The 2 has a value of 200.
The 8 has a value of 80.
The 5 has a value of 5.

Here is 285 in a chart:

Number	Hundreds	Tens	Ones
285	2	8	5

Example 2:

The number 946 written in words is _______ hundred and __________-six.

The 9 has a value of ______.
The 4 has a value of ___.
The 6 has a value of __.

Here is 946 in a chart:

Number	H	T	O
946	9		

Check your answer on the video!

Can you figure out how many three-digit numbers there are?

Complete the chart.

	Number	Hundreds	Tens	Ones
●	672	6	7	2
a	127			
b	249			
c	863			
d	524			
e	780			

SELF CHECK Tick how you feel

Got it!	Need help...	I don't get it
☐	☐	☐

Check your answers
How many did you get correct? ☐

PRACTICE

Write the following numbers in words.

- 657 six hundred and fifty-seven

a 293 ______________________

b 451 ______________________

c 764 ______________________

d 503 ______________________

e 850 ______________________

f 300 ______________________

2 Complete the table.

	Number	Hundreds	Tens	Ones
●	259	2	5	9
a	410			
b	324			
c		5	6	8
d	879			
e		9	0	3

Circle with green all the numbers with 3 hundreds. Use blue for the numbers with 2 tens and red for the numbers with 8 ones.

359	319	218
721	838	422
347	748	362
924	109	300
432	685	426
188	923	878

CATCH UP MATHS YEAR 4 BOOK A © PASCAL PRESS ISBN: 9781925726145

PLACE VALUE TO 1000

The place value is the value of a digit based on where it is in a number.

The place value of the 6 is hundreds because it is in the hundreds place.

The place value of the 4 is ones because it is in the ones place.

The place value of the 2 is tens because it is in the tens place.

Example 1:
395 is a three-digit number that has 3 hundreds, 9 tens and 5 ones.

395 — hundreds place, tens place, ones place

Example 2:
763 is a three-digit number that has __ hundreds, __ tens and __ ones.

763 — hundreds place, tens place, ones place

Check your answer on the video!

Your turn

1 What is the place value of the 8 in these numbers?

- 835 8 hundreds
- **a** 328 8 __________
- **b** 820 8 __________
- **c** 781 8 __________
- **d** 498 8 __________
- **e** 482 8 __________

2 Circle the hundreds green, tens blue and ones red.

936 415 671 472 215

324 716 989 543 537

SELF CHECK Tick how you feel

Got it!	Need help...	I don't get it
☐	☐	☐

Check your answers
How many did you get correct? ☐

PRACTICE

1 Write the numbers.

- 6 hundreds, 4 tens, 2 ones = 642

a one ten, two hundreds, three ones = ______

b seven hundreds, five tens, eight ones = ______

c six tens, nine hundreds = ______

d five ones, five hundreds = ______

2 Circle the numbers with 5 tens.

(257)	553	59	157	95
325	159	125	735	556
452	757	56	345	675

3 Cross out the numbers with 9 hundreds.

~~937~~	795	195	94	987
93	909	993	496	890
98	982	957	973	921

4 What is the place value of the 7 in these numbers?

- 473 7 ______________

a 247 7 ______________

b 479 7 ______________

c 7 7 ______________

d 756 7 ______________

e 875 7 ______________

f 497 7 ______________

g 715 7 ______________

h 17 7 ______________

i 79 7 ______________

j 647 7 ______________

k 709 7 ______________

VALUE AND THREE-DIGIT NUMBERS

The value of a digit is how much it is worth.

Example 1: 724

The value of the 7 is 700.

The value of the 2 is 20.

The value of the 4 is 4.

Example 2: 136

The value of the 1 is ____.

The value of the 3 is ____.

The value of the 6 is ___.

Even though 1 is the smallest number, it has the greatest value as it is in the hundreds place.

Example 3: 859

The value of the 8 is ____.

The value of the 5 is ____.

The value of the 9 is ___.

Your turn

Circle in green the digit in each number with the most value and in red the digit with the least value.

- 592
- **a** 243
- **b** 714
- **c** 139
- **d** 287
- **e** 324
- **f** 903
- **g** 874
- **h** 492
- **i** 736
- **j** 649
- **k** 999
- **l** 711
- **m** 118
- **n** 247

SELF CHECK Tick how you feel

Got it!	Need help...	I don't get it

Check your answers

How many did you get correct?

PRACTICE

1 Write the numbers.

- 2 hundreds, 3 tens, 8 ones = 238

a 3 hundreds, 7 tens, 1 one = ______

b 2 tens, 6 hundreds, 3 ones = ______

c 8 ones, 2 hundreds, 5 tens = ______

d 8 tens, 1 hundred, 4 ones = ______

e 9 hundreds, 9 ones, 8 tens = ______

2 Circle the digits with a value of 6.

26	326	246	624	6
643	62	861	16	746

3 Circle the digits with a value of 70.

78	27	872	79	777
7	127	737	175	474

4 Circle the digits with a value of 800.

832	849	28	8	863
847	189	888	782	148

5 What is the value of the 4 in the following?

- 439 400

a 48 ______

b 14 ______

c 436 ______

d 342 ______

e 40 ______

f 47 ______

g 491 ______

h 143 ______

i 470 ______

j 468 ______

k 4 ______

CATCH UP MATHS YEAR 4 BOOK A © PASCAL PRESS ISBN: 9781925726145

NUMBER EXPANDERS AND THREE-DIGIT NUMBERS

This number expander shows the number 273:

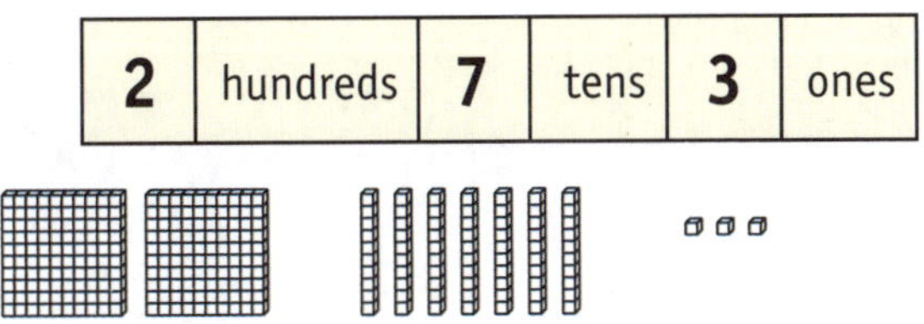

2	hundreds	7	tens	3	ones

If we fold the number expander to make 273 using only tens and ones we have:

2	7	tens	3	ones

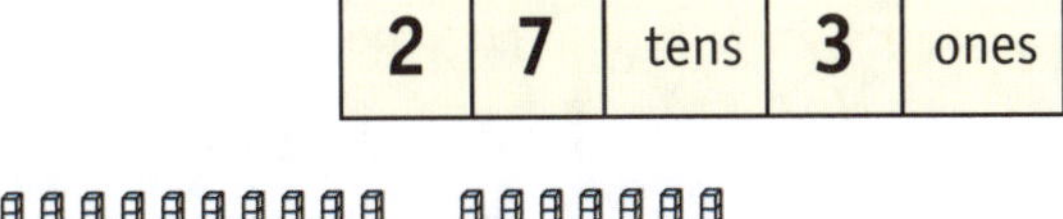

If we fold the number expander again and make 273 using only ones we have:

2	7	3	ones

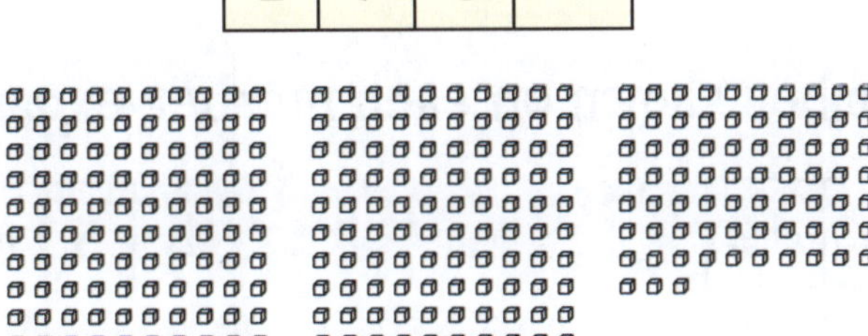

Your turn

Complete the number expanders.

● 872

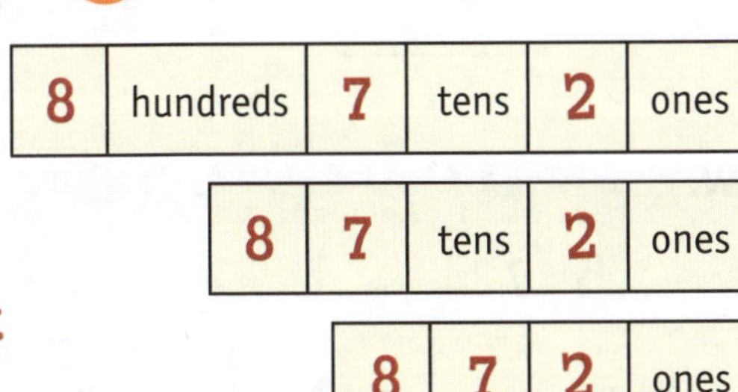

8	hundreds	7	tens	2	ones

8	7	tens	2	ones

8	7	2	ones

a 184

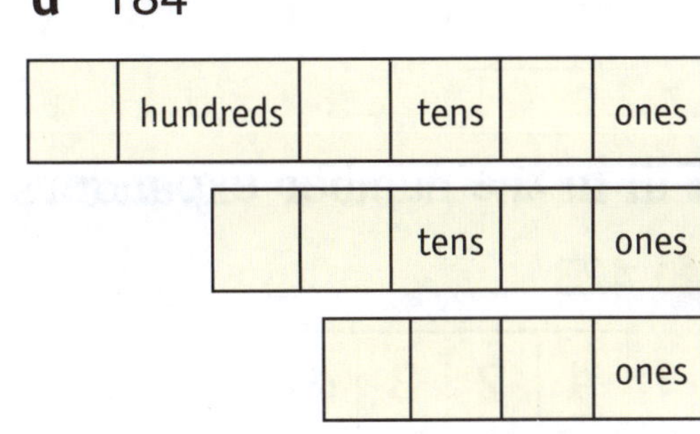

	hundreds		tens		ones

		tens		ones

			ones

SELF CHECK Tick how you feel

Got it!	Need help...	I don't get it
☐	☐	☐

Check your answers
How many did you get correct? ☐

PRACTICE

1 Fill in the number expanders.

520

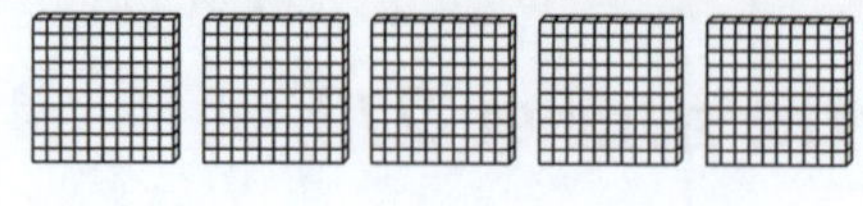

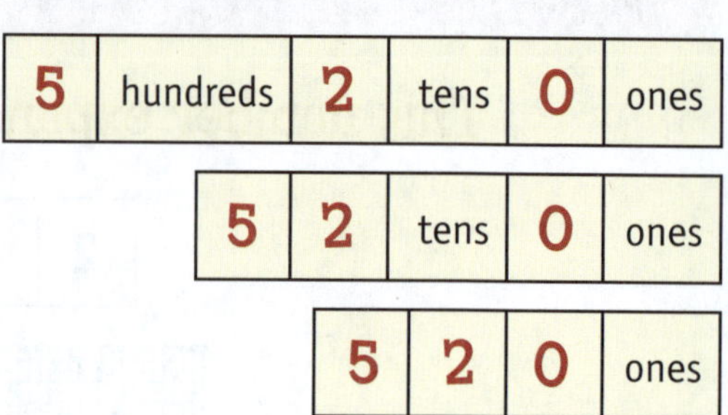

5	hundreds	2	tens	0	ones

5	2	tens	0	ones

5	2	0	ones

a 276

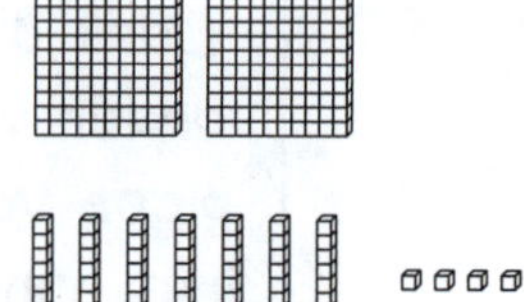

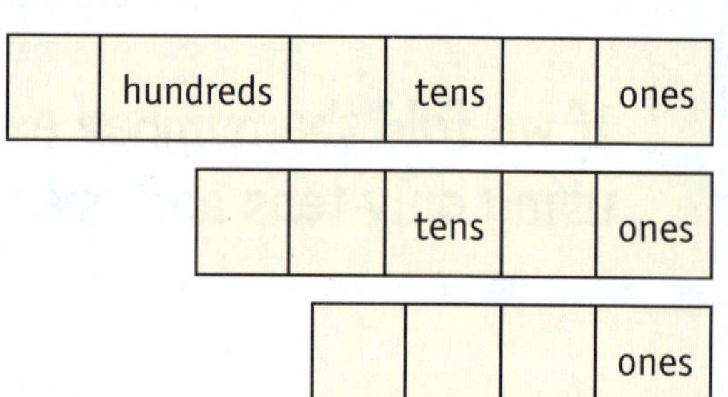

	hundreds		tens		ones

		tens		ones

			ones

b 605

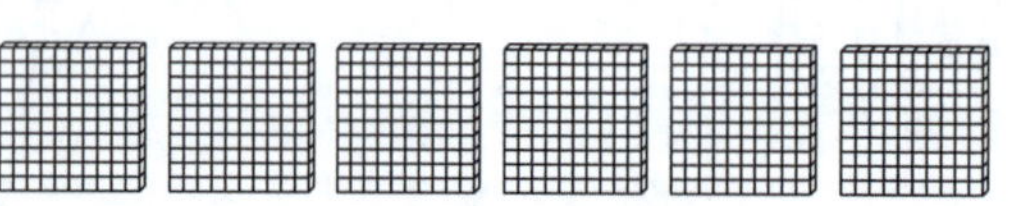

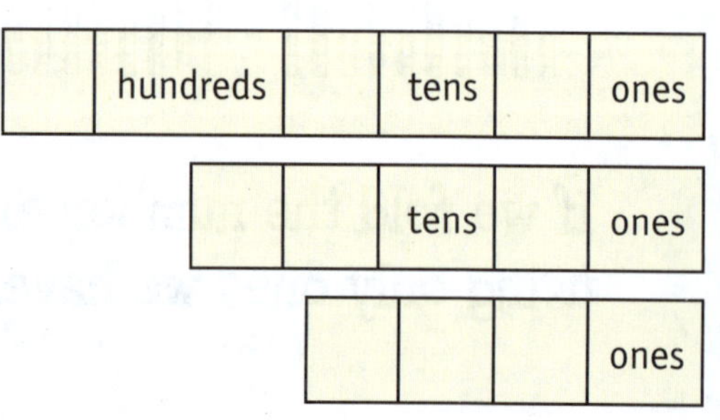

	hundreds		tens		ones

		tens		ones

			ones

2 Write the number shown on each number expander.

7	2	5	ones

725

a

3	2	tens	6	ones

______

b

5	hundreds	0	tens	4	ones

______

c

9	0	6	ones

______

3 Fill in the number expanders below.

423

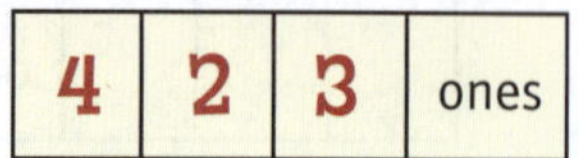

4	2	3	ones

a 342

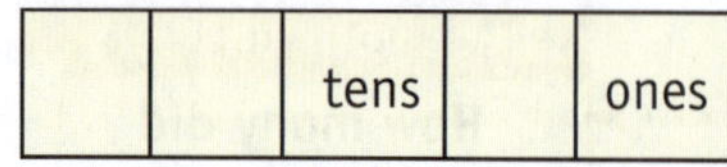

		tens		ones

b 724

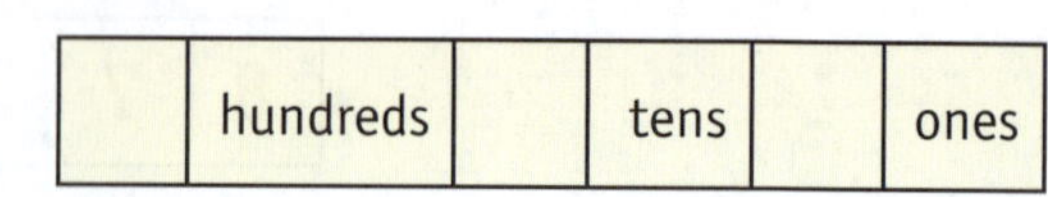

	hundreds		tens		ones

c 649

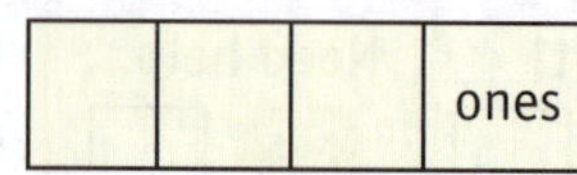

			ones

CATCH UP MATHS YEAR 4 BOOK A © PASCAL PRESS ISBN: 9781925726145

EXPANDED THREE-DIGIT NUMBERS

Expanding a number is writing the number to show the value of each digit.

Example 1: 27 = 20 + 7
(2 tens) (7 ones)

Example 2: 53 = 50 + 3
(5 tens) (3 ones)

Example 3: 198 = 100 + 90 + 8
(__ hundreds) (__ tens) (__ ones)

Check your answer on the video!

Example 4: 346 = _____ + ___ + __
(__ hundreds) (__ tens) (__ ones)

Numbers written like this are in 'expanded form'.

Your turn

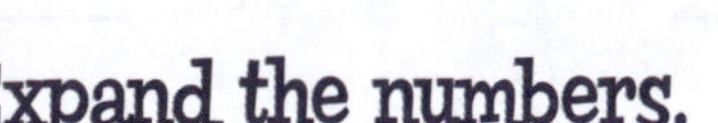

Expand the numbers.

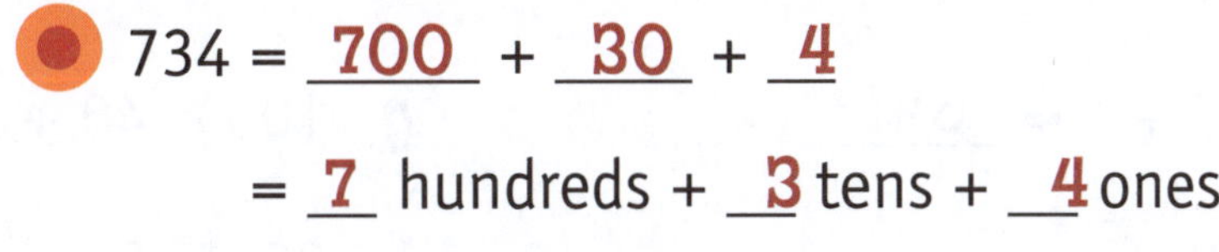

● 734 = 700 + 30 + 4
= 7 hundreds + 3 tens + 4 ones

a 265 = _____ + _____ + __
= __ hundreds + __ tens + __ ones

b 638 = _____ + _____ + __
= __ hundreds + __ tens + __ ones

SELF CHECK Tick how you feel

Got it!	Need help...	I don't get it
☐	☐	☐

Check your answers
How many did you get correct? ☐

PRACTICE

1 Match these.

●	394	800 + 50
a	727	600 + 7
b	850	500 + 80 + 2
c	607	300 + 90 + 4
d	582	400 + 90 + 5
e	500	700 + 20 + 7
f	495	500

2 Write these numbers in expanded form.

● 923 900 + 20 + 3

a 410 ____________

b 359 ____________

c 832 ____________

d 206 ____________

e 300 ____________

f 783 ____________

g 564 ____________

3 Complete.

● 600 + 40 + 1 = 641

a 100 + 90 = ______

b 400 + 30 = ______

c 700 + 40 + 1 = ______

d 200 + 50 + 2 = ______

e 300 + 50 + 5 = ______

f 800 + 2 = ______

g 100 + 40 + 3 = ______

h 900 + 80 + 4 = ______

i 400 + 80 + 7 = ______

j 300 + 70 + 6 = ______

k 600 + 10 + 2 = ______

l 500 + 30 + 8 = ______

MODELLING HUNDREDS

You can model numbers using an abacus and Base 10 blocks.

Example 1:

Here is the number **642** in Base 10 blocks and on the abacus.

6 hundreds **4 tens** **2 ones**

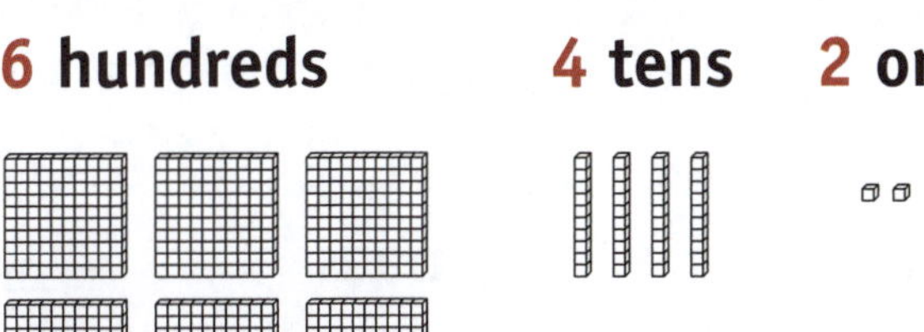

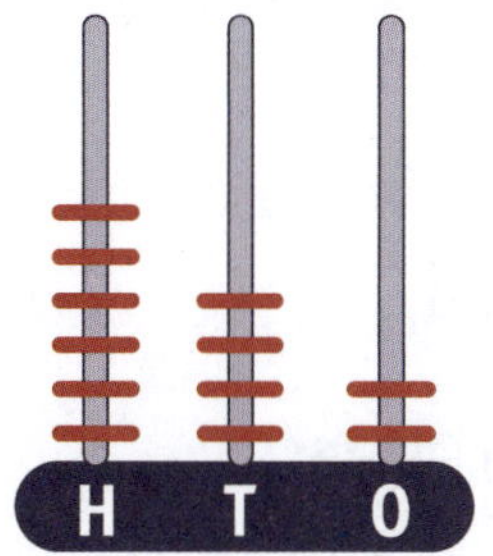

Example 2: Here is the number **207**.

2 hundreds **0 tens** **7 ones**

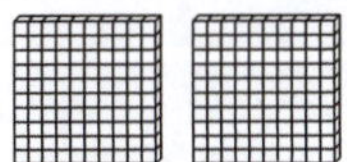

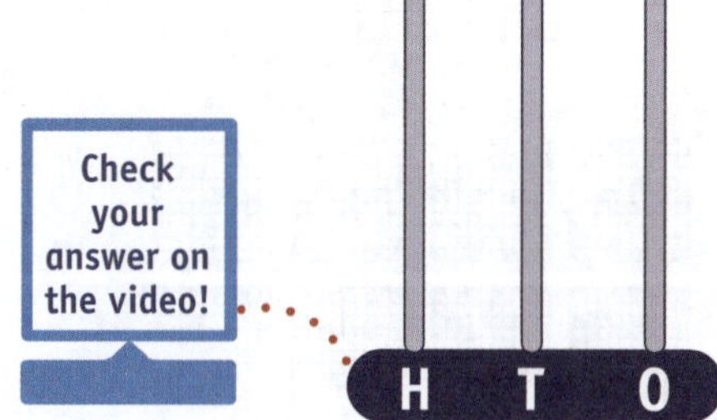

What numbers have been shown below?

● 328

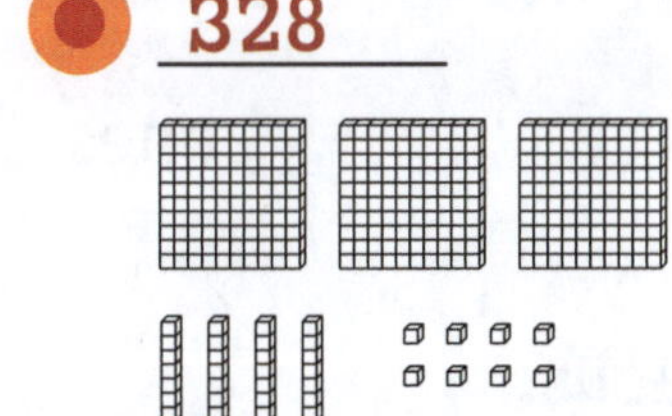

b ________

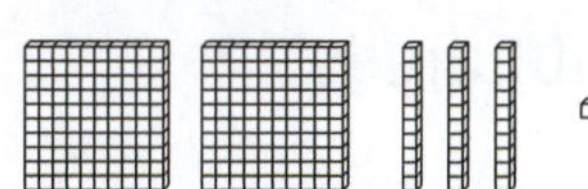

c ________

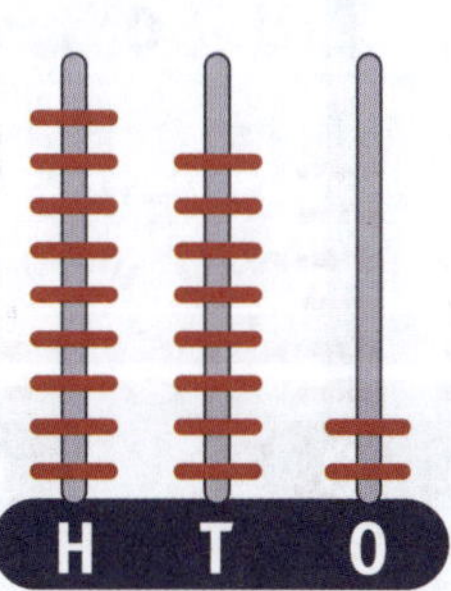

a ________

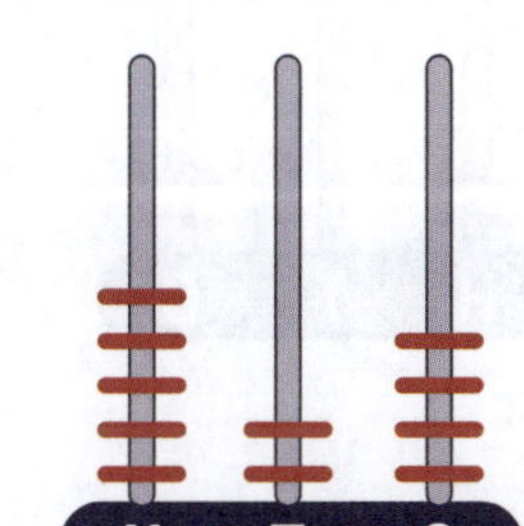

SELF CHECK	Tick how you feel	
Got it! ☐	Need help... ☐	I don't get it ☐

Check your answers

How many did you get correct? ☐

  ISBN: 9781925726145

PRACTICE

1 Write the number shown with the Base 10 blocks.

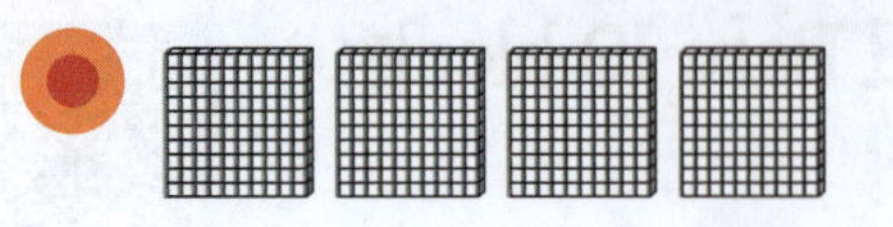

2 hundreds 2 tens 3 ones = 223

a

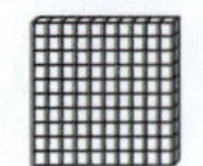
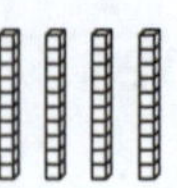

__ hundred __ tens __ ones = ______

b

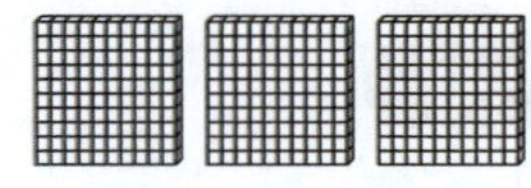

__ hundreds __ ten __ ones = ______

c

__ hundreds __ tens __ ones = ______

d

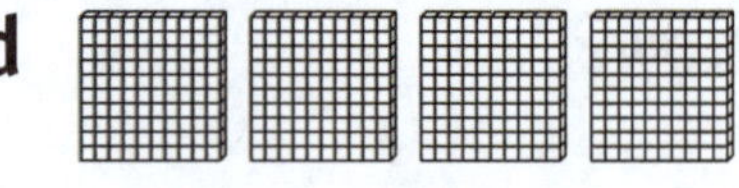

__ hundreds __ tens __ ones = ______

2 Match the number with the correct abacus.

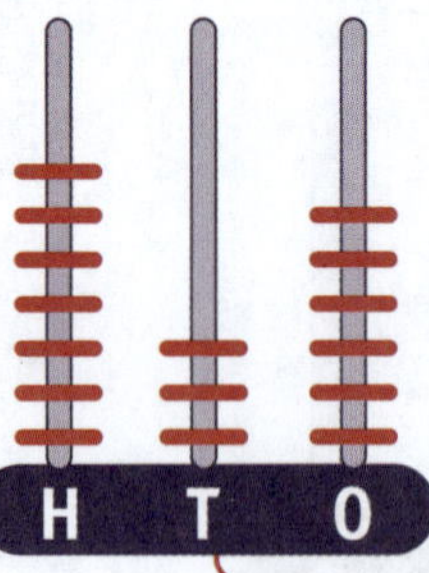

a

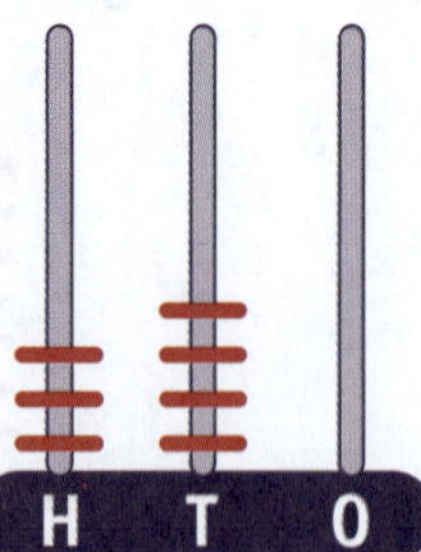

b

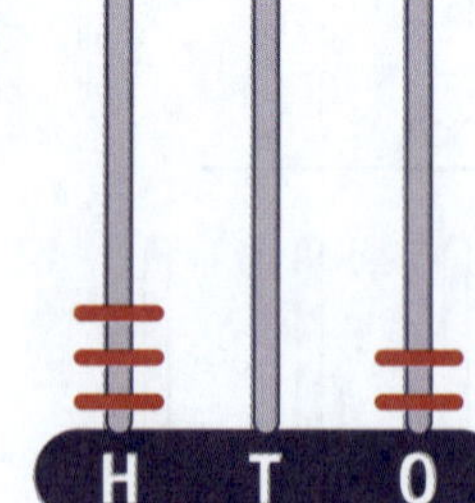

c

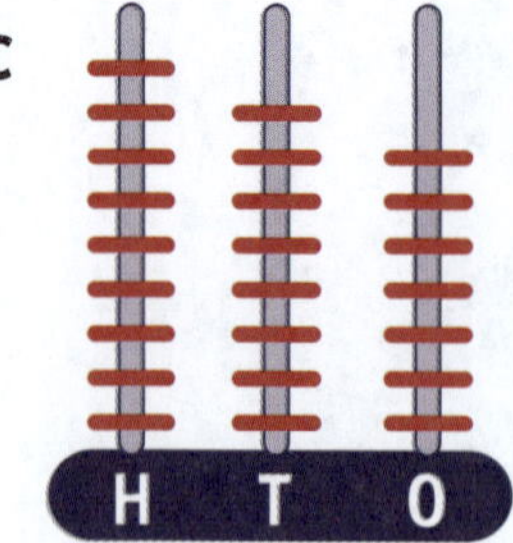

987 302 736 340

CATCH UP MATHS YEAR 4 BOOK A © PASCAL PRESS ISBN: 9781925726145

ORDERING THREE-DIGIT NUMBERS

You can order numbers from smallest to largest or largest to smallest.

These numbers are ordered from smallest to largest:

142, 283, 351, 473.

When numbers are ordered from smallest to largest, it is called ascending order.

These numbers are ordered from largest to smallest:

473, 351, 283, 142.

When numbers are ordered from largest to smallest, it is called descending order.

Remember – 'descending' has a d for down, the numbers are going down.

Example 1:
Circle the smallest number.
147, 263, (109), 634

Example 2:
Circle the largest number.
241, (463), 185, 327

Example 3:
Circle the smallest number.
246, 802, 468, 689

Example 4:
Circle the largest number.
943, 782, 879, 436

1 Order these numbers in ascending order.

- 132, 103, 237, 433
  103, 132, 237, 433

a 243, 147, 824, 763

____

b 431, 454, 371, 124

____

2 Order these numbers in descending order.

- 124, 203, 109, 732
  732, 203, 124, 109

a 513, 315, 153, 335

____

b 532, 235, 105, 205

____

SELF CHECK	Tick how you feel	
Got it! ☐	Need help... ☐	I don't get it ☐

Check your answers
How many did you get correct? ☐

Order these numbers in descending order.

						→					
a	136	215	143	872	602	→					
b	204	142	703	371	411	→					
c	526	341	101	242	711	→					
d	950	133	139	624	524	→					

2 Order these numbers in ascending order.

						→					
a	954	588	741	955	103	→					
b	207	481	111	135	740	→					
c	299	301	531	126	621	→					
d	702	243	145	321	412	→					

3 Write A for ascending or D for descending in the box to describe the order of these numbers.

- 132, 272, 324, 449, 571 **A**
- a 736, 791, 793, 802, 824 ☐
- b 121, 149, 257, 327, 481 ☐
- c 243, 202, 194, 173, 101 ☐
- d 912, 934, 942, 956, 965 ☐
- e 129, 119, 108, 103, 100 ☐
- f 742, 814, 856, 910, 989 ☐
- g 641, 632, 606, 592, 411 ☐
- h 733, 509, 410, 300, 120 ☐
- i 643, 689, 698, 714, 741 ☐

CATCH UP MATHS YEAR 4 BOOK A © PASCAL PRESS ISBN: 9781925726145

COUNTING BY HUNDREDS

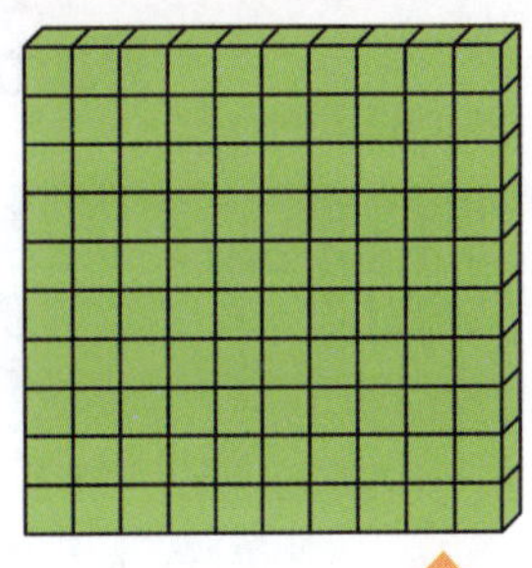

This Base 10 block has 100 squares.

Numbers from 100 to 999 are three-digit numbers, or hundreds.

A number in the hundreds has three digits: 392, 103, 274, 965.

When you count by 100s, you count every 100th number.

Counting forwards by 100s

Example 1:

+100 +100 +100 +100

100, 200, 300, 400, 500

Example 2:

+100 +100 +100 +100

107, 207, 307, ____, ____

Look at the first digit in each number. What do you notice?

Counting backwards by 100s

Example 3:

−100 −100 −100 −100

500, 400, 300, 200, 100

Example 4:

−100 −100 −100 −100

564, 464, 364, ____, ____

1 Circle all the three-digit numbers.

8 243 128 4 62 9 75
920 103 17 81 349

2 Fill in the missing numbers.

a 300, ______, 500, ______

b 700, 600, ______, ______

c 125, 225, ______, ______

d 762, 662, ______, ______

Check your answers
How many did you get correct?

PRACTICE

Match these numbers.

Words	Base 10	Number
● seven hundred and three		300
a nine hundred		642
b six hundred and forty-two		703
c three hundred		900

2 Fill in the missing numbers.

● 442, 542, 642, 742, 842, 942

a 200, 300, 400, ______, ______, ______

b 147, 247, 347, ______, ______, ______

c 400, 500, 600, ______, ______, ______

d 900, 800, 700, ______, ______, ______

e 543, 443, 343, ______, ______, ______

f 771, 671, 571, ______, ______, ______

g 400, ______, 600, ______, 800, ______

h 245, 345, ______, 545, ______, ______

i 378, 478, ______, ______, ______, 878

j 992, ______, 792, 692, ______, ______

CATCH UP MATHS YEAR 4 BOOK A © PASCAL PRESS ISBN: 9781925726145

FOUR-DIGIT NUMBERS

A four-digit number is a number that is made up of four numbers (or digits).

Numbers from 1000 to 9999 are four-digit numbers.

Here are some four-digit numbers:

4873 2972 5097 6203 4200

Example 1: 3946

The number 3946 has four digits. Written in words, it is three thousand, nine hundred and forty-six.

The 3 has a value of 3000.
The 9 has a value of 900.
The 4 has a value of 40.
The 6 has a value of 6.

Here is 3946 in a chart:

Number	Th	H	T	O
3946	3	9	4	6

Check your answer on the video!

Example 2: 2873

The number 2873 in words is

____

____

____

In a number chart it is:

Number	Th	H	T	O
2873				

Complete the chart.

	Number	Th	H	T	O	Words
●	2319	2	3	1	9	two thousand, three hundred and nineteen
a	4692					four thousand, six hundred and ninety-two
b	5300					

SELF CHECK Tick how you feel

Got it!	Need help...	I don't get it
☐	☐	☐

Check your answers
How many did you get correct? ☐

PRACTICE

1 Write these numbers in words.

- 2495 two thousand, four hundred and ninety-five
- **a** 7219 ______
- **b** 3490 ______
- **c** 2005 ______
- **d** 8943 ______

2 Put these numbers in ascending order. Write a 1 in the box under the smallest number, 2 in the box under the next largest number and so on.

●	1429	1325	1703	1421
	3	1	4	2
a	2150	5125	1253	1520
	☐	☐	☐	☐
b	7375	4283	6154	1095
	☐	☐	☐	☐
c	2493	3286	1078	2197
	☐	☐	☐	☐

3 Order these numbers in descending order by putting a 1 in the box with the largest number and a 4 in the box with the smallest number.

●	8621	9235	4396	6981
	2	1	4	3
a	1325	5937	9377	9529
	☐	☐	☐	☐
b	6123	9583	1252	9525
	☐	☐	☐	☐
c	6303	9524	3781	8874
	☐	☐	☐	☐

4 Write these words in numbers.

- three thousand, two hundred and sixteen 3216
- **a** four thousand, nine hundred and twenty-two ______
- **b** eight thousand and three ______
- **c** seven thousand, four hundred and sixty-nine ______

CATCH UP MATHS YEAR 4 BOOK A © PASCAL PRESS ISBN: 9781925726145

PLACE VALUE AND FOUR-DIGIT NUMBERS

The place value is the value of a digit based on where it is in a number.

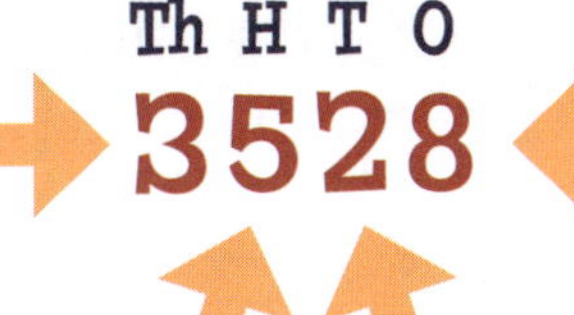

The place value of the 3 is thousands because it is in the thousands place.

The place value of the 8 is ones because it is in the ones place.

The place value of the 5 is hundreds because it is in the hundreds place.

The place value of the 2 is tens because it is in the tens place.

Example 1: 3528 has 3 thousands, 5 hundreds, 2 tens and 8 ones.

Th	H	T	O
3	5	2	8

Example 2:

4930 has __ thousands, __ hundreds, __ tens and __ ones.

Th	H	T	O
4			0

thousands 4930 ones
hundreds tens

Example 3:

8972 has __ thousands, __ hundreds, __ tens and __ ones.

Th	H	T	O
8		7	

Circle the thousands purple, the hundreds green, the tens blue and the ones red.

3921   4374   5500   9746   9525

2125   5251   6303   1809

SELF CHECK Tick how you feel

Got it!	Need help...	I don't get it
 ☐	 ☐	☐

Check your answers
How many did you get correct? ☐

PRACTICE

1 Write the numbers.

- 6 thousands, 3 hundreds, 4 tens, 5 ones = 6345

a 8 hundreds, 1 thousand, 8 tens, 2 ones = ______

b 7 ones, 4 tens, 3 thousands = ______

c 6 thousands, 3 tens = ______

d 8 tens, 2 hundreds, 9 thousands = ______

2 Complete the chart.

	Number	Thousands	Hundreds	Tens	Ones
●	2491	2	4	9	1
a	3269				
b	9038				
c		4	7	3	0
d		5	4	7	6
e		6	9	2	4

3 Circle the numbers below in
red for the numbers with 6 ones
blue for the numbers with 8 tens
green for the numbers with 1 hundred
purple for the numbers with 4 thousands.

4372 4362 6153 1956 6666

1483 2140 8147 3482 4040

3206 2736 3884 4711 1134 3381

4370 5296 8706 9283 4635

1782 8132 4975 4125 3776

CATCH UP MATHS YEAR 4 BOOK A © PASCAL PRESS ISBN: 9781925726145

VALUE AND FOUR-DIGIT NUMBERS

The value of a number is how much it is worth.

	Th	H	T	O	
The value of the 8 is 8000. →	8	6	3	2	← The value of the 2 is 2.

The value of the 6 is 600. ↑ ↑ The value of the 3 is 30.

Example 1: 1745

In 1745, even though 1 is the smallest number, it has the greatest value because it is in the thousands place. The 1 is worth the most. It is worth 1000. The 7 is worth 700, the 4 is worth 40 and the 5 is worth 5.

Example 2: 3472

The 3 is worth the most. It is worth _____.
The 4 is worth _____, the 7 is worth ___ and the 2 is worth __.

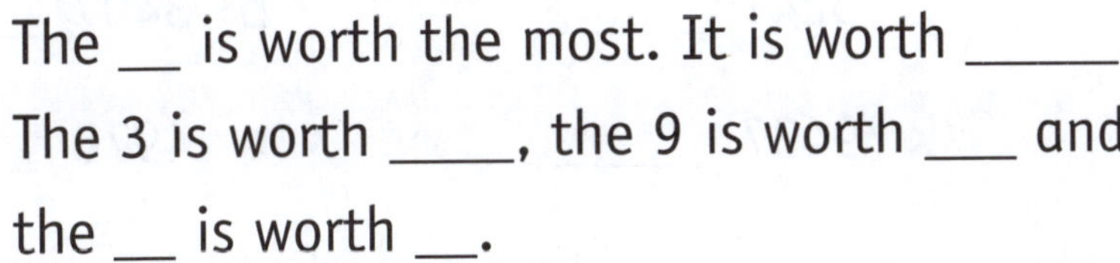

Example 3: 1397

The __ is worth the most. It is worth _____.
The 3 is worth _____, the 9 is worth ___ and the __ is worth __.

Your turn

1 Circle the numbers where the value of the 8 is 8000.

8432 847 18 836 82
78 81 8936

2 For each number, use red to circle the digit worth the most and blue to circle the digit worth the least.

2294 1157 458 3866 5490
8745 8894 6128

Check your answers
How many did you get correct?

PRACTICE

1 Circle the numbers where the value of the 4 is 40.

248   24   463   74   142

640   4925   458   4710   343

2 Circle the numbers where the value of 5 is 500.

536   53   25   524   5382

5920   599   56   507   5

3 What is the value of the 4 in these numbers?

● 842	40	f 849	______	l 4	______
a 429	______	g 74	______	m 4306	______
b 94	______	h 463	______	n 2934	______
c 347	______	i 4739	______	o 8740	______
d 4913	______	j 5241	______	p 3497	______
e 14	______	k 9437	______	q 4978	______

4 What is the value of the underlined number?

● $6\underline{3}8$	30	f $11\underline{2}5$	______	l $2\underline{4}5$	______
a $49\underline{1}$	______	g $\underline{2}543$	______	m $\underline{8}095$	______
b $1\underline{3}57$	______	h $39\underline{3}0$	______	n $6\underline{7}$	______
c $\underline{9}300$	______	i $1\underline{6}31$	______	o $\underline{7}61$	______
d $6\underline{1}31$	______	j $\underline{3}421$	______	p $3\underline{5}$	______
e $62\underline{3}$	______	k $5\underline{9}30$	______	q $1\underline{2}6$	______

CATCH UP MATHS YEAR 4 BOOK A © PASCAL PRESS ISBN: 9781925726145

NUMBER EXPANDERS AND FOUR-DIGIT NUMBERS

The number expander shows 1324.

SCAN to watch video

Example 1:

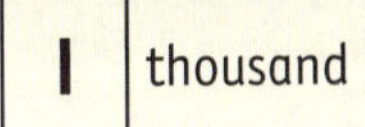
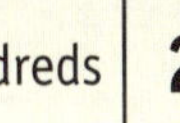

1	thousand	3	hundreds	2	tens	4	ones

= 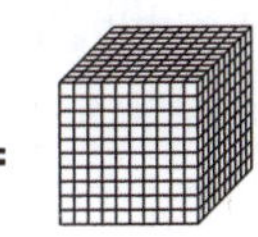 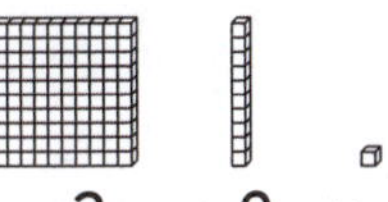

× 1 × 3 × 2 × 4

We can make the same number using hundreds, tens and ones.

1	3	hundreds	2	tens	4	ones

= × 13 × 2 × 4

Here 1324 is made using only tens and ones.

1	3	2	tens	4	ones

= × 132 × 4

Now 1324 is made only using ones.

1	3	2	4	ones

= × 1324

Example 2: Complete the number expanders for 6845.

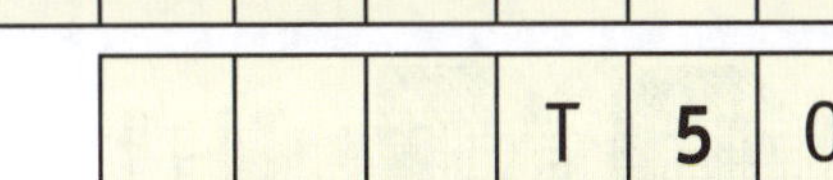
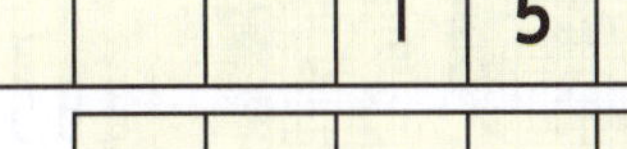

6	Th	8	H	4	T	5	O
	6	8	H	4	T		O
					T	5	O
		6					O

Check your answer on the video!

Fill in the gaps.

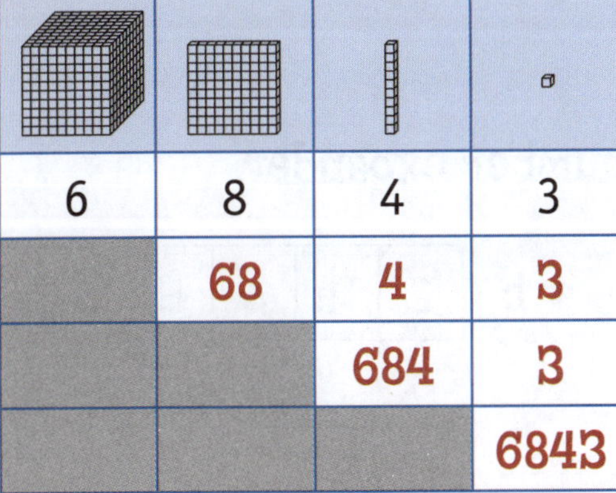

6	8	4	3
	68	4	3
		684	3
			6843

a

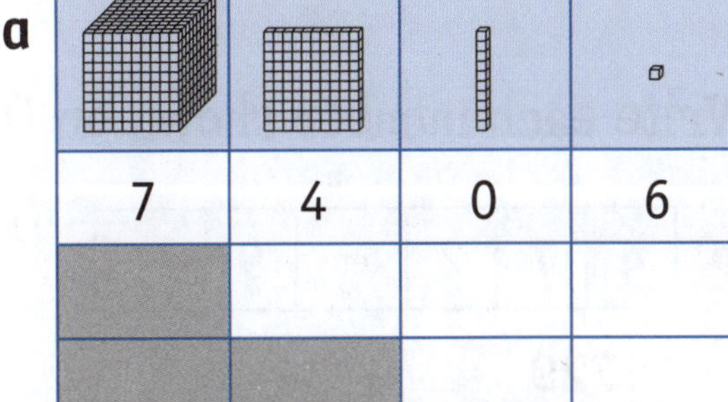

7	4	0	6

SELF CHECK Tick how you feel

Got it!	Need help...	I don't get it
 ☐	 ☐	 ☐

Check your answers
How many did you get correct? ☐

1 Match the blocks with the number expanders.

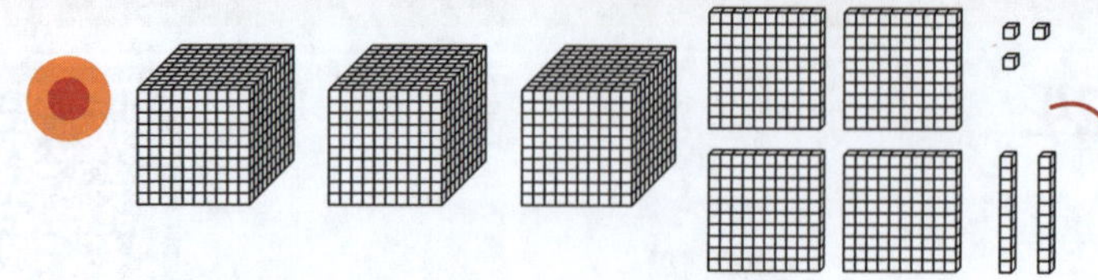

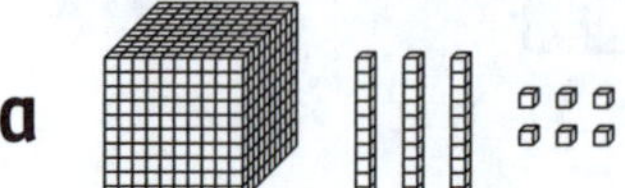

a

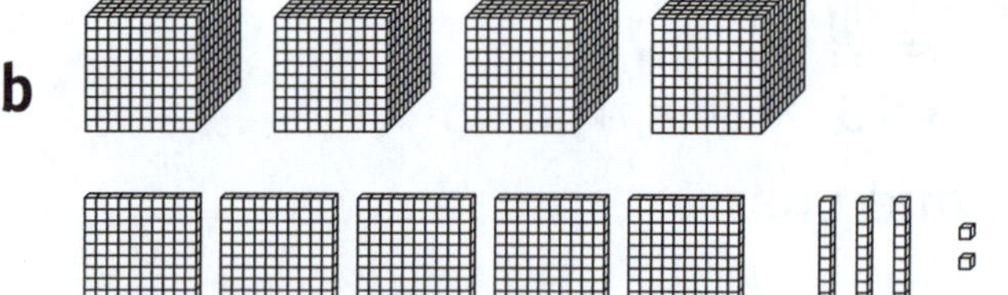

b

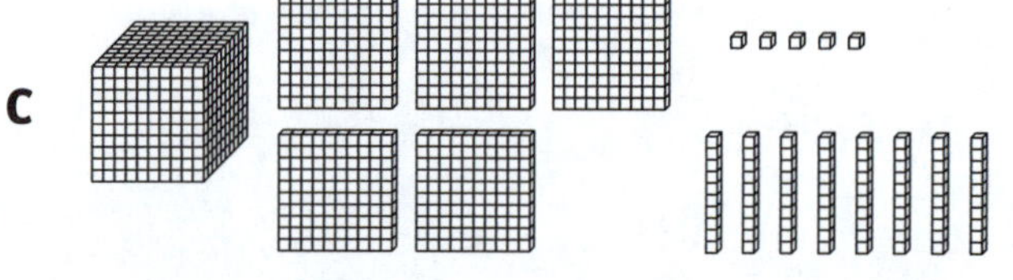

c

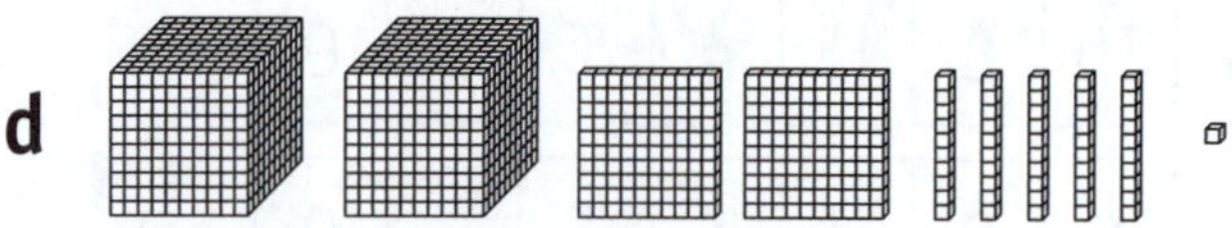

d

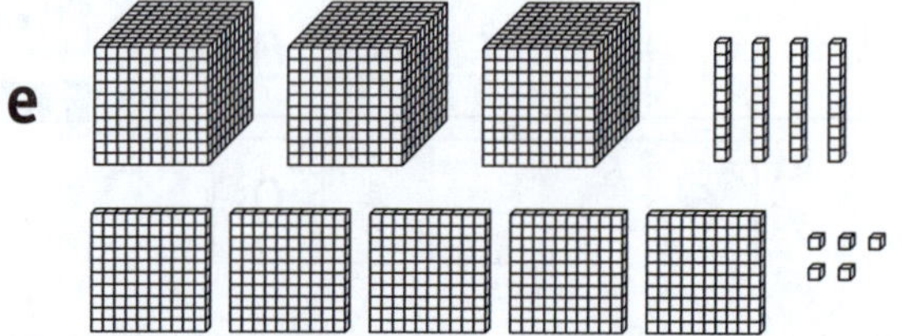

e

2	Th	2	H	5	T	1	0

3	Th	5	H	4	T	5	0

3	Th	4	H	2	T	3	0

4	Th	5	H	3	T	2	0

1	Th	5	H	8	T	5	0

1	Th	0	H	3	T	6	0

2 Write each number shown on the number expander.

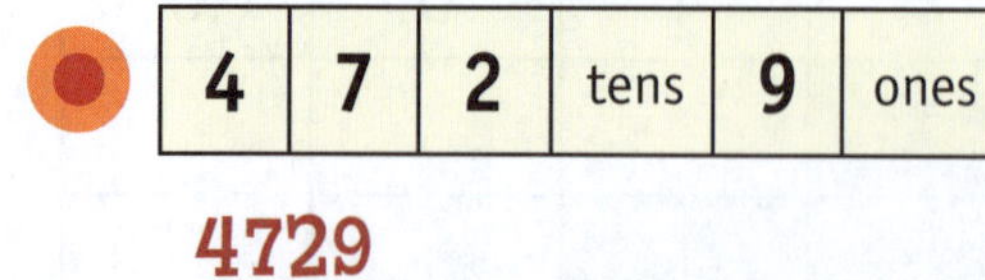

4	7	2	tens	9	ones

4729

a

8	5	3	2	ones

______

b

2	Th	4	H	6	T	8	0

______

c

7	5	H	6	T	4	0

______

CATCH UP MATHS YEAR 4 BOOK A © PASCAL PRESS ISBN: 9781925726145

EXPANDED FOUR-DIGIT NUMBERS

You can show numbers using Base 10 blocks, an abacus, and expanded form.

Example 1:

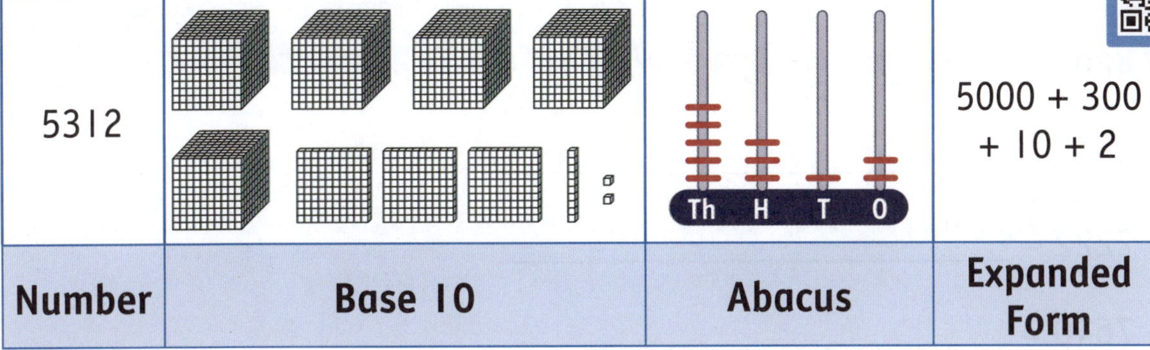

Number	Base 10	Abacus	Expanded Form
5312			5000 + 300 + 10 + 2

Example 2:

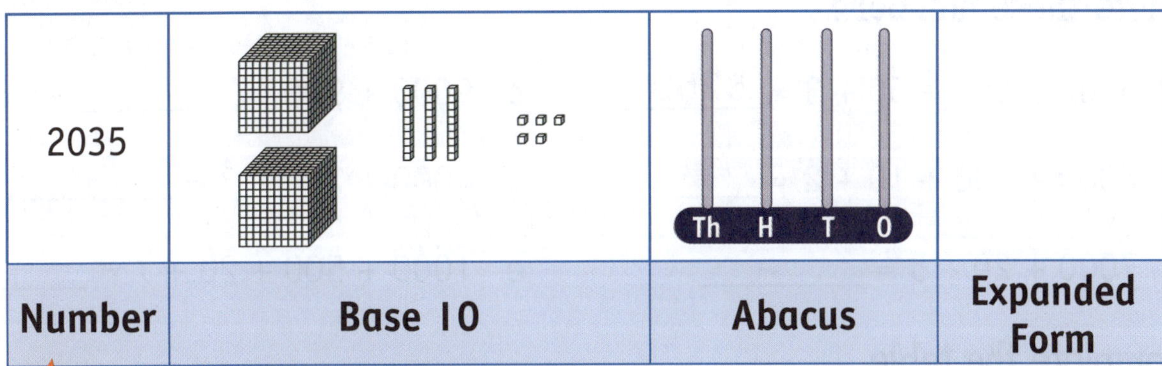

Number	Base 10	Abacus	Expanded Form
2035			

Check your answer on the video!

Your turn Complete the table.

Number	Base 10	Abacus	Expanded Form
1126		Th H T O	
2315		Th H T O	

SELF CHECK Tick how you feel

Got it!	Need help...	I don't get it
☐	☐	☐

Check your answers
How many did you get correct? ☐

PRACTICE

1 Write these four-digit numbers in expanded form.

- 1832 = 1000 + 800 + 30 + 2

a 3627 = ____________________

b 4803 = ____________________

c 5093 = ____________________

d 2222 = ____________________

e 7640 = ____________________

2 Write these numbers.

- 5000 + 200 + 50 + 3 = 5253

a 6000 + 300 + 10 + 2 = ________

b 7000 + 20 + 5 = ________

c 9000 + 900 = ________

d 4000 + 200 + 3 = ________

e 1000 + 500 + 20 + 7 = ________

3 Complete the table.

Number	Base 10	Abacus	Expanded Form
1223		Th H T O	1000 + 200 + 20 + 3
2339		Th H T O	
3459		Th H T O	

CATCH UP MATHS YEAR 4 BOOK A © PASCAL PRESS ISBN: 9781925726145

MODELLING FOUR-DIGIT NUMBERS

These Base 10 blocks and an abacus can be used to model a four-digit number.

Example 1: Here is the four-digit number 2456 in Base 10 blocks and on an abacus.

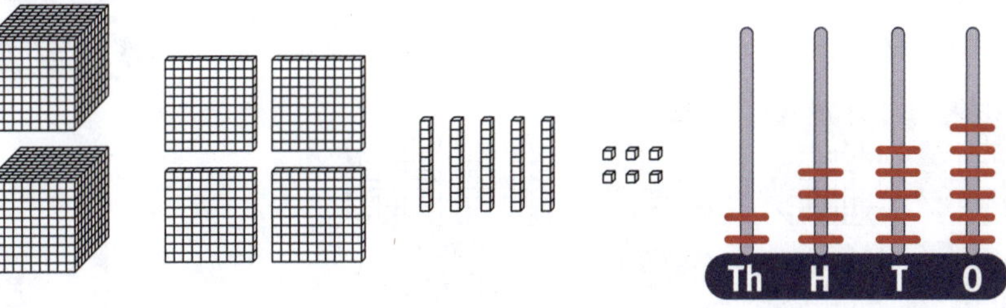

There are lots of different ways to show numbers.

Example 2: Complete the four-digit number 3672 in Base 10 blocks and on the abacus.

Check your answer on the video!

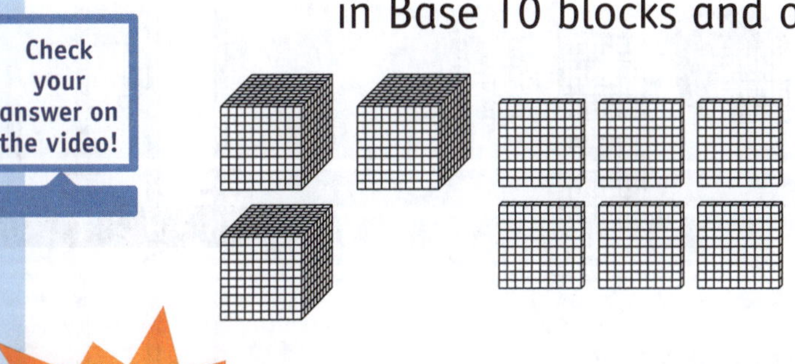

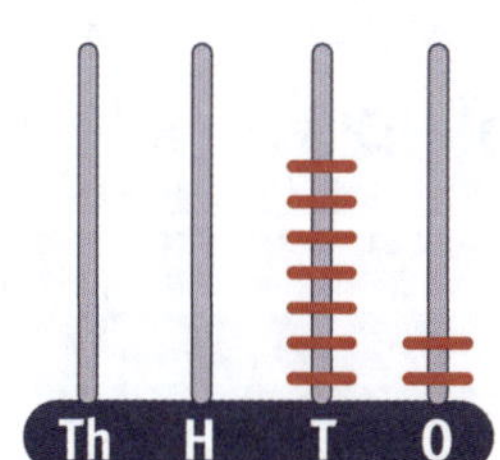

Your turn

What numbers have been modelled below?

2353

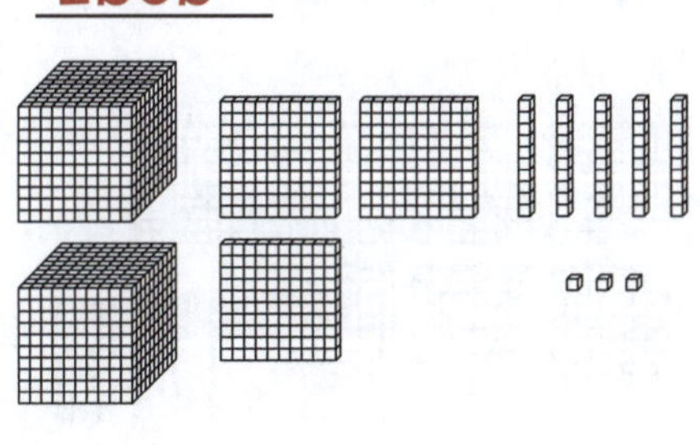

b ______

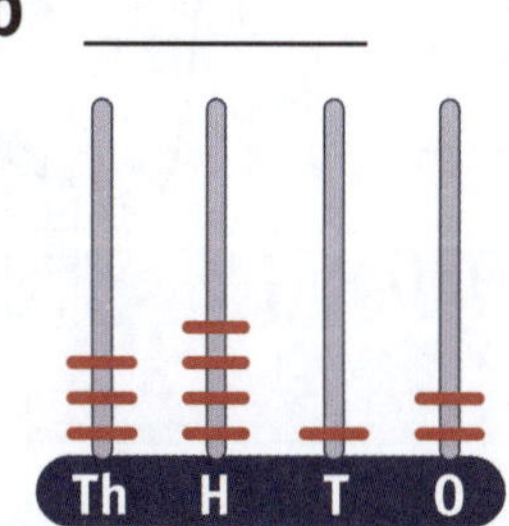

a ______

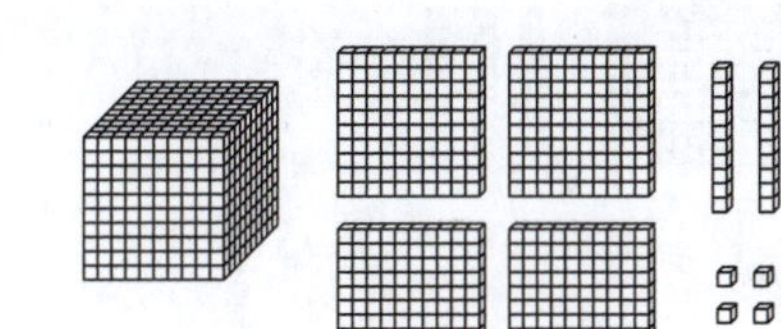

c ______

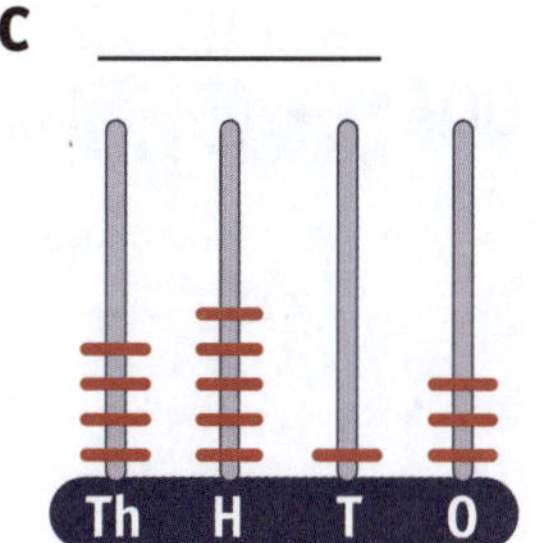

SELF CHECK Tick how you feel

Got it!	Need help...	I don't get it
☐	☐	☐

Check your answers

How many did you get correct?

PRACTICE

Match the numbers with their models.

Number	Base 10	Abacus
3750		Th H T O
a 4213		Th H T O
b 5231		Th H T O
c 1530		Th H T O
d 2030		Th H T O
e 1004		Th H T O
f 6300		Th H T O

CATCH UP MATHS YEAR 4 BOOK A © PASCAL PRESS ISBN: 9781925726145

ORDERING FOUR-DIGIT NUMBERS

Ascending order is when numbers are ordered from smallest to largest. These numbers are in ascending order:
3256, 4815, 7394, 8216, 9018.

Descending order is when numbers are ordered from largest to smallest. Now the numbers are rearranged into descending order:
9018, 8216, 7394, 4815, 3256.

SCAN to watch video

Example 1:

Write these numbers in ascending order.
6258 3724 9215 4711 9283

3724	4711	6258	9215	9283

Example 2:

Write these numbers in descending order.
6258 3724 9215 4711 9283

1 Order these numbers in ascending order.

● 2473, 1215, 3791, 3971 1215, 2473, 3791, 3971

a 5253, 9691, 8873, 3788 ______

b 6740, 2440, 9894, 6440 ______

2 Order these numbers in descending order.

a 2020, 1936, 7791, 3719 ______

b 2904, 1973, 1503, 7179 ______

SELF CHECK Tick how you feel		
Got it! ☐	Need help... ☐	I don't get it ☐

Check your answers
How many did you get correct? ☐

  ISBN: 9781925726145

PRACTICE

1 Order these in ascending order by numbering them from 1 (smallest) to 5 (largest).

●	5217	3284	7190	6177	2849
	3	2	5	4	1
a	9037	9730	9073	9307	9703
b	7213	3127	2137	1723	1327
c	9876	6879	8967	7896	6987
d	2496	6942	6429	6400	6404

2 Order these in descending order by numbering them from 1 (largest) to 5 (smallest).

●	2606	3295	1273	4915	5231
	4	3	5	2	1
a	7312	2137	7321	7123	7231
b	5243	7819	5013	3310	2440
c	7217	9524	1806	2449	9373
d	8532	2964	1425	5400	4500

CATCH UP MATHS YEAR 4 BOOK A © PASCAL PRESS ISBN: 9781925726145

GREATER THAN, EQUAL TO, LESS THAN

SCAN to watch video

>	=	<
the symbol for greater than	the symbol for equal to	the symbol for less than (it looks like an L)

Example 1: 453 [>] 249

Read from left to right:

Is 453 greater than (>) 249? ✓ Yes

Is 453 equal to (=) 249? ✗ No

Is 453 less than (<) 249? ✗ No

Which number does the pointy end point to? That helps you tell the difference between < and >.

Example 2: 1275 [=] 1275

Is 1275 greater than (>) 1275? ✗ No

Is 1275 equal to (=) 1275? ✓ Yes

Is 1275 less than (<) 1275? ✗ No

Example 3: 2410 [ ] 6870

Is 2410 greater than (>) 6870? ______

Is 2410 equal to (=) 6870? ______

Is 2410 less than (<) 6870? ______

Write True or False.

- ● 1010 < 2734 True
- **a** 7145 > 6243 ______
- **b** 1315 = 1315 ______
- **c** 434 > 109 ______
- **d** 6251 < 1290 ______
- **e** 745 > 893 ______

Check your answers

How many did you get correct?

PRACTICE

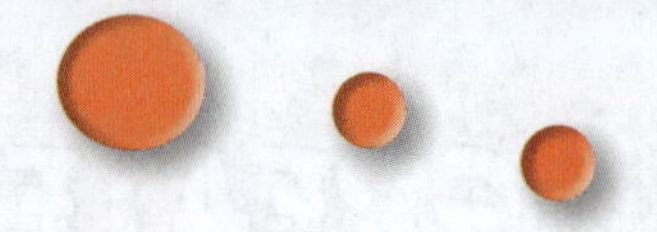

1 Fill in the missing symbols.

●	473	$<$	792
a	129	☐	792
b	925	☐	127
c	103	☐	103
d	461	☐	416
e	382	☐	283
f	545	☐	711
g	813	☐	318
h	207	☐	720

2 Write the words **is greater than**, **is equal to** or **is less than** to make these true.

●	173	is less than	472
a	321		123.
b	841		426.
c	909		836.
d	417		714.
e	523		523.
f	920		902.
g	747		774.
h	209		209.
i	345		304.

3 Circle the numbers that match the descriptions.

●	more than 73	(102) (78) 58 73 (89) (95) 70
a	less than 21	21 73 29 19 16 42 37 18
b	more than 93	93 47 94 16 95 39 98
c	less than 87	87 89 73 17 41 82 9
d	more than 325	129 325 392 145 463 173 497
e	equal to 127	127 721 127 127 371 271 128
f	less than 432	432 317 324 147 214 234 541
g	equal to 607	607 706 176 607 607 67 167

CATCH UP MATHS YEAR 4 BOOK A © PASCAL PRESS ISBN: 9781925726145

4 Fill in the missing symbols.

- 2417 [<] 7412
- a 7600 [ ] 7006
- b 2491 [ ] 2491
- c 8491 [ ] 1459
- d 3857 [ ] 7583
- e 2934 [ ] 4932
- f 5921 [ ] 5912
- g 6030 [ ] 6300
- h 5678 [ ] 8765
- i 9930 [ ] 9033
- j 2011 [ ] 2020
- k 6292 [ ] 6922

5 Write three numbers greater than the number shown.

- 4352 4532 4590 6511
- a 5231 ______ ______ ______
- b 6593 ______ ______ ______
- c 5847 ______ ______ ______
- d 9458 ______ ______ ______

6 Write three numbers less than the number shown.

- 5249 5010 4132 1754
- a 7342 ______ ______ ______
- b 8294 ______ ______ ______
- c 9311 ______ ______ ______
- d 6425 ______ ______ ______

7 Write a number in between.

- 1247 [1389] 1435
- a 3136 [ ] 4425
- b 2634 [ ] 2723
- c 2854 [ ] 5859
- d 935 [ ] 1034
- e 3649 [ ] 3784

8 Write True or False.

●	2432 < 6295	True	e	5740 = 5704	______
a	7143 > 2163	______	f	9386 > 1057	______
b	2409 = 2409	______	g	6322 < 6895	______
c	5382 < 5163	______	h	9855 < 9981	______
d	8740 = 8740	______	i	7249 < 7942	______

9 Circle the numbers that match the descriptions.

● more than 125 — (192) 125 (127) 124 (291) (129)

a less than 626 — 620 662 262 668 266 594 694

b equal to 714 — 714 741 417 174 714 471 147 714

c less than 422 — 422 224 442 424 142 242 422

d equal to 928 — 299 928 829 928 298 928

e more than 1253 — 5231 1352 1253 1235 1325 1523 1253

f more than 9727 — 9727 9729 7729 9772 9727 7927 2779

g less than 3493 — 3493 3439 4393 3394 9433 3499 3493

h equal to 8352 — 5382 2583 8352 8352 3852 2835 8352

10 Circle the correct symbol

●	343 [(>) = <] 249		e	425 [> = <] 462
a	648 [> = <] 846		f	860 [> = <] 796
b	728 [> = <] 320		g	947 [> = <] 780
c	103 [> = <] 103		h	581 [> = <] 343
d	242 [> = <] 585		i	669 [> = <] 468

CATCH UP MATHS YEAR 4 BOOK A © PASCAL PRESS ISBN: 9781925726145

ROUNDING TO 100 AND 1000

Rounding is useful when you need to estimate an answer.

Round down numbers					Round up numbers				
0	1	2	3	4	5	6	7	8	9

Rounding to 100

1 Go to the hundreds column.
2 Write the round up number above the hundreds.
3 Circle the number in the tens column.
4 Is it a round up or round down number?
5 Round your number.

Rounding to 1000

1 Go to the thousands column.
2 Write the round up number above the thousands.
3 Circle the number in the hundreds column.
4 Is it a round up or round down number?
5 Round your number.

If the number is less than 5, you round down.
If the number is 5 or more, you round up.

Example 1:
Round 349 to the nearest 100.

300

Example 2:
Round 582 to the nearest 100.

6 up
582
600

Example 3:
Round 8953 to the nearest 1000. ________

____

Example 4:
Round 21 492 to the nearest 1000. ________

____

Check your answer on the video!

1 Round to the nearest 100.

a 653

____

b 2438

____

2 Round to the nearest 1000.

a 9573

____

b 24 252

____

Check your answers
How many did you get correct?

Round these numbers to the nearest 100.

● 3524 (6 written above the 2, 2 circled) 3500

c 963 ______

f 1590 ______

a 1725 ______

d 1043 ______

g 4923 ______

b 257 ______

e 2597 ______

h 648 ______

Round these numbers to the nearest 1000.

● 7295 (6 written above the 2) ______

b 4935 ______

d 6259 ______

a 3428 ______

c 7135 ______

e 2843 ______

3 Round these numbers to the nearest 100, then 1000.

● 2593
nearest 100 2600
nearest 1000 3000

c 8395
nearest 100 ______
nearest 1000 ______

a 4938
nearest 100 ______
nearest 1000 ______

d 4723
nearest 100 ______
nearest 1000 ______

b 3142
nearest 100 ______
nearest 1000 ______

e 6492
nearest 100 ______
nearest 1000 ______

CATCH UP MATHS YEAR 4 BOOK A © PASCAL PRESS ISBN: 9781925726145

FIVE-DIGIT NUMBERS

A five-digit number is made up of five numbers (or digits). Numbers in the ten thousands have five digits: 10 000 to 99 999.

Here are some five-digit numbers:

39 256 73 824 19 003 24 893

Read the numbers aloud to help you work out how to write them.

Example 1:

23 495 is written as twenty-three thousand, four hundred and ninety-five.

Example 2:

19 053 is written as nineteen thousand and fifty-three.

Example 3: 60 425 is written as sixty ______________,
four ______________ and ___________-_________.

Example 4: 42 950 is written as forty-___________ thousand,
__________ hundred and ___________.

1 Circle the five-digit numbers.

24 957 (circled) **1003 42 056 73 950 26 435**
7632 7458 17 537 835

2 Match the numbers to the words.

●	28 419	seventy-three thousand, four hundred and twenty
a	73 420	twenty-eight thousand, four hundred and nineteen
b	19 596	nineteen thousand, five hundred and ninety-six
c	84 253	eighty four thousand, two hundred and fifty-three

Check your answers
How many did you get correct?

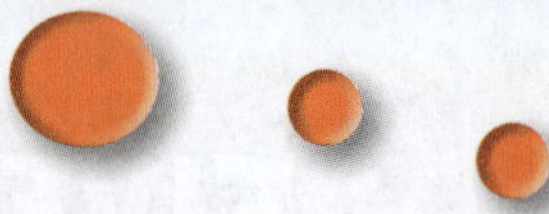

Write these numerals in words.

- 23 976 twenty-three thousand, nine hundred and seventy-six

a 49 350 ______________________

b 57 420 ______________________

c 62 943 ______________________

d 81 462 ______________________

e 93 258 ______________________

f 16 422 ______________________

Match these numbers to the words.

● 47 215	ninety-four thousand, three hundred and eighty-two
a 53 847	forty-seven thousand, two hundred and fifteen
b 71 329	sixty-two thousand, four hundred and seventy-one
c 94 382	seventy-six thousand, two hundred and eighteen
d 81 432	seventy-one thousand, three hundred and twenty-nine
e 76 218	thirty-one thousand, two hundred and forty
f 62 471	fifty-three thousand, eight hundred and forty-seven
g 31 240	eighty-one thousand, four hundred and thirty-two

CATCH UP MATHS YEAR 4 BOOK A © PASCAL PRESS ISBN: 9781925726145

PLACE VALUE TO 100 000

The place value is the value of a digit based on where it is in a number.

37 289 is a five-digit number.

The place value of the 3 is ten thousands because it is in the ten thousands place.

The place value of the 7 is thousands because it is in the thousands place.

The place value of the 2 is hundreds because it is in the hundreds place.

The place value of the 8 is tens because it is in the tens place.

The place value of the 9 is ones because it is in the ones place.

Example 1:

37 289 has 3 ten thousands, 7 thousands, 2 hundreds, 8 tens and 9 ones.

This is 37 289 in a place value chart:

TT	Th	H	T	O
3	7	2	8	9

Example 2: 47 568 has __ ten thousands, __ thousands, 5 __________, 6 tens and 8 __________.

This is 47 568 in a place value chart:

TT	Th	H	T	O

Your turn

Circle the ten thousands orange, the thousands purple, the hundreds green, the tens blue and the ones red.

73 608   89 243   63 825   74 103

61 000   47 090   32 361   24 140

Check your answers
How many did you get correct?

PRACTICE

1 Write these numbers.

- 27 thousands, 6 tens, 2 hundreds **27 260**

a 2 ten thousands, 3 tens, 4 hundreds, 1 ones ________

b 94 thousands, 6 ones, 7 hundreds, 3 tens ________

c 5 tens, 6 hundreds, 5 ones, 35 thousands ________

d 6 ten thousands, 8 ones, 2 hundreds, 5 thousands, 1 tens ________

2 Complete the chart.

	Number	Ten thousands	Thousands	Hundreds	Tens	Ones
●	34 398	**3**	**4**	**3**	**9**	**8**
a	67 138					
b	15 840					
c		7	1	4	5	9
d		2	5	0	4	3
e	75 458					

3 Write the numbers in the correct place on the chart.

		Ten thousands	Thousands	Hundreds	Tens	Ones
●	60 000	**60 000**				
a	5000					
b	2					
c	10					
d	80 000					
e	5					
f	80					

CATCH UP MATHS YEAR 4 BOOK A © PASCAL PRESS ISBN: 9781925726145

THE VALUE OF NUMBERS TO 100 000

The value of a number is how much it is worth.

Example 1:

H T O

692

The value of the 6 is 600.

The value of the 2 is 2.

The value of the 9 is 90.

Example 2:

Th H T O

3685

The value of the 3 is 3000.

The value of the 5 is 5.

The value of the 6 is 600.

The value of the 8 is 80.

Example 3:

TT Th H T O

24 769

The value of the 2 is ________.

The value of the 9 is ___.

The value of the 4 is ________.

The value of the 6 is ____.

The value of the 7 is _______.

Circle in green the digit with the most value and in red the digit with the least value.

- 28 491   10 398   24 937   82 358

90 009   12 683   73 469   44 444

35 293   47 247   64 293

SELF CHECK Tick how you feel

Got it!	Need help...	I don't get it
☐	☐	☐

Check your answers

How many did you get correct? ☐

PRACTICE

1 Write numbers using these digits to make a number with:

3 5 8 7 6

- 6 hundreds 35 687
- **a** 3 ten thousands ________
- **b** 8 hundreds ________
- **c** 7 thousands ________
- **d** 6 ones ________
- **e** 5 ten thousands ________
- **f** 3 tens ________
- **g** 5 ones ________
- **h** 7 hundreds ________
- **i** 7 ten thousands ________

2 Write the missing numbers to make 32 487.

- 32 thousands + 487
- **a** 3 ten thousands + ________
- **b** 324 hundreds + ________
- **c** 32 487 ones + ________
- **d** 3248 tens + ________

3 What is the value of the underlined digit?

- 24 <u>7</u>31 = 700
- **a** <u>3</u>8 420 = ________
- **b** 16 2<u>1</u>6 = ________
- **c** <u>4</u>1 249 = ________
- **d** 7<u>2</u> 542 = ________
- **e** 31 <u>0</u>53 = ________
- **f** 17 45<u>8</u> = ________
- **g** <u>9</u>1 004 = ________
- **h** 85 3<u>7</u>2 = ________
- **i** 21 <u>4</u>35 = ________
- **j** 41 27<u>3</u> = ________
- **k** 8<u>1</u> 431 = ________
- **l** 57 <u>6</u>31 = ________
- **m** 24 6<u>2</u>2 = ________
- **n** <u>8</u>1 030 = ________
- **o** 6<u>0</u> 239 = ________
- **p** <u>4</u>8 346 = ________
- **q** 3<u>4</u> 189 = ________
- **r** 97 2<u>5</u>7 = ________
- **s** 6<u>7</u> 201 = ________

  ISBN: 9781925726145

NUMBER EXPANDERS AND FIVE-DIGIT NUMBERS

The number expander shows 36 425.

3	ten thousands	6	thousands	4	hundreds	2	tens	5	ones

SCAN to watch video

We can show 36 425 using only thousands, hundreds, tens and ones:

3	6	thousands	4	hundreds	2	tens	5	ones

Or using only hundreds, tens and ones:

3	6	4	hundreds	2	tens	5	ones

Or only tens and ones:

3	6	4	2	tens	5	ones

And finally, 36 425 using only ones:

3	6	4	2	5	ones

Example 2: Show 14 935 on the number expanders.

	ten thousands		thousands		hundreds		tens		ones

		thousands		hundreds		tens		ones

			hundreds		tens		ones

				tens		ones

					ones

There are lots of different ways to show numbers.

Your turn

Complete the number expanders.

7	TT	3	Th	4	H	6	T	3	O

		Th		H		T		O

			H		T		O

				T		O

					O

Check your answer on the video!

SELF CHECK Tick how you feel

Got it!	Need help...	I don't get it
☐	☐	☐

Check your answers

How many did you get correct? ☐

PRACTICE

1 Write the number shown on these number expanders.

●

6	ten thousands	3	thousands	4	hundreds	7	tens	2	ones

63 472

a

4	0	thousands	2	hundreds	8	tens	3	ones

____

b

2	7	5	hundreds	3	tens	9	ones

____

c

7	4	6	8	tens	1	ones

____

d

9	0	7	2	4	ones

____

e

8	3	thousands	7	hundreds	4	tens	5	ones

____

2 Fill in the numbers on the number expanders.

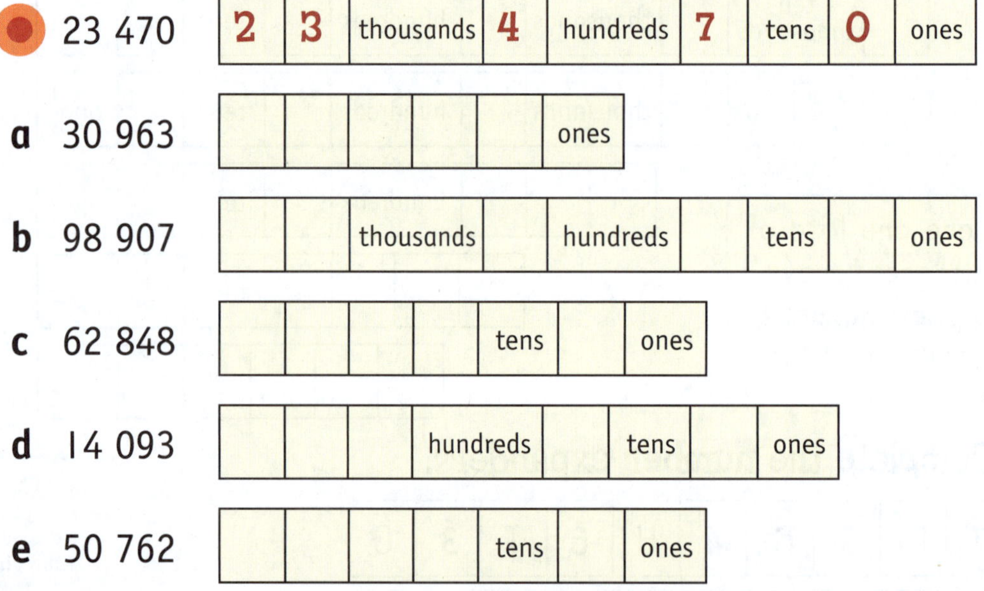

● 23 470

2	3	thousands	4	hundreds	7	tens	0	ones

a 30 963

					ones

b 98 907

		thousands		hundreds		tens		ones

c 62 848

				tens		ones

d 14 093

			hundreds		tens		ones

e 50 762

				tens		ones

3 What are the following numbers?

● 32 thousands and 64 ones ________

a 526 hundreds 9 tens 2 ones ________

b 2 ten thousands, 3 hundreds, 6 tens ________

c 28 thousands, 2 hundreds, 5 tens, 3 ones ________

CATCH UP MATHS YEAR 4 BOOK A © PASCAL PRESS ISBN: 9781925726145

EXPANDED FIVE-DIGIT NUMBERS

Expanding a number is writing the number to show the value of each digit.

You can work out what a number is by looking at its expanded form.

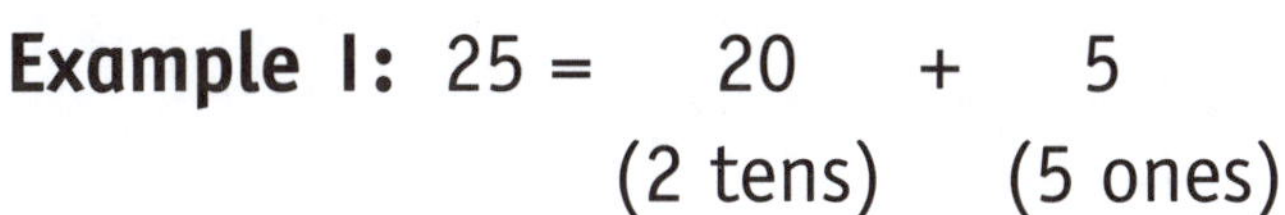

Example 1: 25 = 20 + 5
(2 tens) (5 ones)

Example 2: 179 = 100 + 70 + 9
(1 hundred) (7 tens) (9 ones)

Example 3:
2843 = 2000 + 800 + 40 + 3
(__ thousands) (__ hundreds) (__ tens) (__ ones)

Example 4:
69 512 = __________ + _______ + ______ + ___ + __
(__ ten thousands) (__ thousands) (__ hundreds) (__ tens) (__ ones)

Expand the following numbers.

● 236 = 200 + 30 + 6
= 2 hundreds + 3 tens + 6 ones

a 57 = ___ + __
= __________ + ________

b 159 = _____ + ___ + __
= ______________ + __________ + ________

Check your answers
How many did you get correct?

PRACTICE

1 Expand the following numbers.

- 285 = 200 + 80 + 5

a 58 = ______

b 726 = ______

c 64 = ______

d 5 = ______

e 9373 = ______

f 9771 = ______

g 95 256 = ______

h 30 357 = ______

i 128 = ______

j 72 421 = ______

2 Match these numbers with their expanded form.

• 62 473	20 000 + 5000 + 200 + 90 + 5
a 25 295	10 000 + 400 + 90 + 3
b 68 490	60 000 + 8000 + 400 + 90 + 3
c 74 002	80 000 + 9000 + 700 + 3
d 10 493	10 000 + 8000 + 50 + 6
e 18056	60 000 + 2000 + 400 + 70 + 3
f 89 703	70 000 + 4000 + 2

CATCH UP MATHS YEAR 4 BOOK A © PASCAL PRESS ISBN: 9781925726145

MODELLING FIVE-DIGIT NUMBERS

These Base 10 blocks and an abacus can be used to model a five-digit number.

Example 1: Here is the five-digit number 15 372 in Base 10 blocks and on an abacus.

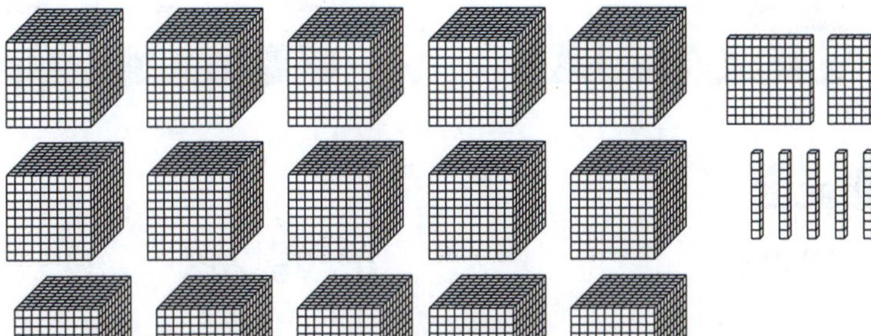

TT Th H T O

Check your answer on the video!

Example 2: Write the number shown on the abacus.

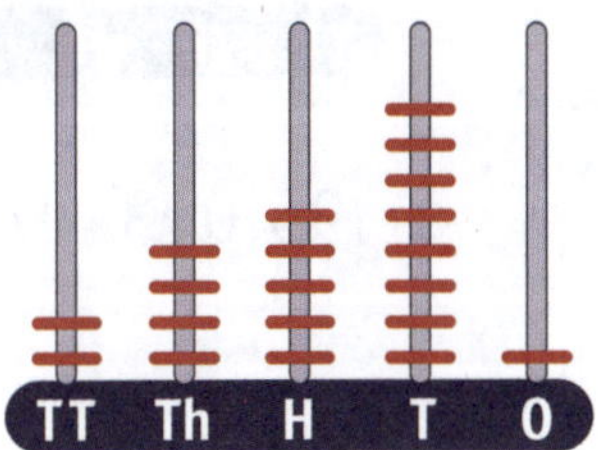

Example 3: Write the number.

Number				
	24	0	3	5

Complete the table.

	Number				
●	30 256	**30**	**2**	**5**	**6**
a	72 345				
b		63	1	6	2
c		21	5	1	9
d	32 481				

SELF CHECK Tick how you feel

Got it!	Need help...	I don't get it
☐	☐	☐

Check your answers

How many did you get correct? ☐

1 What numbers have been shown on the abacus?

● 53 632

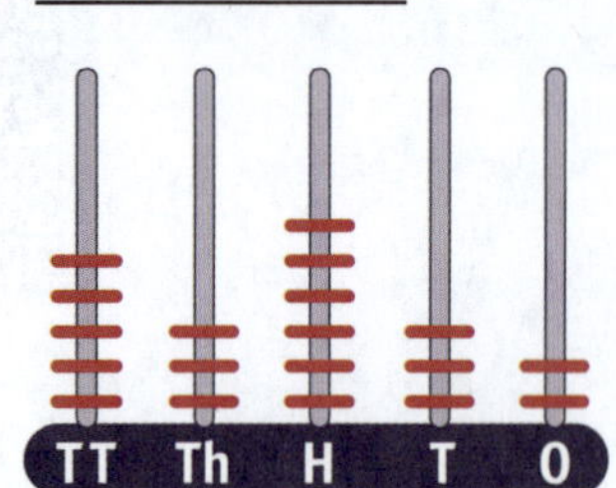

a ______

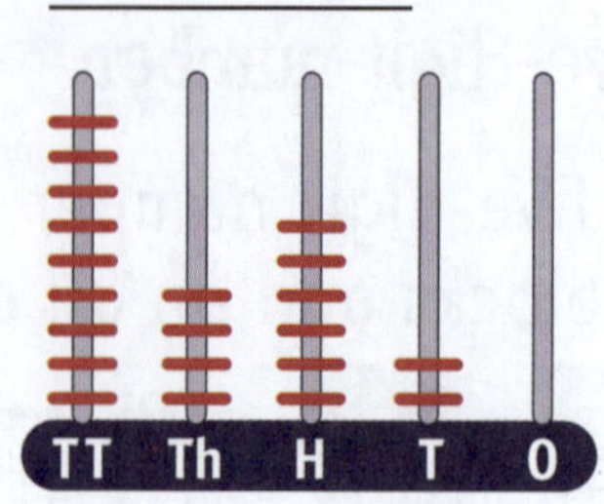

b ______

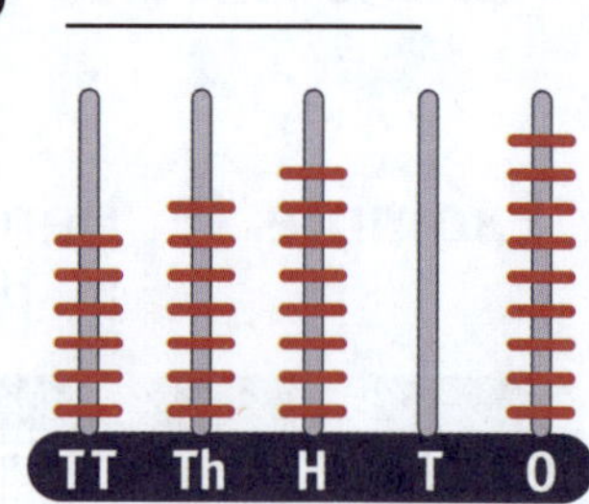

2 Show each number on the abacus.

● 17 295

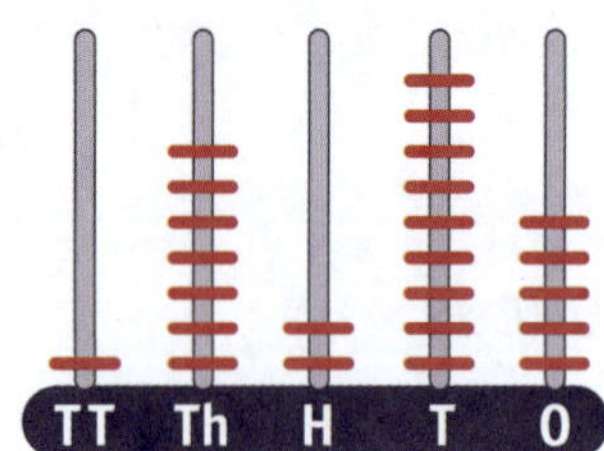

b 98 962

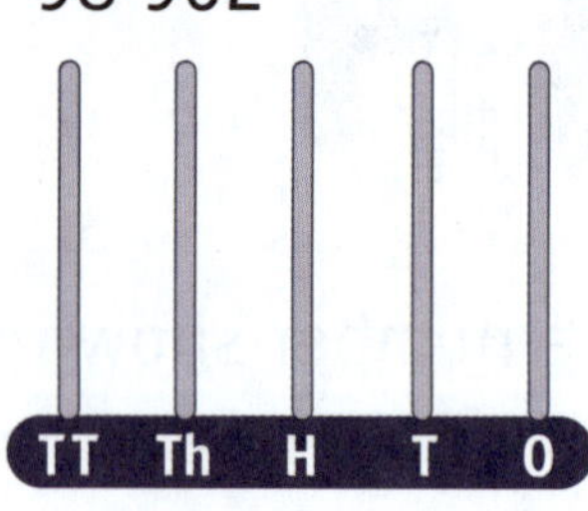

d 95 243

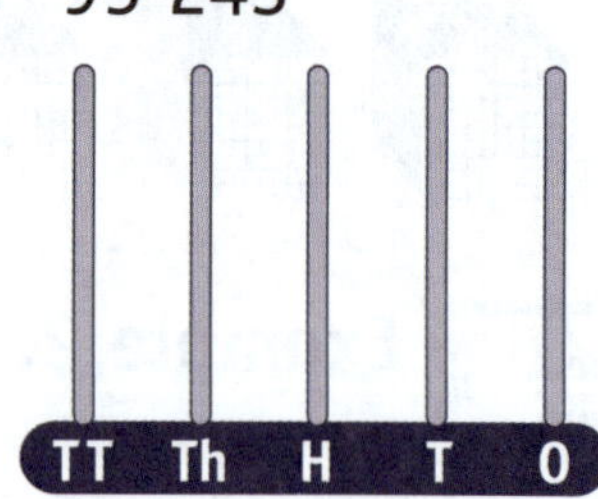

a 54 372

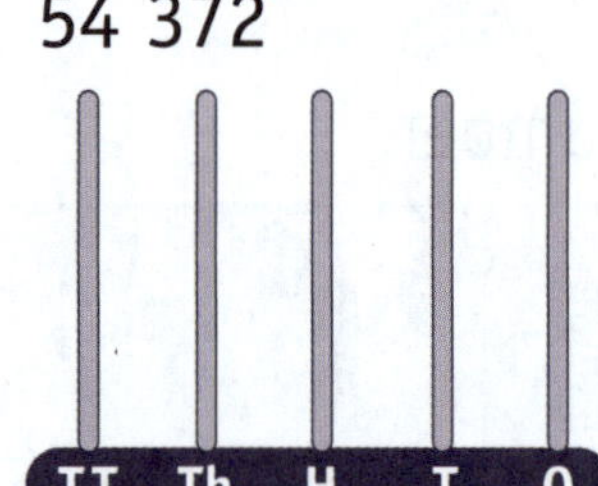

c 24 409

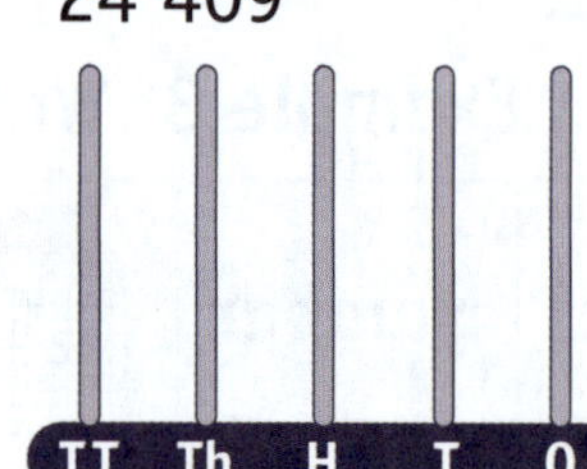

e 95 013

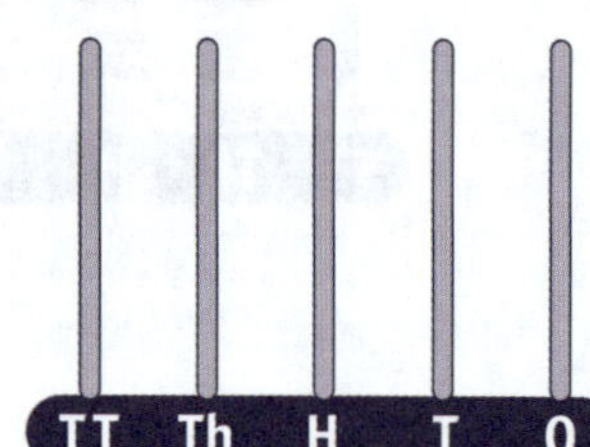

3 Complete the table.

	Number	(thousands)	(hundreds)	(tens)	(ones)
●	32 035	32	0	3	5
a	64 043				
b	59 321				
c	95 296				
d	12 395				
e	83 125				
f	29 791				
g	69 614				

CATCH UP MATHS YEAR 4 BOOK A © PASCAL PRESS ISBN: 9781925726145

ORDERING FIVE-DIGIT NUMBERS

You can order numbers in ascending order (smallest to largest) or descending order (largest to smallest).

These numbers are in descending order:

57 529, 43 290, 37 411, 22 115

These numbers are in ascending order:

12 537, 29 385, 31 624, 49 240

Example 1: Write these numbers in ascending order.
43 456, 21 437, 32 459, 98 436

21 437	32 459	43 456	98 436
↑ smallest number			↑ largest number

Example 2: Write these numbers in descending order.
46 202, 74 016, 85 134, 23 579

↑ largest number			↑ smallest number

Write ascending or descending.

- 12 579, 19 257, 19 752, 21 957 ascending

a 29 357, 27 235, 25 223, 23 522 __________

b 57 950, 57 590, 55 790, 55 097 __________

c 14 379, 14 973, 17 913, 19 713 __________

d 62 521, 62 522, 65 252, 65 522 __________

e 79 179, 79 971, 97 791, 99 791 __________

SELF CHECK Tick how you feel		
Got it! ☐	Need help... ☐	I don't get it ☐

Check your answers
How many did you get correct? ☐

PRACTICE

Arrange these numbers into ascending order.

- 14 253, 35 241, 24 351, 42 531

  14 253, 24 351, 35 241, 42 531

a 62 435, 27 942, 13 578, 84 259

____

b 13 524, 31 524, 21 531, 25 311

____

c 75 426, 62 457, 57 462, 72 624

____

d 58 385, 49 326, 85 358, 62 572

____

e 27 345, 54 372, 42 573, 35 472

____

Put these numbers in descending order. Write a 1 under the largest number and a 5 under the smallest.

•	24 382	29 375	21 637	73 621	53 762
	4	3	5	1	2
a	54 927	72 945	25 947	45 729	79 542
	☐	☐	☐	☐	☐
b	36 438	38 463	46 843	84 643	64 368
	☐	☐	☐	☐	☐
c	14 936	63 941	41 369	34 961	16 943
	☐	☐	☐	☐	☐
d	52 934	45 923	93 254	39 524	69 245
	☐	☐	☐	☐	☐

CATCH UP MATHS YEAR 4 BOOK A © PASCAL PRESS ISBN: 9781925726145

GREATER THAN, EQUAL TO, LESS THAN: NUMBERS TO 100 000

> greater than	= equal to	< less than

Example 1: 13 157 [<] 23 546

Read from left to right. Ask yourself:

Is 13 157 greater than (>) 23 546? ✗ No

Is 13 157 equal to (=) 23 546? ✗ No

Is 13 157 less than (<) 23 546? ✓ Yes

Example 2: 34 957 [=] 34 957

Is 34 957 greater than (>) 34 957? ✗ No

Is 34 957 equal to (=) 34 957? ✓ Yes

Is 34 957 less than (<) 34 957? ✗ No

Example 3: 49 520 [ ] 15 490

Is 49 520 greater than (>) 15 490? ______

Is 49 520 equal to (=) 15 490? ______

Is 49 520 less than (<) 15 490? ______

Your turn

Answer Yes or No.

● Is 65 249 < 57 957? No

a Is 14 573 > 24 297? ______

b Is 36 429 < 59 419? ______

c Is 41 000 = 41 000? ______

SELF CHECK Tick how you feel

Got it!	Need help...	I don't get it
 ☐	 ☐	☐

Check your answers

How many did you get correct? ☐

PRACTICE

1 Insert the correct symbol: > < or =

- 24 579 [<] 25 479
- a 32 003 [ ] 33 200
- b 54 920 [ ] 54 920
- c 14 019 [ ] 41 910
- d 37 249 [ ] 94 375
- e 93 002 [ ] 92 030
- f 72 490 [ ] 74 290
- g 49 273 [ ] 49 723
- h 94 875 [ ] 95 874
- i 61 439 [ ] 16 439
- j 22 044 [ ] 22 440
- k 16 134 [ ] 16 134
- l 83 270 [ ] 83 720
- m 72 467 [ ] 72 467

2 Circle the correct answer.

- 35 284 is less than / is equal to / (is greater than) 34 582.
- a 72 145 is less than / is equal to / is greater than 75 214.
- b 42 957 is less than / is equal to / is greater than 24 957.
- c 53 240 is less than / is equal to / is greater than 53 420.
- d 17 490 is less than / is equal to / is greater than 17 490.

3 Write a number to make the following true.

- 43 245 > 15 256
- a 72 193 < ________
- b 28 057 = ________
- c 41 347 > ________
- d 32 090 > ________
- e 16 325 < ________
- f 43 495 > ________
- g 22 043 = ________
- h 52 949 < ________
- i 49 563 < ________
- j 79 345 < ________
- k 81 090 < ________

CATCH UP MATHS YEAR 4 BOOK A © PASCAL PRESS ISBN: 9781925726145

LARGEST AND SMALLEST FIVE-DIGIT NUMBERS

Here are five digits: 7 3 8 4 1

I can move them around to make the smallest number possible.

SCAN to watch video

I put the 1 in first position as it has the least value.
Then the 3, then the 4, followed by the 7 and then 8.

The smallest five-digit number I can make is: 1 3 4 7 8

To make the largest number possible, put the 8 first as it has the greatest value. Next is 7, then 4 followed by 3 and then 1.

The largest five-digit number I can make is: 8 7 4 3 1

Example 1: Make the largest possible number and the smallest possible number using these digits: 4 5 9 0 6

9 6 5 4 0 — largest number

4 0 5 6 9 — smallest number

You can't start a 5-digit number with zero!

Example 2: Make the largest possible number and the smallest possible number using these digits: 6 4 7 8 9

9 __ __ __ __ — largest number

4 __ __ __ __ — smallest number

Check your answer on the video!

Your turn

Write the smallest number you can using these digits.

- ● 4 3 — 34
- a 7 2 ______
- b 0 3 1 ______
- c 7 5 7 ______
- d 4 9 3 2 ______
- e 4 7 8 5 3 ______

SELF CHECK Tick how you feel

Got it!	Need help...	I don't get it
☐	☐	☐

Check your answers
How many did you get correct? ☐

PRACTICE

1 Write the smallest number you can using these digits.

- ● 7, 6, 4, 5, 1 — 14 567
- a 2, 8, 4, 7, 5 ______
- b 9, 3, 2, 4, 0 ______
- c 0, 8, 9, 4, 1 ______
- d 3, 8, 4, 2, 1 ______
- e 7, 2, 1, 4, 5 ______
- f 1, 2, 4, 5, 3 ______
- g 5, 7, 9, 8, 2 ______
- h 8, 7, 6, 4, 1 ______
- i 7, 4, 0, 3, 6 ______

2 Write the largest five-digit number you can using these digits.

- ● 5, 9, 2, 4, 1 — 95 421
- a 3, 7, 2, 4, 4 ______
- b 1, 2, 7, 8, 8 ______
- c 5, 9, 0, 3, 7 ______
- d 6, 4, 9, 0, 3 ______
- e 8, 4, 2, 4, 0 ______
- f 5, 4, 3, 2, 8 ______
- g 2, 3, 5, 9, 1 ______
- h 8, 7, 4, 0, 2 ______
- i 4, 7, 9, 1, 1 ______

3 Complete the table.

	Digits	Smallest number	Largest number
●	1 5 7 9 0	10 579	97 510
a	3 2 4 8 5		
b	1 6 5 6 3		
c	7 4 0 9 8		
d	0 2 4 6 8		
e	1 3 7 5 9		
f	7 7 4 9 3		
g	9 9 8 9 5		

CATCH UP MATHS YEAR 4 BOOK A © PASCAL PRESS ISBN: 9781925726145

ODD AND EVEN NUMBERS TO 100 000

Odd numbers end in 1, 3, 5, 7, 9.	Even numbers end in 2, 4, 6, 8, 0.

Example 1: Is the number 29 837 odd or even?

Look at the last digit: **29 837**

7 is an odd number so 29 837 is an odd number.

Example 2: Is the number 46 498 odd or even?

Look at the last digit.

8 is an even number so 46 498 is an even number.

Check your answer on the video!

Example 3: Is the number 72 941 odd or even? ___________

It doesn't matter how big the number is – always look at the ones place to find out if it is odd or even.

Circle all the even five-digit numbers.
Cross out all the odd five-digit numbers.

~~74 953~~ 34 686 24 985 59 992

41 323 52 430 62 431

75 999 42 034 57 597 95 018

SELF CHECK Tick how you feel		
Got it! ☐	Need help... ☐	I don't get it ☐

Check your answers

How many did you get correct? ☐

  ISBN: 9781925726145

PRACTICE

1 Circle the even numbers with green and the odd numbers with red.

● 25 924	**d** 24 980	**h** 53 255
a 72 411	**e** 55 148	**i** 43 342
b 35 136	**f** 13 389	**j** 34 913
c 47 327	**g** 13 579	**k** 24 680

2 Write the next four even numbers.

● 24 920, 24 922, 24 924, 24 926, 24 928

a 16 432, ______, ______, ______, ______

b 36 588, ______, ______, ______, ______

c 72 054, ______, ______, ______, ______

d 58 600, ______, ______, ______, ______

e 90 020, ______, ______, ______, ______

3 Write the next four odd numbers.

● 26 483, 26 485, 26 487, 26 489, 26 491

a 59 599, ______, ______, ______, ______

b 61 111, ______, ______, ______, ______

c 72 463, ______, ______, ______, ______

d 90 007, ______, ______, ______, ______

e 10 035, ______, ______, ______, ______

f 52 331, ______, ______, ______, ______

g 38 419, ______, ______, ______, ______

CATCH UP MATHS YEAR 4 BOOK A © PASCAL PRESS ISBN: 9781925726145

ROUNDING TO 10 000

Rounding is useful when you need to estimate an answer.

Round down numbers					Round up numbers				
0	1	2	3	4	5	6	7	8	9

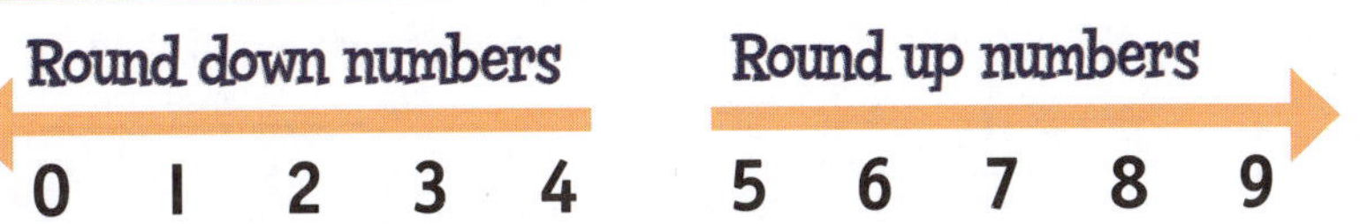

Rounding to 10 000

1 Go to the ten thousands column.

2 Write the round up number above the ten thousands.

3 Circle the number in the thousands column.

4 Is it a round up or round down number?

5 Round your number.

Example 1:
Round 24 358 to the nearest 10 000.

3
24 358
down
20 000

Example 2:
Round 36 529 to the nearest 10 000.

4 up
36 529
40 000

Example 3:
Round 15 372 to the nearest 10 000. ________ ________

Example 4:
Round 89 647 to the nearest 10 000. ________ ________

Check your answer on the video!

Round these number to the nearest 10 000.

● 24 597

3
24 597
down
20 000

a 37 924 ________ ________

b 52 599 ________ ________

c 75 023 ________ ________

SELF CHECK Tick how you feel

Got it!	Need help...	I don't get it
 ☐	 ☐	☐

Check your answers
How many did you get correct? ☐

PRACTICE

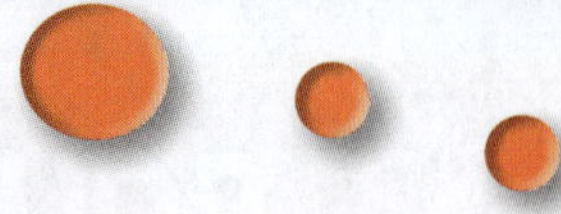

1 Round these numbers to the nearest 10 000.

- 45 926 (5 circled) — 50 000
- **a** 36 439 ________
- **b** 72 493 ________
- **c** 55 386 ________
- **d** 91 254 ________
- **e** 82 649 ________
- **f** 79 501 ________
- **g** 86 255 ________
- **h** 47 153 ________

2 Round the following.

- 62 510
  - nearest 10: 62 510
  - nearest 100: 62 600
  - nearest 1000: 63 000
  - nearest 10 000: 60 000

a 37 502
- nearest 10 ________
- nearest 100 ________
- nearest 1000 ________
- nearest 10 000 ________

b 71 539
- nearest 10 ________
- nearest 100 ________
- nearest 1000 ________
- nearest 10 000 ________

c 84 965
- nearest 10 ________
- nearest 100 ________
- nearest 1000 ________
- nearest 10 000 ________

d 26 608
- nearest 10 ________
- nearest 100 ________
- nearest 1000 ________
- nearest 10 000 ________

e 30 429
- nearest 10 ________
- nearest 100 ________
- nearest 1000 ________
- nearest 10 000 ________

CATCH UP MATHS YEAR 4 BOOK A © PASCAL PRESS ISBN: 9781925726145

WHOLE NUMBERS REVIEW

1 Write the following numbers in words.

a 675 ______

b 429 ______

c 3056 ______

d 2350 ______

e 7438 ______

f 8503 ______

g 26 590

______

h 37 429

______

2 Complete the table.

	Number	Ten thousands	Thousands	Hundreds	Tens	Ones
a	79					
b	838					
c	903					
d	1 430					
e	2 574					
f	3 827					
g	12 507					
h	35 639					
i	40 256					
j	57 007					

REVIEW

3 Circle the numbers with eight hundreds.

836 4798 3863 12 484 983 842 29 836

5815 183 98 28 3683 19 837 2862

4 Circle the numbers with four thousands.

24 293 94 47 473 2414 4684 8436

2594 4004 499 94 362 4937 34 876

5 What is the place value of the 6 in these numbers?

a 625 __________

b 1629 __________

c 6384 __________

d 26 378 __________

e 64 321 __________

f 69 342 __________

g 16 373 __________

h 47 624 __________

i 56 290 __________

j 97 635 __________

k 24 376 __________

l 16 __________

m 25 960 __________

n 3640 __________

o 74 362 __________

6 Write the numbers.

a 6 ten thousands, 3 hundreds, 4 ones = __________

b 8 thousands, 2 ten thousands, 1 ones, 3 tens = __________

c 7 tens, 4 hundreds, 8 ones = __________

d 5 thousands, 7 tens = __________

e 8 ten thousands, 6 tens, 3 ones = __________

f 3 thousands, 5 ten thousands, 2 ones = __________

g 8 ones, 6 tens, 4 ten thousands = __________

h 9 tens, 7 ten thousands, 5 hundreds = __________

CATCH UP MATHS YEAR 4 BOOK A © PASCAL PRESS ISBN: 9781925726145

7 What is the value of the 2 in these numbers?

a 423 ________ g 32 ________ m 72 ________

b 732 ________ h 24 358 ________ n 239 ________

c 729 ________ i 52 436 ________ o 2993 ________

d 12 ________ j 57 230 ________ p 23 491 ________

e 2495 ________ k 42 135 ________ q 62 410 ________

f 3245 ________ l 29 693 ________ r 43 216 ________

8 Fill in the number expanders.

a 2759

		thousands		hundreds		tens		ones

b 55 329

		thousands		hundreds		tens		ones

c 3248

					ones

d 46 819

	ten thousands		thousands		hundreds		tens		ones

e 72 385

			hundreds		tens		ones

f 1037

		thousands		hundreds		tens		ones

9 Complete these number expanders.

a 17 439

	TT		Th		H		T		O
			Th		H		T		O
					H		T		O
							T		O
									O

REVIEW

b 98 431

	TT		Th		H		T		O
			Th		H		T		O
					H		T		O
							T		O
									O

Write in expanded form.

a 632 = ______________________

b 207 = ______________________

c 910 = ______________________

d 1 438 = ______________________

e 3 589 = ______________________

f 7 136 = ______________________

g 23 458 = ______________________

h 62 410 = ______________________

i 90 007 = ______________________

j 42 070 = ______________________

Show these numbers below.

a 3942

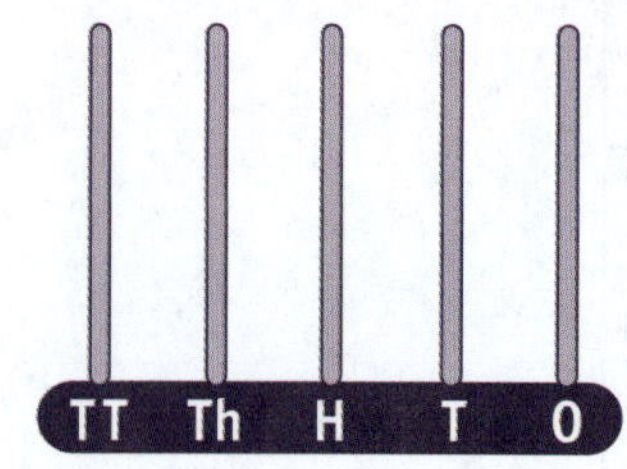

b 64 952

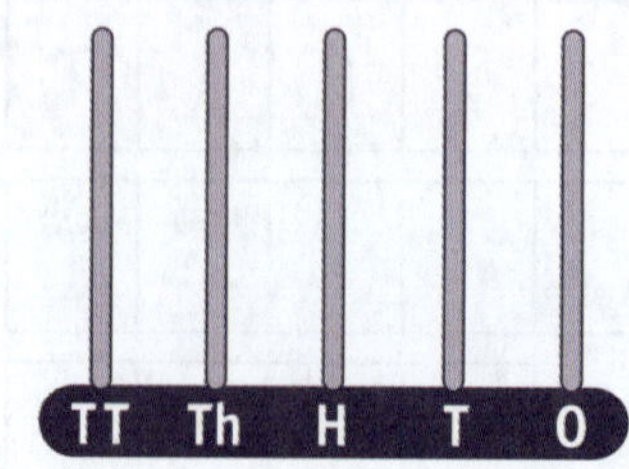

c 32 015

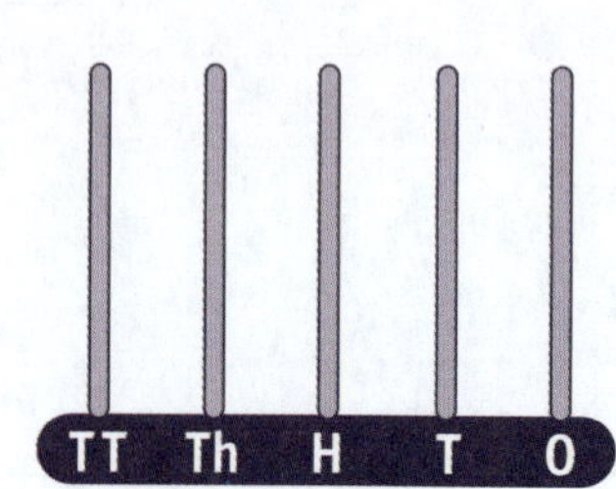

CATCH UP MATHS YEAR 4 BOOK A © PASCAL PRESS ISBN: 9781925726145

d 7495

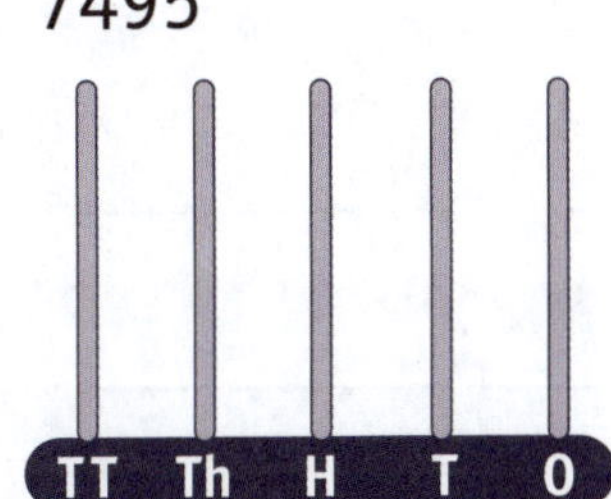

e 8973

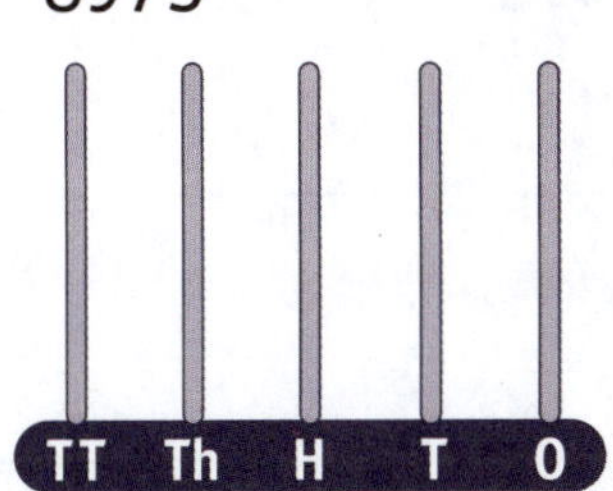

f 90 379

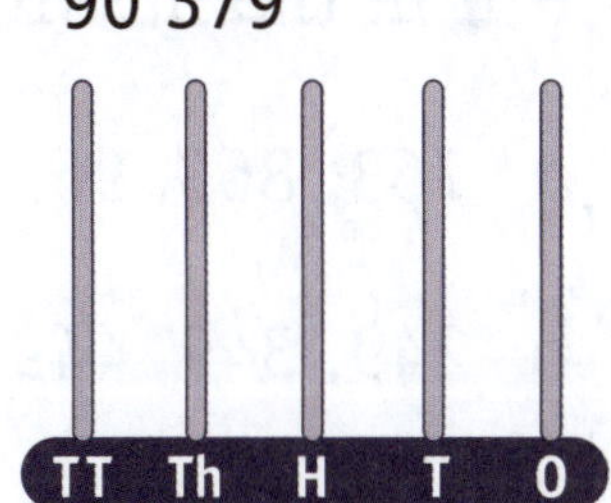

12 Write these numbers in ascending order.

a 325, 632, 149, 836, 473

____

b 1432, 248, 632, 2593, 3025

____

c 631, 136, 316, 361, 103

____

d 13 524, 27 626, 10 425, 35 927, 41 341

____

13 Write these numbers in descending order.

a 763, 1249, 62 538, 34 273, 8439

____

b 42 438, 84 327, 74 283, 14 327, 32 495

____

c 28 434, 44 382, 34 284, 38 428, 84 823

____

d 63 294, 49 236, 36 492, 24 936, 34 692

____

REVIEW

14 Fill in the missing numbers.

a 763, 863, 963, ________, ________, ________, ________

b 249, 349, 449, ________, ________, ________, ________

c 932, 832, 732, ________, ________, ________, ________

d 1147, 1047, 947, ________, ________, ________, ________

15 Write the correct symbol: > < or =

a 347 ☐ 293

b 1250 ☐ 1205

c 7646 ☐ 7646

d 24 937 ☐ 29 473

e 8932 ☐ 2938

f 63 241 ☐ 61 243

g 11 345 ☐ 54 311

h 76 201 ☐ 72 106

i 13 254 ☐ 13 254

j 92 374 ☐ 97 234

k 84 921 ☐ 81 924

l 74 328 ☐ 83 274

16 Round the numbers to complete the table.

	Number	Nearest 10	Nearest 100	Nearest 1000	Nearest 10 000
a	13 427				
b	49 238				
c	56 502				
d	60 348				
e	99 037				
f	74 295				
g	86 697				

CATCH UP MATHS YEAR 4 BOOK A © PASCAL PRESS ISBN: 9781925726145

17 Write the largest and smallest numbers using the digits.

	Digits	Smallest number	Largest number
a	6 2 4		
b	7 3 2		
c	8 6 3		
d	1 3 8 7		
e	9 4 8 5		
f	4 5 6 0		
g	5 6 2 0 3		
h	7 1 0 4 9		
i	8 5 6 7 4		

18 Circle all the odd numbers with red and even numbers with green.

4 1343 243 73 248 16 741 51 424 58

634 8915 2462 21 18 71245 838 6410

257 3 25 245 93 45 933 93 001 7

19 Circle all the one-digit numbers with red
two-digit numbers with blue
three-digit numbers with green
four-digit numbers with purple
five-digit numbers with orange.

27 492 9 24 293 41 321 8211 35 3524

82 2 7363 78 920 67 341 13 8 164

40 000 613 3711 7 4940 98 710 103

ADDING THREE OR MORE SINGLE-DIGIT NUMBERS

A quick and easy method of adding three numbers is to first find the numbers that add to 10.

Example 1:

2 + 3 + 8

= 2 + 8 + 3

= 10 + 3

= 13

Example 2:

7 + 3 + 5

= 7 + 3 + 5

= 10 + 5

= 15

Check your answer on the video!

Example 3:

1 + 7 + 9

= 1 + __ + 7

= ___ + 7

= ___

Example 4:

4 + 3 + 6

= 4 + 6 + __

= ___ + __

= ___

Your turn

Add by first finding the numbers that add to 10.

7 + 2 + 8

= 2 + 8 + 7

= 10 + 7

= 17

a 6 + 4 + 5

= ___

= ___

= ___

b 1 + 8 + 9

= ___

= ___

= ___

c 6 + 3 + 7

= ___

= ___

= ___

d 5 + 5 + 4

= ___

= ___

= ___

e 1 + 9 + 1

= ___

= ___

= ___

SELF CHECK Tick how you feel

Got it!	Need help...	I don't get it

Check your answers

How many did you get correct?

CATCH UP MATHS YEAR 4 BOOK A © PASCAL PRESS ISBN: 9781925726145

PRACTICE

1 Add these numbers by first finding the numbers that add to 10.

● 9 + 3 + 1

= 9 + 1 + 3

= 10 + 3

= 13

a 2 + 7 + 8

= ______

= ______

= ______

b 4 + 8 + 6

= ______

= ______

= ______

c 3 + 8 + 7

= ______

= ______

= ______

d 9 + 7 + 1

= ______

= ______

= ______

e 5 + 9 + 5

= ______

= ______

= ______

2 Now try these additions with four numbers.

● 4 + 3 + 6 + 2

= 4 + 6 + 3 + 2

= 10 + 5

= 15

a 6 + 5 + 2 + 5

= ______

= ______

= ______

b 9 + 4 + 1 + 3

= ______

= ______

= ______

c 5 + 6 + 5 + 4

= ______

= ______

= ______

d 2 + 3 + 4 + 8

= ______

= ______

= ______

e 1 + 4 + 6 + 4

= ______

= ______

= ______

RELATING ADDITION TO SUBTRACTION

Addition and subtraction are opposite actions.

A fact family is a group of facts about three numbers that are related to each other.

Here are four facts about the three numbers 4, 6 and 10:

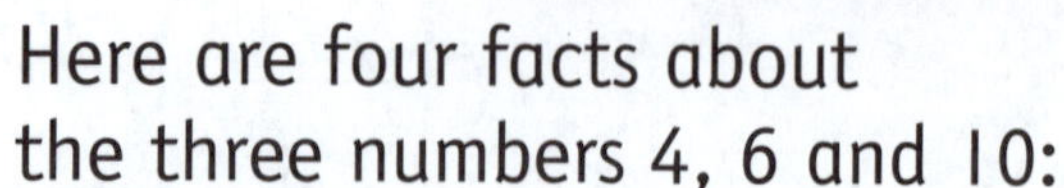

Fact 1	4 + 6 = 10	**Addition facts**
Fact 2	6 + 4 = 10	
Fact 3	10 − 6 = 4	**Subtraction facts**
Fact 4	10 − 4 = 6	

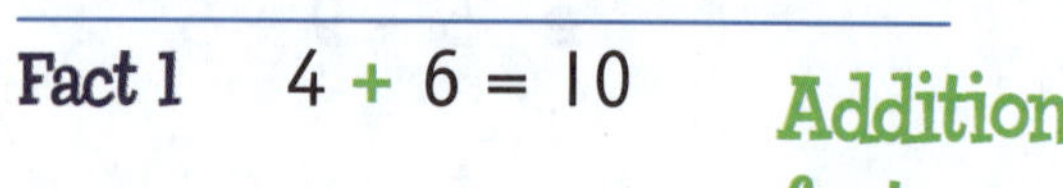

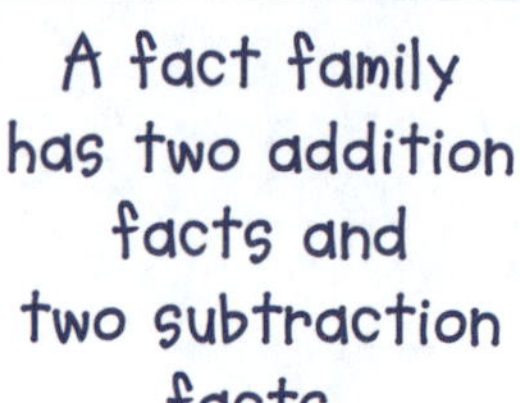

Example 1:
Write the fact family.

(7) (13) (20)

Fact 1 7 + 13 = 20

Fact 2 13 + 7 = 20

Fact 3 20 − 13 = 7

Fact 4 20 − 7 = 13

Example 2:
Write the fact family.

Fact 1 11 + ___ = 17

Fact 2 ___ + 11 = 17

Fact 3 17 − ___ = 11

Fact 4 17 − ___ = 6

Your turn

Write the fact family for each group of numbers.

Fact 1 16 + 4 = 20

Fact 2 4 + 16 = 20

Fact 3 20 − 4 = 16

Fact 4 20 − 16 = 4

a (5) (3) (8)

Fact 1 ___ + ___ = ___

Fact 2 ___ + ___ = ___

Fact 3 ___ − ___ = ___

Fact 4 ___ − ___ = ___

Check your answers
How many did you get correct?

CATCH UP MATHS YEAR 4 BOOK A © PASCAL PRESS ISBN: 9781925726145

PRACTICE

1 Fill in the missing numbers.

● 16 + 4 = 20
20 − 4 = 16

a 9 + 11 = 20
20 − ___ = 9

b 7 + 3 = 10
10 − ___ = 7

c 15 + 5 = 20
20 − ___ = 15

d 19 + 1 = 20
20 − 1 = ___

e 5 + 5 = 10
___ − 5 = 5

2 Complete these facts.

● 14 + 3 = 17
3 + 14 = 17
17 − 3 = 14
17 − 14 = 3

a 16 + 8 = 24
___ + 16 = 24
___ − 8 = 16
24 − ___ = 8

b 18 + 7 = 25
___ + 18 = 25
25 − ___ = 18
___ − 18 = 7

c 13 + 6 = 19
6 + ___ = 19
19 − 13 = ___
19 − ___ = 13

d 17 + 9 = 26
___ + 17 = 26
26 − 17 = ___
26 − 9 = ___

e 15 + 6 = 21
6 + ___ = 21
21 − ___ = 15
21 − 15 = ___

f 12 + 4 = 16
___ + ___ = ___
___ − ___ = ___
___ − ___ = ___

g 19 + 4 = 23
___ + ___ = ___
___ − ___ = ___
___ − ___ = ___

h 11 + 9 = 20
___ + ___ = ___
___ − ___ = ___
___ − ___ = ___

3 Write the fact family for each set of numbers.

● 5 6 11
5 + 6 = 11
6 + 5 = 11
11 − 6 = 5
11 − 5 = 6

a 3 9 12

◯ + ◯ = ◯
◯ + ◯ = ◯
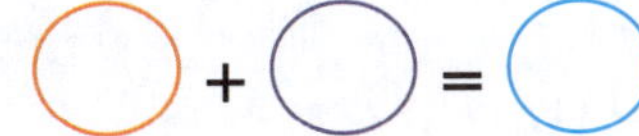
◯ − ◯ = ◯

◯ − ◯ = ◯

b 2 18 20

◯ + ◯ = ◯

◯ + ◯ = ◯

◯ − ◯ = ◯

◯ − ◯ = ◯

c 2 15 17

___ + ___ = ___
___ + ___ = ___
___ − ___ = ___
___ − ___ = ___

e 8 16 24

___ + ___ = ___
___ + ___ = ___
___ − ___ = ___
___ − ___ = ___

g 15 33 48

___ + ___ = ___
___ + ___ = ___
___ − ___ = ___
___ − ___ = ___

d 7 11 18

___ + ___ = ___
___ + ___ = ___
___ − ___ = ___
___ − ___ = ___

f 14 22 36

___ + ___ = ___
___ + ___ = ___
___ − ___ = ___
___ − ___ = ___

h 6 46 52

___ + ___ = ___
___ + ___ = ___
___ − ___ = ___
___ − ___ = ___

4 Now write the fact family for each of these sets of numbers.

24 16 40

16 + 24 = 40
24 + 16 = 40
40 − 24 = 16
40 − 16 = 24

b 15 13 28

___ + ___ = ___
___ + ___ = ___
___ − ___ = ___
___ − ___ = ___

a 19 18 37

___ + ___ = ___
___ + ___ = ___
___ − ___ = ___
___ − ___ = ___

c 15 41 56

___ + ___ = ___
___ + ___ = ___
___ − ___ = ___
___ − ___ = ___

CATCH UP MATHS YEAR 4 BOOK A © PASCAL PRESS ISBN: 9781925726145

d 59 18 41

☐ + ☐ = ☐
☐ + ☐ = ☐
☐ − ☐ = ☐
☐ − ☐ = ☐

e 22 73 51

☐ + ☐ = ☐
☐ + ☐ = ☐
☐ − ☐ = ☐
☐ − ☐ = ☐

f 36 69 33

☐ + ☐ = ☐
☐ + ☐ = ☐
☐ − ☐ = ☐
☐ − ☐ = ☐

g 73 52 21

☐ + ☐ = ☐
☐ + ☐ = ☐
☐ − ☐ = ☐
☐ − ☐ = ☐

h 63 21 84

☐ + ☐ = ☐
☐ + ☐ = ☐
☐ − ☐ = ☐
☐ − ☐ = ☐

i 26 152 126

☐ + ☐ = ☐
☐ + ☐ = ☐
☐ − ☐ = ☐
☐ − ☐ = ☐

j 115 247 132

☐ + ☐ = ☐
☐ + ☐ = ☐
☐ − ☐ = ☐
☐ − ☐ = ☐

k 365 203 162

☐ + ☐ = ☐
☐ + ☐ = ☐
☐ − ☐ = ☐
☐ − ☐ = ☐

ADDITION WITHOUT TRADING

Addition is when two or more numbers are combined to make one larger number.

Example 1: Add.

	Tens	Ones
	2	4
+	1	5
	3	**9**

Example 2: Add.

	Tens	Ones
	3	5
+	2	3
	5	**8**

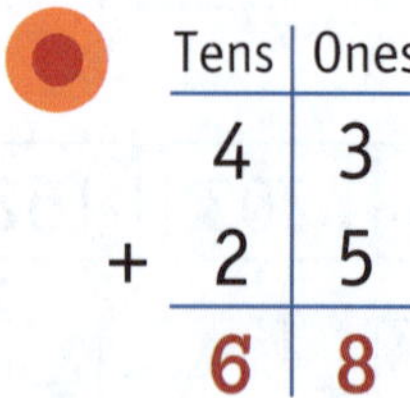

Example 3: Add.

	Tens	Ones
	6	2
+	3	5

Check your answer on the video!

Example 4: Add.

	Tens	Ones
	8	4
+	1	3

Solve.

●

	Tens	Ones
	4	3
+	2	5
	6	**8**

a

	Tens	Ones
	5	2
+	4	0

b

	Tens	Ones
	7	3
+	1	5

c

	Tens	Ones
	6	1
+	2	2

d

	Tens	Ones
	8	6
+	1	1

e

	Tens	Ones
	7	5
+	2	3

SELF CHECK Tick how you feel

Got it!	Need help...	I don't get it
 ☐	 ☐	 ☐

Check your answers

How many did you get correct? ☐

CATCH UP MATHS YEAR 4 BOOK A © PASCAL PRESS ISBN: 9781925726145

1 Solve these additions.

Example

	Tens	Ones
	5	3
+	4	1
	9	4

a

	Tens	Ones
	1	5
+	2	4

b

	Tens	Ones
	3	2
+	2	5

c

	Tens	Ones
	6	8
+	2	1

d

	Tens	Ones
	7	5
+	1	1

e

	Tens	Ones
	8	3
+		4

f

	Tens	Ones
	5	6
+	4	2

g

	Tens	Ones
	2	8
+	5	1

2 Complete the additions.

	Tens	Ones
	4	2
+	5	6
	9	8

a

	Tens	Ones
	3	5
+	5	1

b

	Tens	Ones
	2	4
+	5	5

c

	Tens	Ones
	3	7
+	4	2

d

	Tens	Ones
	4	3
+	5	3

e

	Tens	Ones
	5	6
+	2	1

f

	Tens	Ones
	4	7
+	4	1

g

	Tens	Ones
	3	0
+	6	5

h

	Tens	Ones
	4	7
+	3	2

i

	Tens	Ones
	3	7
+	1	0

j

	Tens	Ones
	6	3
+	2	5

k

	Tens	Ones
	5	3
+	3	5

l

	Tens	Ones
	3	1
+	2	1

m

	Tens	Ones
	5	4
+	4	3

n

	Tens	Ones
	7	3
+	1	3

o

	Tens	Ones
	4	5
+	5	0

p

	Tens	Ones
	3	6
+	3	1

q

	Tens	Ones
	4	7
+	3	2

r

	Tens	Ones
	5	9
+	4	0

s

	Tens	Ones
	9	3
+		2

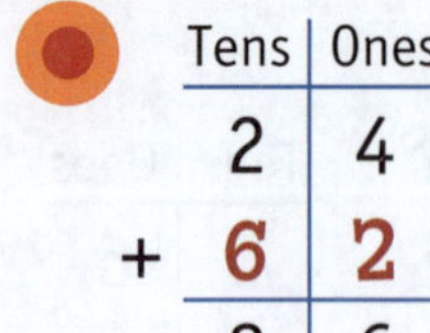

3 Fill in the missing numbers.

● (example)

	Tens	Ones
	2	4
+	6	2
	8	6

a

	Tens	Ones
	4	1
+		
	9	3

b

	Tens	Ones
	6	8
+		
	9	9

c

	Tens	Ones
	5	3
+		
	9	8

d

	Tens	Ones
	7	2
+		
	9	5

e

	Tens	Ones
	2	1
+		
	4	1

f

	Tens	Ones
	3	0
+		
	5	7

g

	Tens	Ones
	4	9
+		
	6	9

4 Round each number to the nearest 10 to estimate an answer for these number sentences.

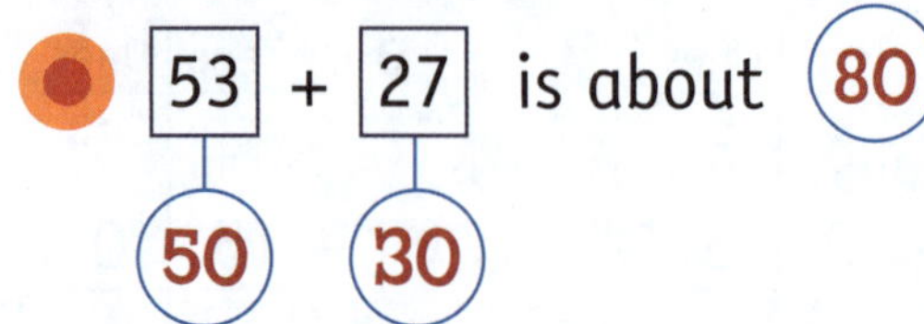

● 53 + 27 is about 80 (50, 30)

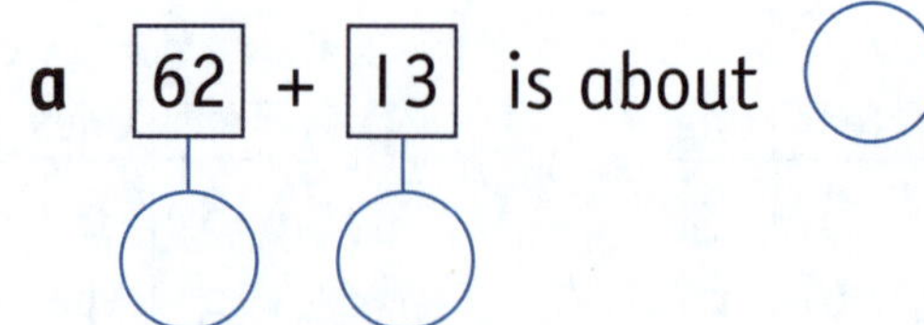

a 62 + 13 is about ◯

b 78 + 24 is about ◯

c 81 + 19 is about ◯

d 73 + 24 is about ◯

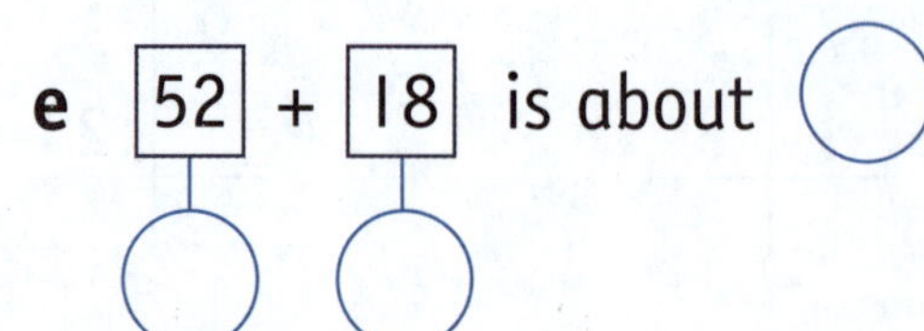

e 52 + 18 is about ◯

f 72 + 15 is about ◯

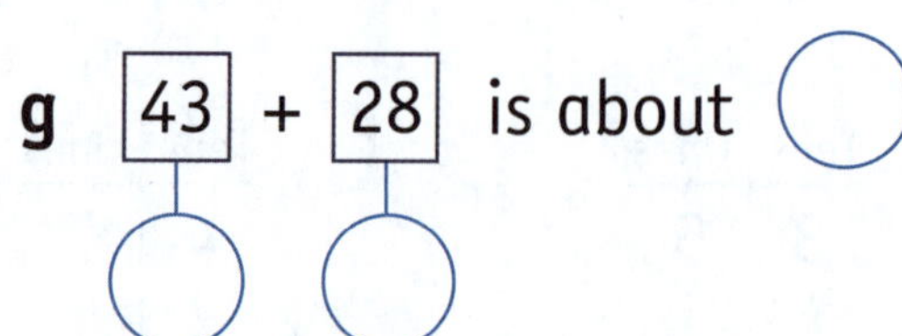

g 43 + 28 is about ◯

h 56 + 31 is about ◯

i 36 + 42 is about ◯

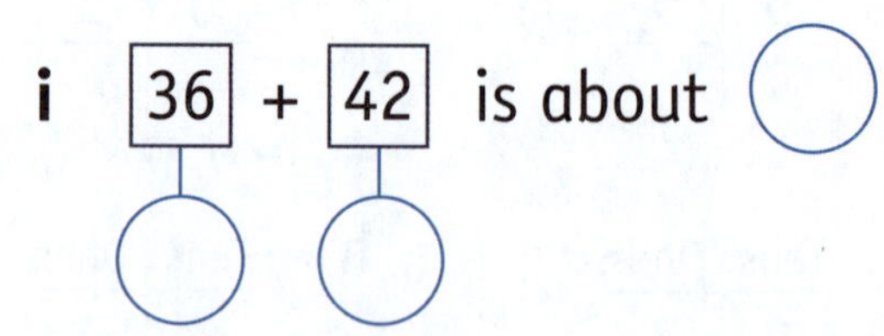

j 41 + 17 is about ◯

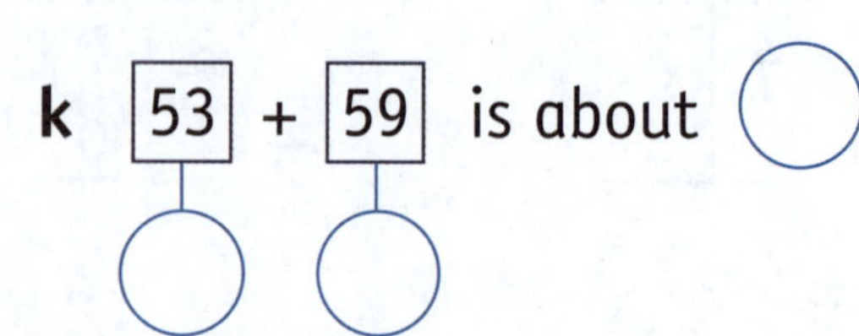

k 53 + 59 is about ◯

CATCH UP MATHS YEAR 4 BOOK A © PASCAL PRESS ISBN: 9781925726145

5 Solve these three-digit additions.

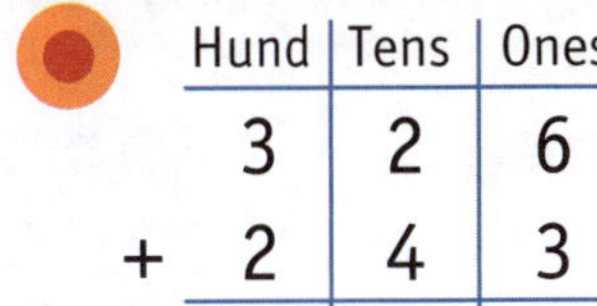

	Hund	Tens	Ones
	3	2	6
+	2	4	3
	5	6	9

a

	Hund	Tens	Ones
	8	1	5
+	1	6	3

b

	Hund	Tens	Ones
	2	2	7
+	7	5	2

c

	Hund	Tens	Ones
	1	7	0
+	5	1	5

d

	Hund	Tens	Ones
	6	4	2
+	3	4	5

e

	Hund	Tens	Ones
	7	2	3
+	2	5	6

f

	Hund	Tens	Ones
	1	7	4
+	5	2	0

g

	Hund	Tens	Ones
	1	3	8
+	4	2	1

h

	Hund	Tens	Ones
	1	3	6
+	8	5	2

6 Fill in the missing numbers.

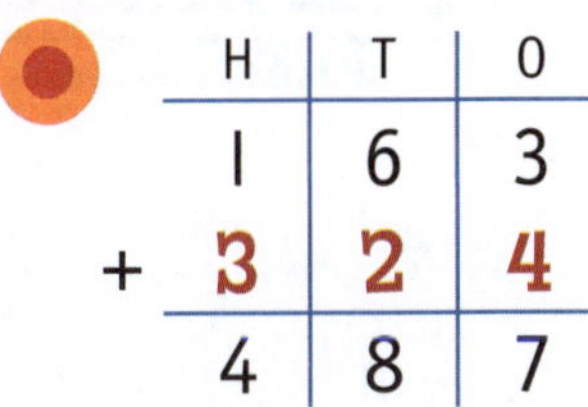

	H	T	O
	1	6	3
+	3	2	4
	4	8	7

a

	H	T	O
	3	1	0
+			
	5	9	0

b

	H	T	O
	6	2	4
+			
	8	5	7

c

	H	T	O
	5	2	3
+			
	7	5	4

d

	H	T	O
	4	3	8
+			
	5	5	9

e

	H	T	O
	4	2	6
+			
	9	9	8

7 Round each number to the nearest 10 to estimate an answer for these number sentences.

137 + 28 is about 170
140 30

a 256 + 17 is about ___

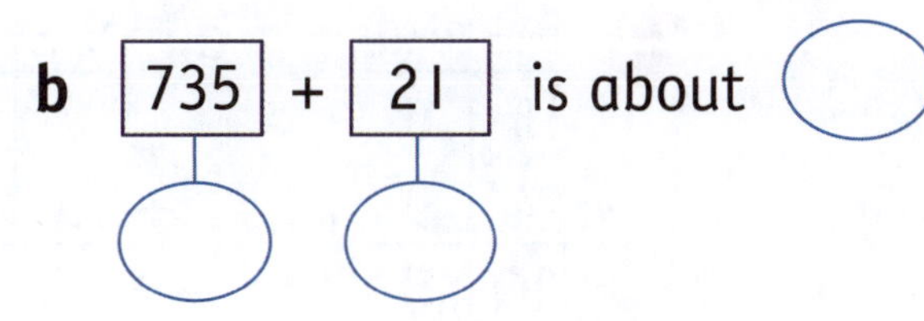

b 735 + 21 is about ___

c 589 + 12 is about ___

ADDITION WITH TRADING

Addition is when two or more numbers are combined to make one larger number.

In vertical addition, sometimes the sum of two digits in the same place value column is 10 or more.

Because there is only space for one digit, the 10 has to be traded with (or carried over to) the next column.

Example 1: Add.

	Tens	Ones
	[1] 7	5
+	5	8
	13	**3**

Example 2: Add.

	Tens	Ones
	[1] 3	7
+	5	9
	9	**8**

Example 3: Add.

	Tens	Ones
	[1] 1	5
+	7	6
		1

Example 4: Add.

	Tens	Ones
	2	9
+	8	3

Your turn

Complete these additions.

Example:

	Tens	Ones
	[1] 5	6
+	4	9
	10	**5**

a

	Tens	Ones
	7	8
+	9	5

b

	Tens	Ones
	5	7
+	3	4

c

	Tens	Ones
	6	2
+	4	8

SELF CHECK Tick how you feel

Got it!	Need help...	I don't get it
☐	☐	☐

Check your answers

How many did you get correct? ☐

CATCH UP MATHS YEAR 4 BOOK A © PASCAL PRESS ISBN: 9781925726145

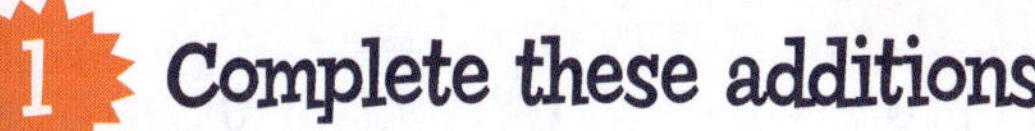

PRACTICE

1 Complete these additions.

Example

	Tens	Ones
	[1]5	5
+	4	9
	10	4

a

	Tens	Ones
	6	9
+	4	5

b

	Tens	Ones
	8	7
+	3	6

c

	Tens	Ones
	4	7
+	3	9

d

	Tens	Ones
	7	2
+	1	9

e

	Tens	Ones
	2	6
+	3	5

f

	Tens	Ones
	8	4
+	5	9

g

	Tens	Ones
	9	3
+	4	8

2 Now try these.

Example

	T	O
	[1]1	5
+	6	5
	8	0

a

	T	O
	1	4
+	5	6

b

	T	O
	9	7
+	2	3

c

	T	O
	8	8
+	3	9

d

	T	O
	3	6
+	9	7

e

	T	O
	7	2
+	6	8

f

	T	O
	5	4
+	8	9

g

	T	O
	3	9
+	7	8

h

	T	O
	4	8
+	9	8

i

	T	O
	8	9
+	4	9

j

	T	O
	3	9
+	7	2

k

	T	O
	3	8
+	9	4

l

	T	O
	5	9
+	2	5

m

	T	O
	8	3
+	1	7

n

	T	O
	4	9
+	2	5

o

	T	O
	5	5
+	4	8

p

	T	O
	3	6
+	8	7

q

	T	O
	9	4
+	8	8

r

	T	O
	5	4
+	2	8

s

	T	O
	1	7
+	6	7

  ISBN: 9781925726145

3 Fill in the missing numbers.

Example

	T	O
	¹5	6
+	3	4
	9	0

a

	T	O
	7	7
+		
	11	1

b

	T	O
	8	4
+		
	13	1

c

	T	O
	9	6
+		
	15	0

d

	T	O
	5	8
+		
	10	7

e

	T	O
	9	2
+		
	13	0

f

	T	O
	8	6
+		
	13	1

g

	T	O
	7	4
+		
	14	1

h

	T	O
	4	8
+		
	13	4

i

	T	O
	3	7
+		
	6	2

j

	T	O
	5	8
+		
	11	7

k

	T	O
	2	9
+		
	4	2

l

	T	O
	1	9
+		
	3	4

m

	T	O
	3	5
+		
	9	2

n

	T	O
	4	7
+		
	14	0

o

	T	O
	5	9
+		
	12	4

p

	T	O
	9	9
+		
	14	5

q

	T	O
	3	4
+		
	9	2

r

	T	O
	5	6
+		
	11	1

s

	T	O
	2	8
+		
	6	4

4 Complete these three-digit additions.

Example

	H	T	O
	¹4	¹7	3
+	4	2	8
	9	0	1

a

	H	T	O
	4	8	5
+	2	7	6

b

	H	T	O
	5	3	8
+	4	5	2

c

	H	T	O
	2	8	7
+	1	3	9

d

	H	T	O
	4	9	3
+	3	5	8

e

	H	T	O
	2	3	7
+	4	5	9

CATCH UP MATHS YEAR 4 BOOK A © PASCAL PRESS ISBN: 9781925726145

f

	H	T	O
	8	4	5
+	1	3	6

g

	H	T	O
	2	5	6
+	6	8	9

h

	H	T	O
	1	5	3
+	3	2	9

i

	H	T	O
	2	5	8
+	3	1	4

j

	H	T	O
	6	2	4
+	3	5	6

k

	H	T	O
	7	5	8
+	1	4	3

5 Fill in the missing numbers.

●

	H	T	O
	5	[1]4	7
+	3	0	7
	8	5	4

a

	H	T	O
	6	4	3
+			
	8	5	2

b

	H	T	O
	8	2	6
+		3	
	9	6	1

c

	H	T	O
	7	4	4
+		1	
	9	6	2

d

	H	T	O
	5	2	9
+	2		
	7	4	7

e

	H	T	O
	3	8	9
+		1	
	6	0	8

f

	H	T	O
	4	2	8
+		4	
	9	7	6

g

	H	T	O
	1	9	4
+		6	
	9	6	3

h

	H	T	O
	2	6	8
+	3		
	6	3	7

i

	H	T	O
	6	3	2
+			8
	9	3	0

j

	H	T	O
			9
+	1	3	7
	2	4	6

k

	H	T	O
		8	
+	6	5	7
	8	4	3

l

	H	T	O
			8
+	1	4	7
	6	7	5

m

	H	T	O
	8	6	8
+		0	
	9	7	4

n

	H	T	O
	3		
+	5	1	7
	8	8	4

JUMP STRATEGY TO SOLVE ADDITION

You use a number line to solve addition with the jump strategy. Start with the bigger number and jump up the number line.

Example 1: 38 + (47) = ____

start number

38
3 tens 8 ones

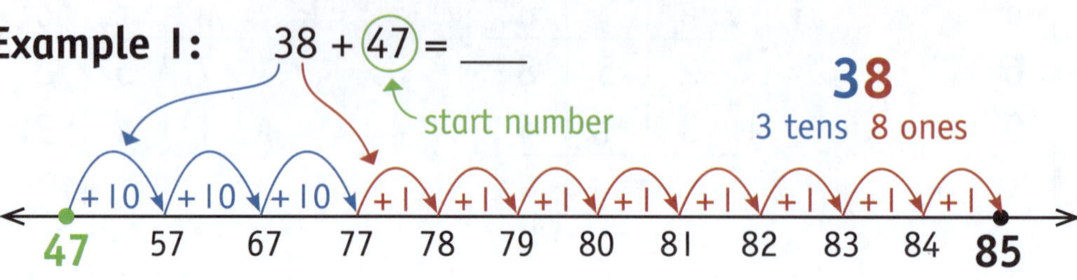

38
3 big 'tens' jumps ↗ ↖ 8 little 'ones' jumps

So, 38 + 47 = 85

Example 2: (43) + 24 = ____

start number

24
2 tens 4 ones

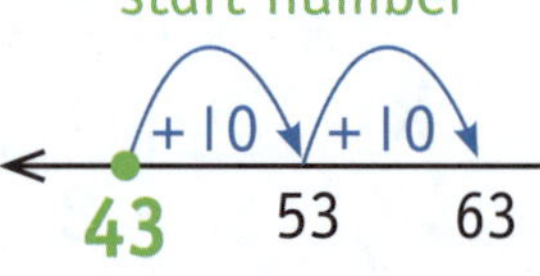

+10 +10
43 53 63

24
__ big 'tens' jumps ↗ ↖ __ little 'ones' jumps

So, 43 + 24 = ____

Your turn

Use the number lines to solve these additions.

● 58 + 34 = 92

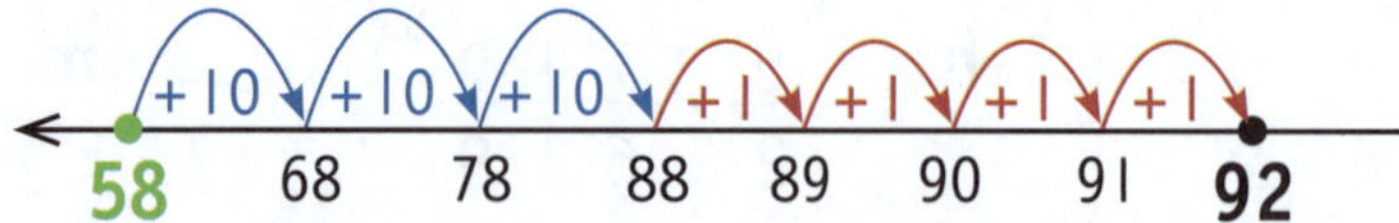

a 62 + 32 = ____

SELF CHECK Tick how you feel

Got it!	Need help...	I don't get it
 ☐	 ☐	 ☐

Check your answers
How many did you get correct? ☐

CATCH UP MATHS YEAR 4 BOOK A © PASCAL PRESS ISBN: 9781925726145

PRACTICE

Solve these additions using the number lines.

● 72 + 44 = 116

a 38 + 25 = ____

b 26 + 59 = ____

c 38 + 33 = ____

d 74 + 26 = ____

e 93 + 25 = ____

f 43 + 28 = ____

g 63 + 75 = ____

h 72 + 15 = ____

JUMP STRATEGY WITH TENS AND HUNDREDS

The jump strategy can also be used to solve addition with bigger numbers.

SCAN to watch video

Example 1: What is 734 + 65?

734 + 65

start number · 6 big 'tens' jumps · 5 little 'ones' jumps

65: 6 tens 5 ones

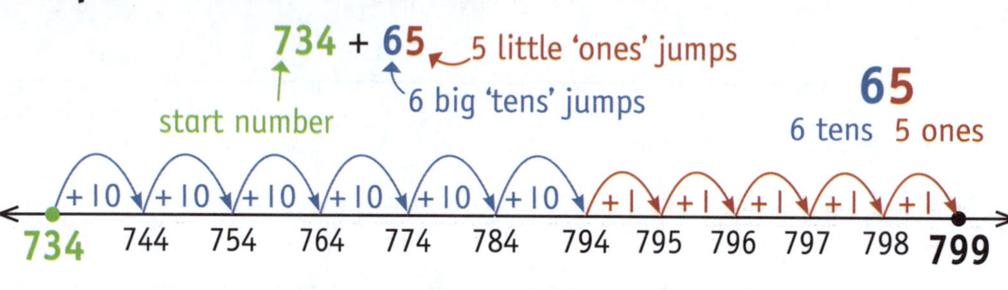

So, 734 + 65 = 799

Example 2: What is 584 + 43?

584 + 43

start number

43: 4 tens 3 ones

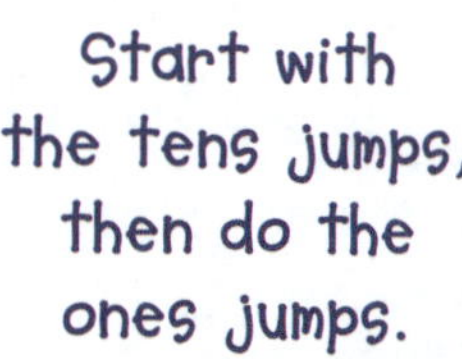

+10 +10 +10 +10

584 594 604 614 624

So, 584 + 43 = ______

Use the number lines to solve these additions.

- 538 + 43 = 581

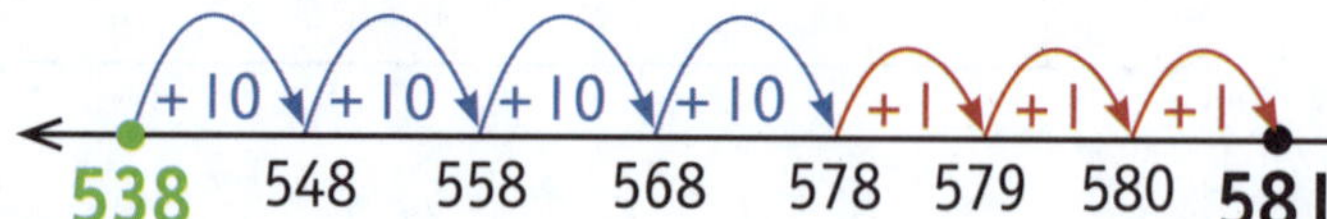

a 647 + 48 = ______

Check your answers

How many did you get correct?

CATCH UP MATHS YEAR 4 BOOK A © PASCAL PRESS ISBN: 9781925726145

PRACTICE

1 Using the number lines, solve the additions.

623 + 24 = 647

a 741 + 52 = ____

b 331 + 54 = ____

c 439 + 13 = ____

d 234 + 73 = ____

e 914 + 58 = ____

f 412 + 65 = ____

g 562 + 38 = ____

h 621 + 55 = ____

JUMP STRATEGY WITH THREE- AND FOUR-DIGIT NUMBERS

You can use the jump strategy to solve the addition of numbers in the hundreds and thousands.

Example 1: 285 + 134 = 419

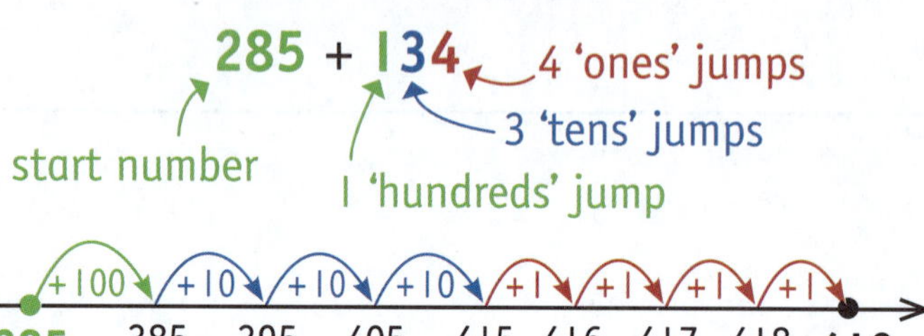

Start with the hundreds jumps, then do the tens jumps, then the ones jumps.

Example 2: 5381 + 562 = 5943

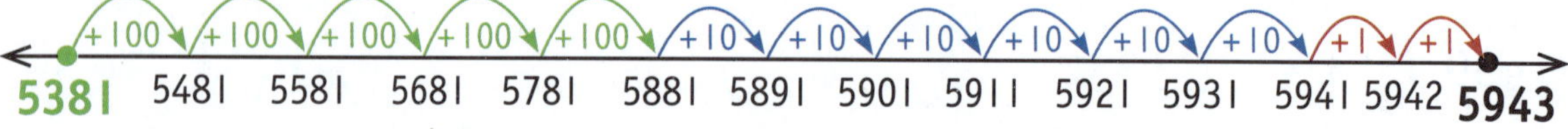

Example 3: 4325 + 321 = ______

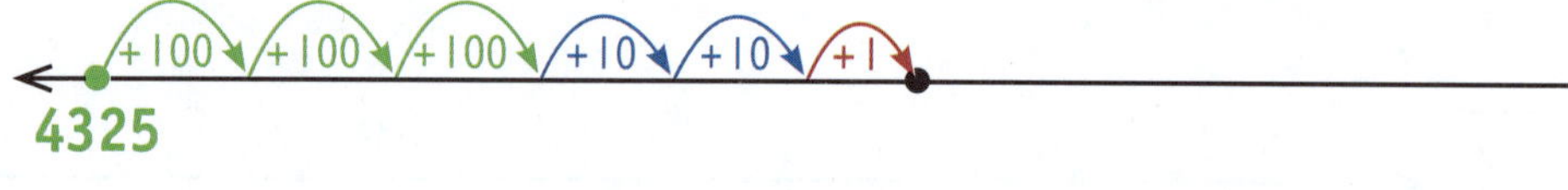

Solve the additions using the number lines.

● 495 + 173 = 668

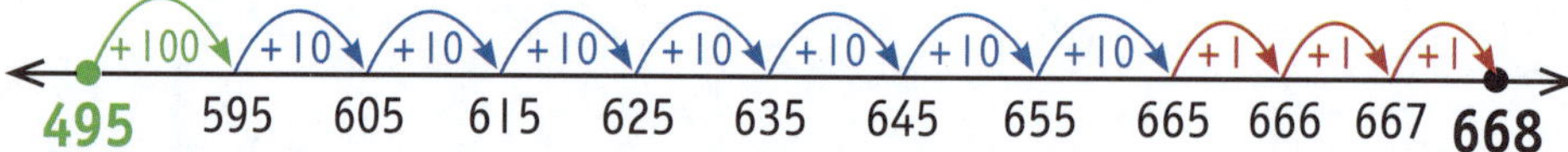

a 379 + 248 = ______

b 4386 + 341 = ______

SELF CHECK Tick how you feel		
Got it! ☐	Need help... ☐	I don't get it ☐

Check your answers
How many did you get correct? ☐

CATCH UP MATHS YEAR 4 BOOK A © PASCAL PRESS ISBN: 9781925726145

PRACTICE

1 Solve using the number lines

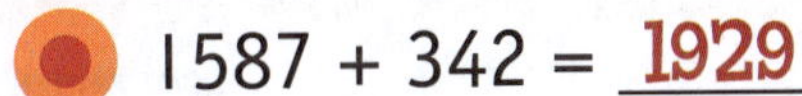

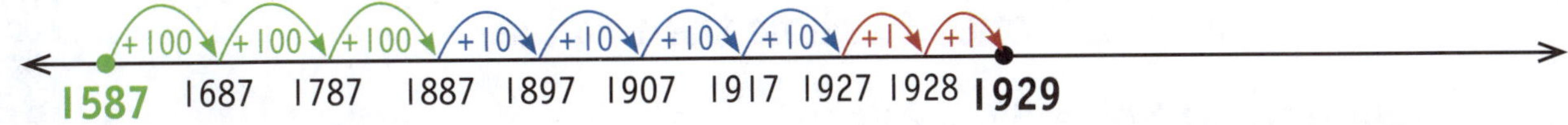

a 343 + 146 = ______

b 726 + 238 = ______

c 563 + 372 = ______

d 832 + 442 = ______

e 2364 + 128 = ______

f 5632 + 253 = ______

g 7520 + 346 = ______

h 6735 + 464 = ______

SPLIT STRATEGY TO ADD TWO-DIGIT NUMBERS

The split strategy splits the numbers into tens and ones to show their value.

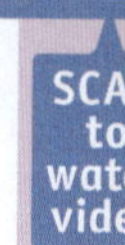

Example 1: Solve 57 + 21.

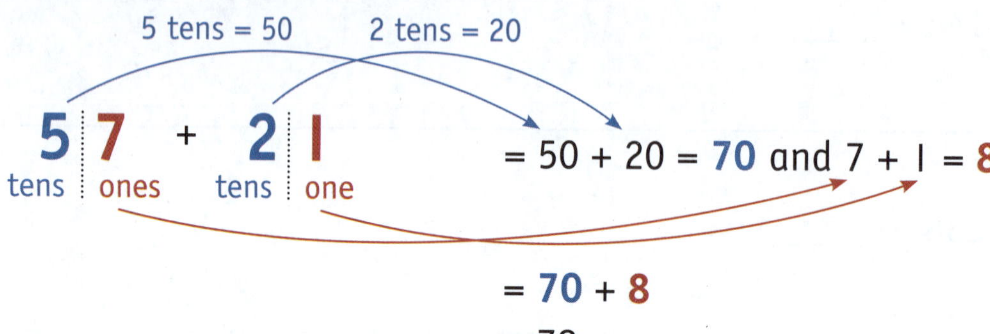

= 50 + 20 = **70** and 7 + 1 = **8**

= **70** + **8**

= 78

Example 2: Solve 38 + 41.

3 tens | 8 ones + 4 tens | 1 one

= 30 + 40 = **70** and 8 + 1 = **9**

= **70** + **9**

= 79

Example 3: Solve 64 + 35.

6 tens | 4 ones + 3 tens | 5 ones

= ___ + ___ = **90** and __ + __ = **9**

= ___ + ___

= ___

Solve the additions using the split strategy.

● 43 + 26 = 40 + 20 = 60 and 3 + 6 = 9

= 60 + 9

= 69

a 52 + 33 = ___ + ___ = ___ and __ + __ = __

= ___ + __

= ___

SELF CHECK Tick how you feel

Got it!	Need help...	I don't get it
☐	☐	☐

Check your answers

How many did you get correct? ☐

CATCH UP MATHS YEAR 4 BOOK A © PASCAL PRESS ISBN: 9781925726145

1 Solve the additions using the split strategy.

36 + 23 = 30 + 20 = 50 and 6 + 3 = 9
= 50 + 9
= 59

a 58 + 11 = ___ + ___ = ___ and __ + __ = __
= ___ + __
= ___

b 63 + 24 = ___ + ___ = ___ and __ + __ = __
= ___ + __
= ___

c 81 + 14 = ___ + ___ = ___ and __ + __ = __
= ___ + __
= ___

d 75 + 23 = ___ + ___ = ___ and __ + __ = __
= ___ + __
= ___

e 81 + 15 = ___ + ___ = ___ and __ + __ = __
= ___ + __
= ___

f 43 + 25 = ___ + ___ = ___ and __ + __ = __
= ___ + __
= ___

g 52 + 36 = ___ + ___ = ___ and __ + __ = __
= ___ + __
= ___

SPLIT STRATEGY TO ADD THREE-DIGIT NUMBERS

You can use the split strategy to add larger numbers.

Example 1:

712 + 63 = 700 and 10 + 60 = 70 and 2 + 3 = 5

= 700 + 70 + 5

= 775

Example 2:

543 + 51 = 500 and 40 + 50 = 90 and 3 + 1 = 4

= 500 + 90 + 4

= 594

Example 3:

832 + 43 = ____ and 30 + ___ = ___ and 2 + __ = __

= ____ + ___ + __

= ____

Solve these addition problems using the split strategy.

- 257 + 42 = 200 and 50 + 40 = 90 and 7 + 2 = 9

= 200 + 90 + 9

= 299

a 314 + 35 = ____ and ___ + ___ = ___ and __ + __ = __

= ____ + ___ + __

= ____

b 715 + 23 = ____ and ___ + ___ = ___ and __ + __ = __

= ____ + ___ + __

= ____

SELF CHECK Tick how you feel		
Got it! ☐	Need help... ☐	I don't get it ☐

Check your answers
How many did you get correct? ☐

PRACTICE

1 **Solve these addition problems using the split strategy.**

526 + 31 = 500 and 20 + 30 = 50 and 6 + 1 = 7
= 500 + 50 + 7
= 557

a 432 + 43 = ____ and ___ + ___ = ___ and __ + __ = __
= ____ + ___ + __
= ____

b 472 + 17 = ____ and ___ + ___ = ___ and __ + __ = __
= ____ + ___ + __
= ____

c 541 + 35 = ____ and ___ + ___ = ___ and __ + __ = __
= ____ + ___ + __
= ____

d 632 + 47 = ____ and ___ + ___ = ___ and __ + __ = __
= ____ + ___ + __
= ____

e 706 + 52 = ____ and ___ + ___ = ___ and __ + __ = __
= ____ + ___ + __
= ____

f 902 + 83 = ____ and ___ + ___ = ___ and __ + __ = __
= ____ + ___ + __
= ____

SPLIT STRATEGY WITH THREE- AND FOUR-DIGIT NUMBERS

Even when adding three- and four-digit numbers, the split strategy can be used to work out the answer.

Example 1: 352 + 45

= 300 and 50 + 40 = 90 and 2 + 5 = 7

= 300 + 90 + 7

= 397

Example 2: 4532 + 354

= 4000 and 500 + 300 = 800 and 30 + 50 = 80 and 2 + 4 = 6

= 4000 + 800 + 80 + 6

= 4886

Example 3: 5283 + 114

= _____ and 200 + ____ = ____ and 80 + ___ = ___ and 3 + __ = __

= _____ + _____ + ___ + __

= _____

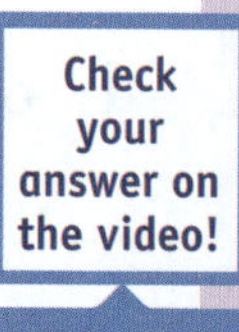

Your turn

Solve this addition problem using the split strategy.

5672 + 213

= _____ and ____ + ____ = ____ and ___ + ___ = ___ and __ + __ = __

= _____ + _____ + ___ + __

= _____

SELF CHECK Tick how you feel

Got it!	Need help...	I don't get it
 ☐	 ☐	 ☐

Check your answers

How many did you get correct? ☐

CATCH UP MATHS YEAR 4 BOOK A © PASCAL PRESS ISBN: 9781925726145

1 Solve these using the split strategy.

632 + 421

= 600 + 400 = 1000 and 30 + 20 = 50 and 2 + 1 = 3

= 1000 + 50 + 3

= 1053

a 541 + 33

= ______ and ____ + ____ = ____ and ___ + ___ = ___ and __ + __ = __

= ______ + ______ + ___ + __

= ______

b 727 + 32

= ______ and ____ + ____ = ____ and ___ + ___ = ___ and __ + __ = __

= ______ + ______ + ___ + __

= ______

c 1352 + 423

= ______ and ____ + ____ = ____ and ___ + ___ = ___ and __ + __ = __

= ______ + ______ + ___ + __

= ______

d 2468 + 221

= ______ and ____ + ____ = ____ and ___ + ___ = ___ and __ + __ = __

= ______ + ______ + ___ + __

= ______

e 8435 + 243

= ______ and ____ + ____ = ____ and ___ + ___ = ___ and __ + __ = __

= ______ + ______ + ___ + __

= ______

ROUNDING MONEY

The smallest Australian coin is 5 cents. So if the total cost of something purchased with cash ends in a number that is not 0 or 5, the cents are rounded.

Example 1: Round $15.32 to the nearest 5 cents.

Look at the last number.

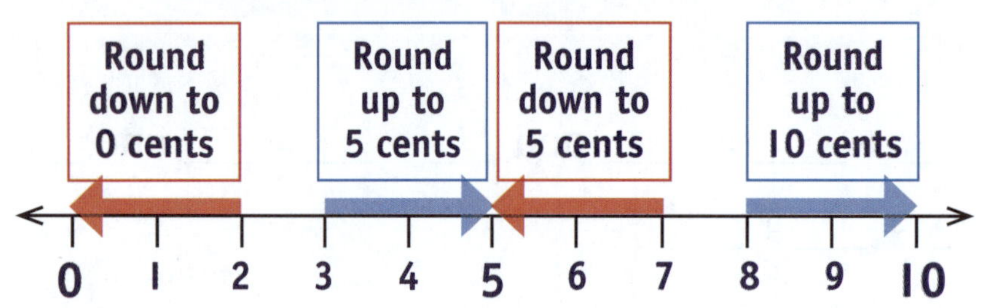

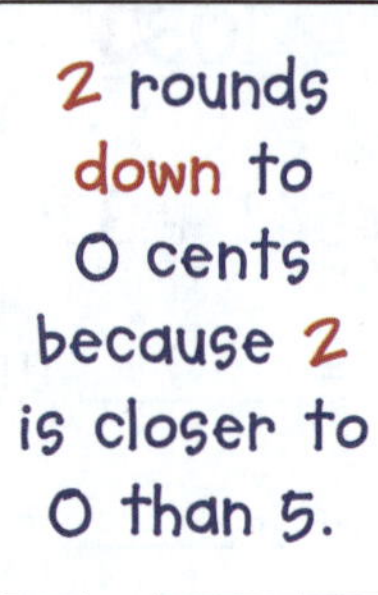

$15.32 rounded to the nearest 5 cents is $15.30.

Example 2: Round $24.87 to the nearest 5 cents.

Look at the last number.

The number 7 rounds down to 5 cents because 7 is closer to 5 than 10.

$24.87 rounded to the nearest 5 cents is $24.85.

Example 3: Round $42.33 to the nearest 5 cents.

$42.3(3) ← rounds up

____

Example 4: Round $67.46 to the nearest 5 cents.

$67.4(6) ←

____

Your turn

Round these amounts to the nearest 5 cents.

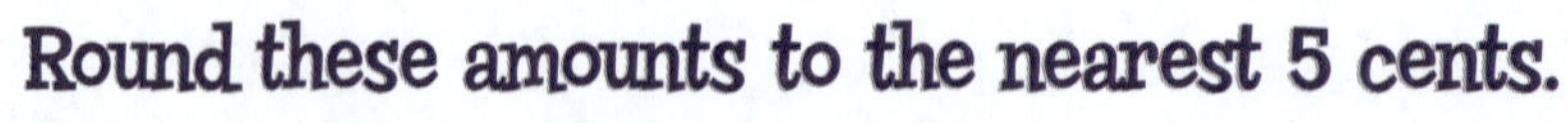

● $4.19	a $10.21	b $18.49	c $21.16
$4.20	________	________	________

Check your answers
How many did you get correct?

CATCH UP MATHS YEAR 4 BOOK A © PASCAL PRESS ISBN: 9781925726145

PRACTICE

1 Round these amounts to the nearest 5 cents.

●	$4.26 $4.25	d	$3.28	h	$9.89	l	$6.42
a	$7.27	e	$1.21	i	$4.69	m	$1.07
b	$9.09	f	$4.43	j	$3.78	n	$5.56
c	$4.16	g	$8.67	k	$2.54	o	$7.91

2 Round to the nearest 5 cents. Circle the correct answer.

●	$15.18	$15.20	$15.15	$15.10
a	$24.72	$24.75	$24.70	$24.80
b	$36.54	$36.50	$36.60	$36.55
c	$78.66	$78.65	$78.60	$78.70
d	$81.83	$81.80	$81.85	$81.90
e	$52.98	$52.90	$52.95	$53.00
f	$135.71	$135.70	$135.75	$135.80
g	$411.27	$411.20	$411.25	$411.30
h	$536.77	$536.70	$536.75	$536.80
i	$2742.44	$2742.40	$2742.45	$2742.50
j	$1352.52	$1352.50	$1352.55	$1352.60

Work out the change from $10.00 for each cost.

	Cost	Rounded to nearest 5 cents	Change from $10.00
●	$2.38	$2.40	$7.60
a	$5.17		
b	$7.49		
c	$8.63		
d	$4.88		
e	$3.31		
f	$0.78		
g	$1.26		
h	$4.61		
i	$6.52		
j	$9.57		

4 Complete the table.

	Price	Nearest 5 cents	Amount given	Change given
●	$2.58	$2.60	$10.00	$7.40
a	$0.76		$2.00	
b	$3.24		$5.00	
c	$15.49		$20.00	
d	$12.63		$20.00	
e	$7.32		$10.00	
f	$3.41		$5.00	
g	$1.26		$2.00	
h	$14.84		$20.00	
i	$3.76		$10.00	
j	$8.52		$20.00	

CATCH UP MATHS YEAR 4 BOOK A © PASCAL PRESS ISBN: 9781925726145

5 Solve these money problems.

- Magenta bought 3 pencils for $1.98 each. She gave $10.00.
How much change did she receive? (Round to the nearest 5 cents.)

$1.99 + $1.99 + $1.99 = $5.97

Change = $4.05

a Allan bought 4 pens for $2.47 each. He gave $20.00.
How much change did he receive? (Round to the nearest 5 cents.)

____

____

b Jade bought 3 books for $3.99 each. She gave $20.00.
How much change did she receive? (Round to the nearest 5 cents.)

____

____

c Fabio bought 6 peaches for 77 cents each. He gave $10.00.
How much change did he receive? (Round to the nearest 5 cents.)

____

____

6 At Eddy's shop every item costs $2.45. Complete the table.

	How many items bought	Total cost	Amount given	Change given
●	2	$4.90	$10.00	$5.10
a	1	$	$10.00	$
b	3	$	$10.00	$
c	4	$	$20.00	$
d	1	$	$5.00	$
e	2	$	$20.00	$
f	3	$	$20.00	$
g	5	$	$20.00	$
h	1	$	$20.00	$

ADDITION REVIEW

1 Solve these additions.

a 2 + 8 + 7

= ____________

= ________

= _____

b 3 + 6 + 4

= ____________

= ________

= _____

c 1 + 6 + 9

= ____________

= ________

= _____

d 1 + 7 + 3 + 8

= ________________

= ___________

= _____

e 9 + 4 + 1 + 4

= ________________

= ___________

= _____

f 6 + 1 + 3 + 4

= ________________

= ___________

= _____

2 Fill in the missing numbers.

a 14 + 8 = 22

___ + 14 = 22

22 − ___ = 14

22 − ___ = 8

b 18 + 9 = 27

9 + ___ = 27

27 − ___ = 18

27 − 18 = ___

c 16 + 7 = 23

7 + ___ = 23

23 − 16 = ___

23 − ___ = 16

d 19 + 5 = 24

___ + ___ = ___

___ − ___ = ___

___ − ___ = ___

e 16 + 9 = 25

___ + ___ = ___

___ − ___ = ___

___ − ___ = ___

f 17 + 4 = 21

___ + ___ = ___

___ − ___ = ___

___ − ___ = ___

CATCH UP MATHS YEAR 4 BOOK A © PASCAL PRESS ISBN: 9781925726145

3 Write the fact family for each set of numbers.

a (12) (3) (15)

◯ + ◯ = ◯
◯ + ◯ = ◯
◯ − ◯ = ◯
◯ − ◯ = ◯

b (32) (20) (52)

◯ + ◯ = ◯
◯ + ◯ = ◯
◯ − ◯ = ◯
◯ − ◯ = ◯

c (27) (46) (73)

◯ + ◯ = ◯
◯ + ◯ = ◯
◯ − ◯ = ◯
◯ − ◯ = ◯

4 Write the fact family for each set of numbers.

a [40] [23] [17]

☐ + ☐ = ☐
☐ + ☐ = ☐
☐ − ☐ = ☐
☐ − ☐ = ☐

b [15] [19] [34]

☐ + ☐ = ☐
☐ + ☐ = ☐
☐ − ☐ = ☐
☐ − ☐ = ☐

c [123] [254] [377]

☐ + ☐ = ☐
☐ + ☐ = ☐
☐ − ☐ = ☐
☐ − ☐ = ☐

d [116] [246] [362]

☐ + ☐ = ☐
☐ + ☐ = ☐
☐ − ☐ = ☐
☐ − ☐ = ☐

5 Solve these additions.

a

	T	O
	5	7
+	4	2

b

	T	O
	6	3
+	3	1

c

	T	O
	5	8
+	4	1

REVIEW

d

T	O
7	0
+ 2	9

f

H	T	O
5	1	7
+ 3	8	1

h

H	T	O
6	0	8
+ 3	9	1

e

H	T	O
3	0	8
+ 4	5	1

g

H	T	O
4	2	3
+ 4	3	5

i

H	T	O
2	7	3
+ 6	2	4

6 Add these numbers.

a

T	O
3	8
+ 6	9

b

T	O
7	6
+ 8	5

c

T	O
8	2
+ 5	9

d

H	T	O
3	7	2
+ 1	1	8

e

H	T	O
5	4	1
+ 2	3	9

f

H	T	O
6	7	5
+ 3	9	5

7 Fill in the missing numbers.

a

T	O
3	5
+	
5	9

b

T	O
4	2
+	
7	8

c

T	O
3	9
+	
8	4

d

T	O
7	6
+	
10	8

e

H	T	O
7	2	4
+		
9	8	6

f

H	T	O
5	9	7
+		
9	1	2

g

H	T	O
2	6	8
+		
7	9	5

h

H	T	O
+ 4	8	5
9	2	3

i

H	T	O
6	8	7
+	4	
8	3	6

CATCH UP MATHS YEAR 4 BOOK A © PASCAL PRESS ISBN: 9781925726145

8 Round each number to the nearest 10 to estimate an answer for each of these number sentences.

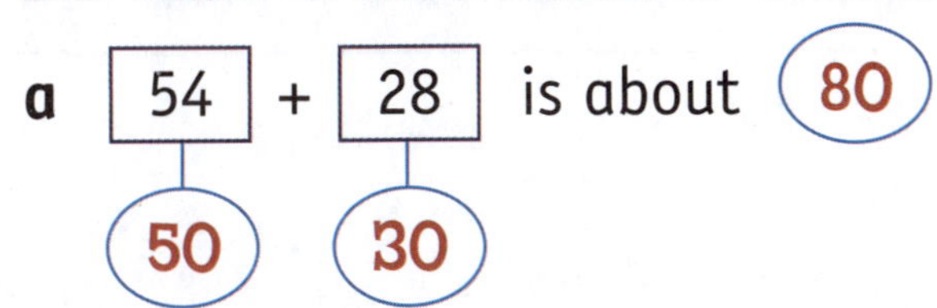

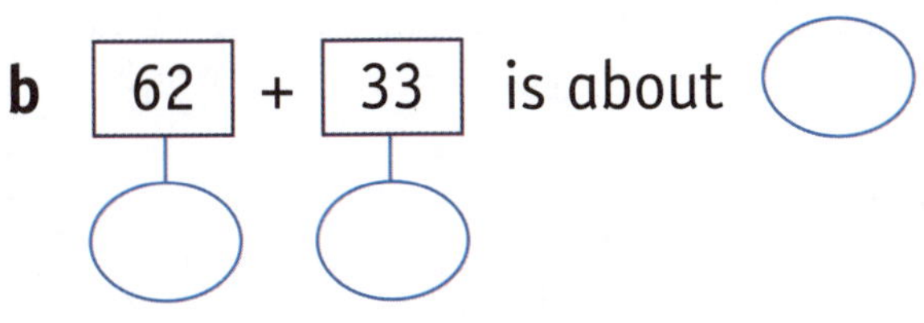

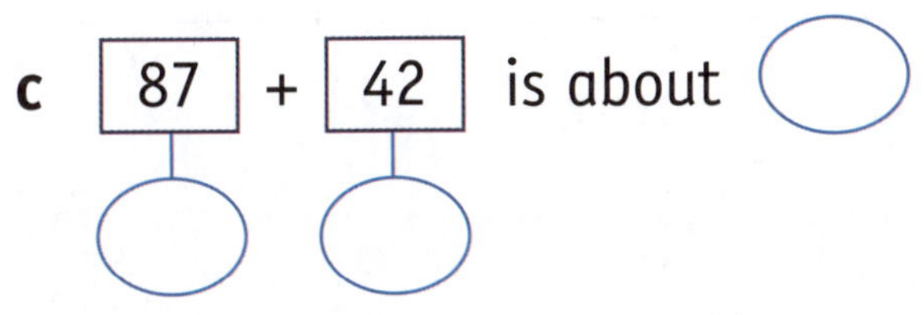

d 37 + 42 is about

e 25 + 81 is about

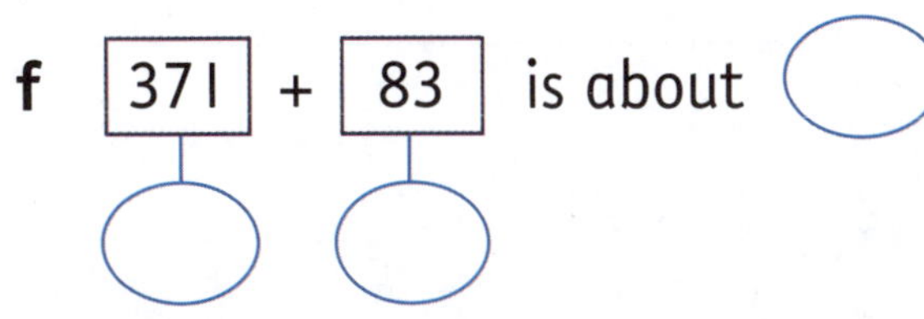

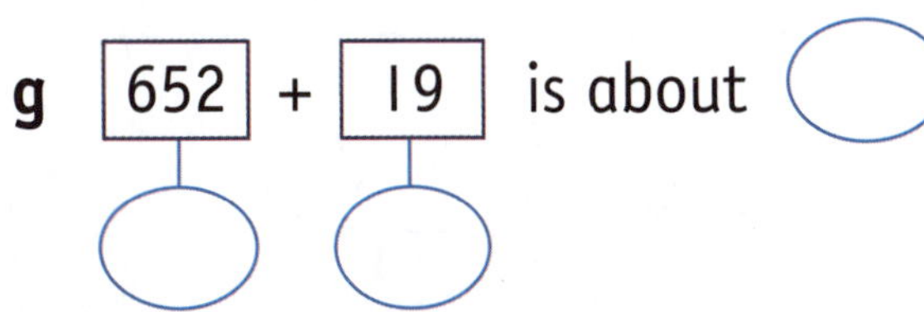

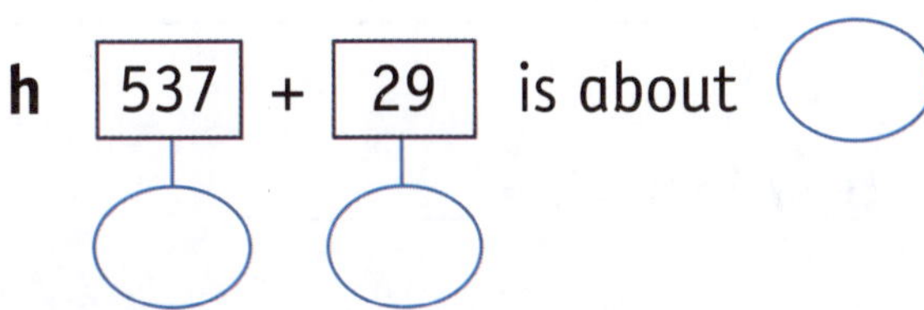

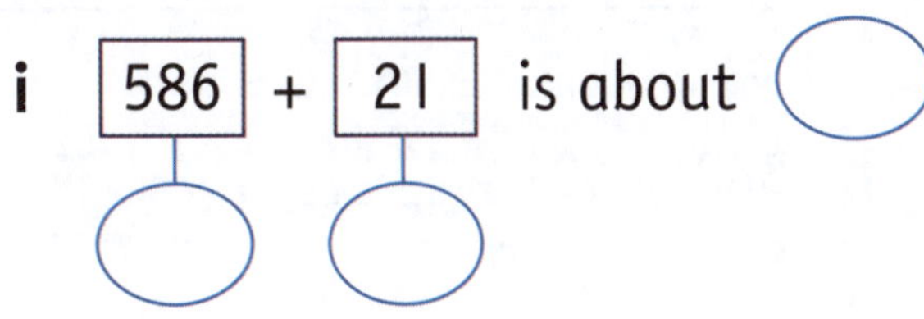

j 746 + 52 is about

9 Use the number lines to solve these additions.

a 48 + 25 = _____

b 21 + 59 = _____

c 83 + 42 = _____

REVIEW

d 47 + 12 = ____

e 736 + 21 = ____

f 548 + 33 = ____

g 746 + 43 = ____

h 583 + 421 = ____

i 1973 + 124 = ____

j 3524 + 153 = ____

k 5246 + 262 = ____

l 32 + 151 = ____

m 23 + 233 = ______

n 16 + 248 = ______

o 103 + 501 = ______

p 216 + 210 = ______

q 4932 + 145 = ______

r 6090 + 403 = ______

10 **Use the split strategy to perform these additions.**

a 54 + 23 = 50 + 20 = 70 and 4 + 3 = 7

= 70 + 7

= 77

b 25 + 63 = ___ + ___ = ___ and __ + __ = __

= ___ + __

= ___

REVIEW

c 42 + 34 = ___ + ___ = ___ and __ + __ = __

= ___ + __

= ___

d 713 + 44 = ____ and ___ + ___ = ___ and __ + __ = __

= ____ + ___ + __

= ____

e 824 + 32 = ____ and ___ + ___ = ___ and __ + __ = __

= ____ + ___ + __

= ____

f 642 + 16 = ____ and ___ + ___ = ___ and __ + __ = __

= ____ + ___ + __

= ____

g 4281 + 213

= _____ and ____ + ____ = ____ and ___ + ___ = ___ and __ + __ = __

= _____ + _____ + ___ + __

= _____

h 2812 + 147

= _____ and ____ + ____ = ____ and ___ + ___ = ___ and __ + __ = __

= _____ + _____ + ___ + __

= _____

11 Round to the nearest 5 cents.

a $4.17 _______

b $2.51 _______

c $6.49 _______

d $43.76 _______

e $89.89 _______

f $142.01 _______

CATCH UP MATHS YEAR 4 BOOK A © PASCAL PRESS ISBN: 9781925726145

12 Complete the table.

	Cost	Nearest 5 cents	Amount given	Change
a	$0.66	$	$2.00	$
b	$1.52	$	$2.00	$
c	$3.64	$	$5.00	$
d	$2.18	$	$5.00	$
e	$0.79	$	$5.00	$
f	$1.83	$	$10.00	$
g	$5.92	$	$10.00	$
h	$7.34	$	$10.00	$
i	$12.14	$	$20.00	$
j	$15.81	$	$20.00	$
k	$17.39	$	$20.00	$

13 Work out the answers to these money problems.
Round each answer to the nearest 5 cents.

a Noah bought 3 pencils for $2.99 each. He gave $10.00.
How much change did he get?

____

____

b Gabrielle bought 3 magazines for $5.98 each. She gave $20.00.
How much change did she get?

____

____

c Dom bought 5 oranges for 67 cents each. She gave $10.00.
How much change did Dom receive?

____

____

JUMP STRATEGY TO SOLVE SUBTRACTION

The jump strategy and a number line can be used to solve subtraction.

SCAN to watch video

Example 1:

Solve 58 – 34 using the number line and the jump strategy.

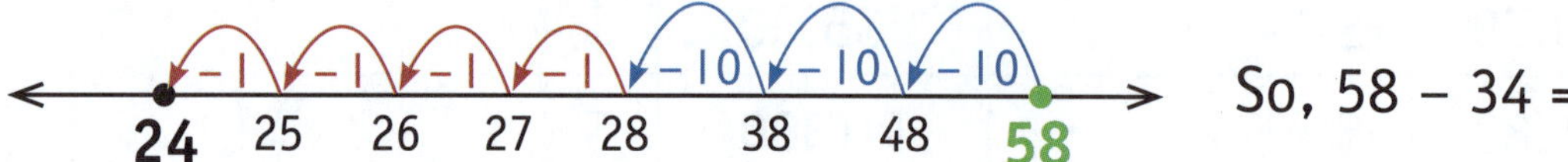

So, 58 – 34 = 24

Check your answer on the video!

Example 2:

Solve 67 – 35 using the number line and the jump strategy.

67 – 35
5 'ones' jumps
3 'tens' jumps
Start with the bigger number.

67

So, 67 – 35 = ___

Your turn

Use the number line to work out the answer.

● 84 – 21 = 63

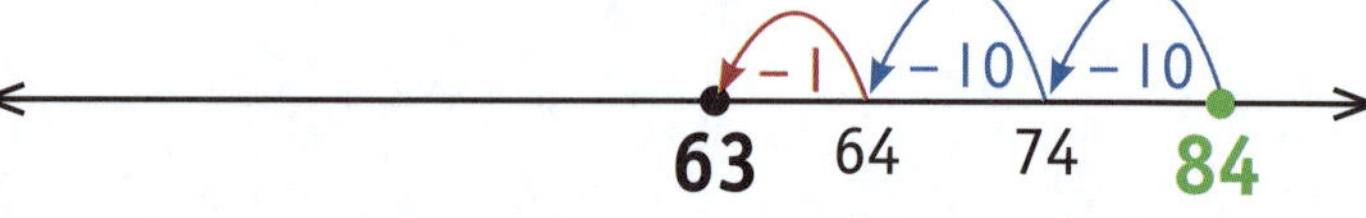

a 93 – 42 = ___

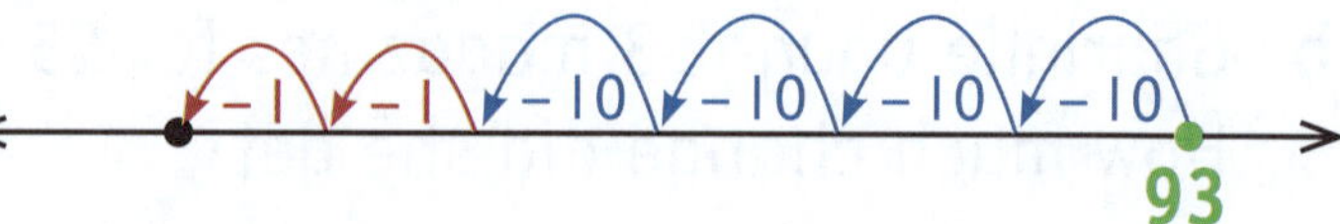

b 56 – 15 = ___

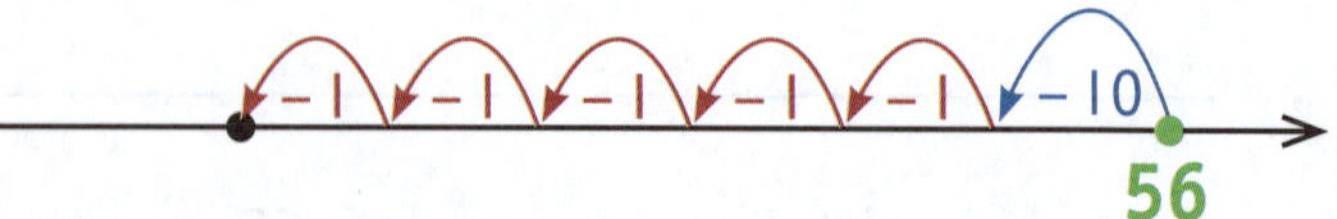

Check your answers

How many did you get correct?

CATCH UP MATHS YEAR 4 BOOK A © PASCAL PRESS ISBN: 9781925726145

PRACTICE

1 Find the answer using the jump strategy and the number line.

64 – 23 = 41

a 53 – 41 = ___

b 72 – 52 = ___

c 87 – 34 = ___

d 33 – 12 = ___

e 96 – 35 = ___

f 85 – 10 = ___

g 74 – 32 = ___

h 99 – 84 = ___

JUMP STRATEGY WITH THREE-DIGIT NUMBERS

The number line and jump strategy can be used to solve subtraction of three-digit numbers.

Example 1: Solve 457 − 125.

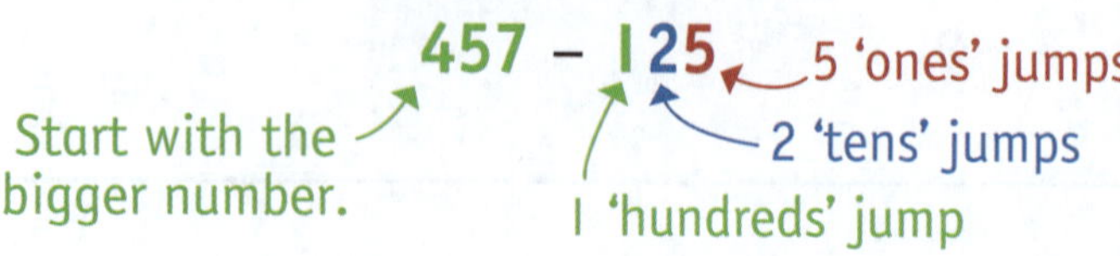

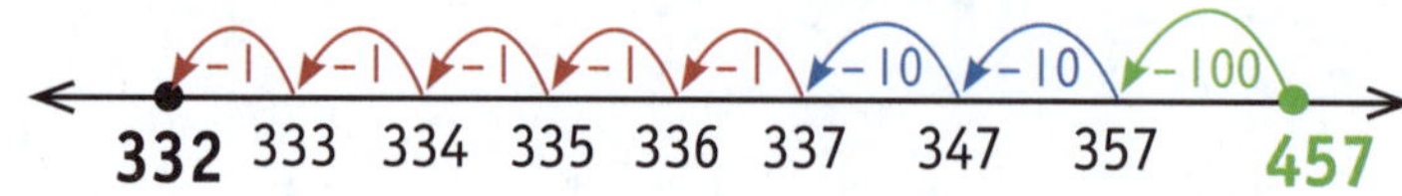

So, 457 − 125 = 332

Example 2: Solve 524 − 213.

524 − 213

Start with the bigger number.

__ 'ones' jumps

__ 'tens' jump

__ 'hundreds' jumps

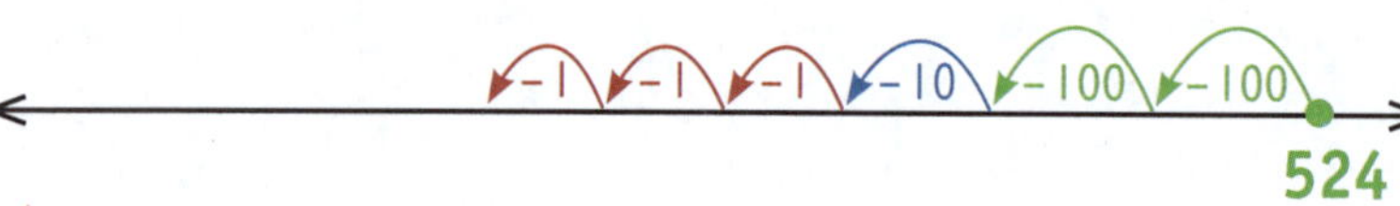

So, 524 − 213 = ____

Solve the subtractions using the jump strategy and the number line.

● 637 − 234 = 403

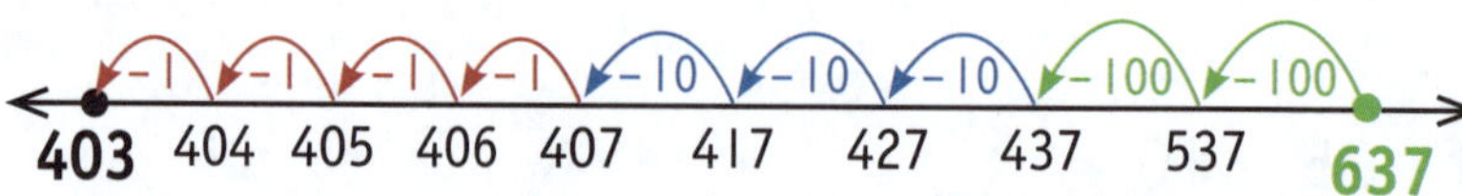

a 492 − 31 = ____

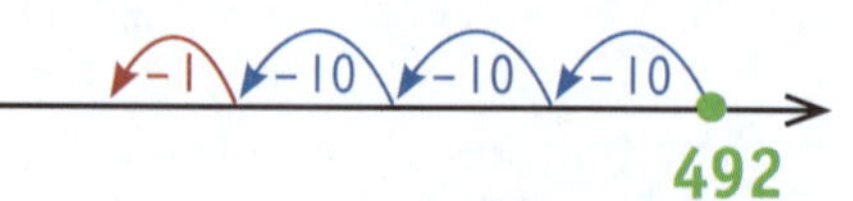

b 589 − 224 = ____

−1 −1 −1 −1 −10 −10 −100 −100

589

SELF CHECK Tick how you feel

Got it!	Need help...	I don't get it
 ☐	 ☐	 ☐

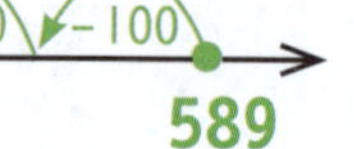

How many did you get correct?

CATCH UP MATHS YEAR 4 BOOK A © PASCAL PRESS ISBN: 9781925726145

PRACTICE

1 Find the answer using the jump strategy and the number line.

- 524 − 13 = 511

a 637 − 24 = ______

b 849 − 36 = ______

c 935 − 24 = ______

d 496 − 44 = ______

e 899 − 152 = ______

f 756 − 631 = ______

g 657 − 215 = ______

h 591 − 141 = ___

JUMP STRATEGY WITH LARGER NUMBERS

The jump strategy can be used to solve the subtraction of larger numbers.

Below are examples where the jump strategy is used with three- and four-digit numbers.

SCAN to watch video

Example 1: 5345 – 214 = 5131

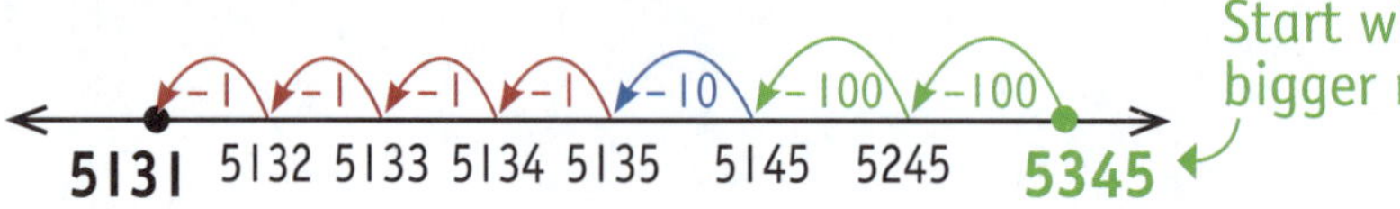

Example 2: 6896 – 2324 = 4572

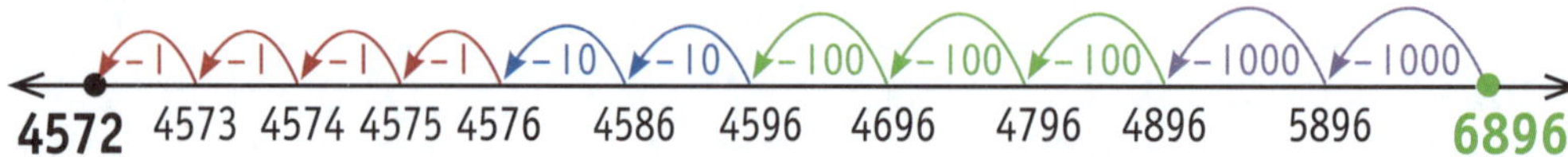

Example 3: 8497 – 1403 = ______

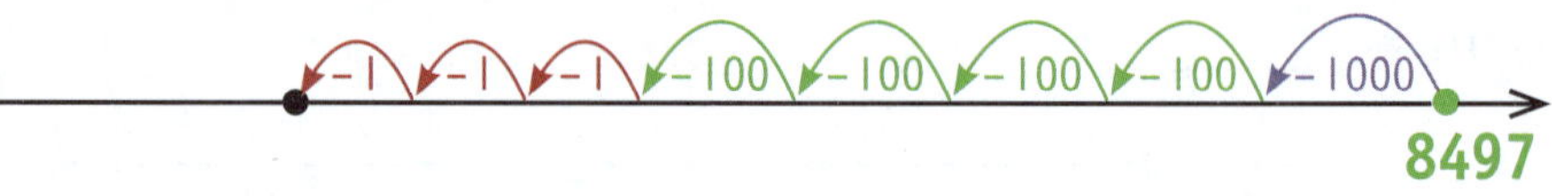

Solve using the jump strategy and the number line.

● 4834 – 2311 = 2523

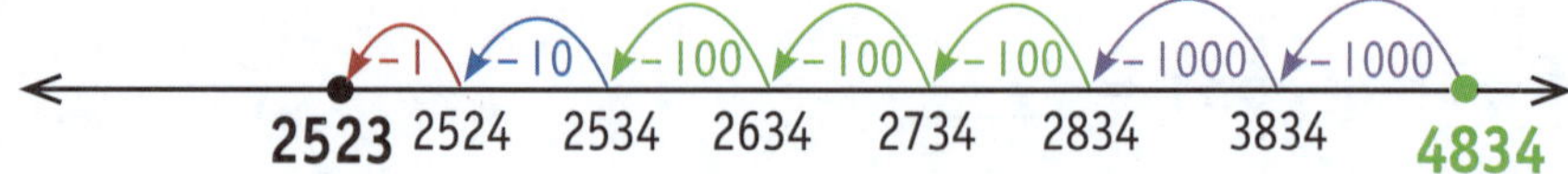

a 5863 – 412 = ______

b 6975 – 1323 = ______

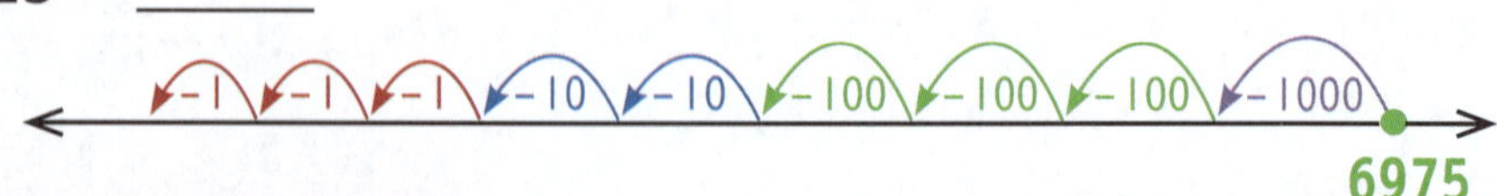

SELF CHECK Tick how you feel

Got it!	Need help...	I don't get it
☐	☐	☐

Check your answers

How many did you get correct?

CATCH UP MATHS YEAR 4 BOOK A © PASCAL PRESS ISBN: 9781925726145

PRACTICE

1 Find the answer using the jump strategy and the number line.

5698 − 243 = 5455

a 6824 − 513 = ________

b 8278 − 143 = ________

c 9825 − 602 = ________

d 5945 − 531 = ________

e 1809 − 1304 = ________

f 5400 − 2522 = ________

g 3980 − 2314 = ________

h 4308 − 1436 = ________

SUBTRACTION USING THE SPLIT STRATEGY

When solving subtraction using the split strategy, we 'split' the numbers into their values.

Example 1:

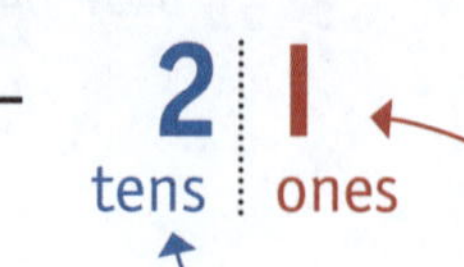

5 (tens)	3 (ones)	–	2 (tens)	1 (ones)

The 5 is worth 50 as it is in the tens column.

The 3 is worth 3 as it is in the ones column.

The 2 is worth 20 as it is in the tens column.

The 1 is worth 1 as it is in the ones column.

50 – 20 = 30

3 – 1 = 2

Here, we add the tens and the ones. → 30 + 2 = 32

First add the tens, then add the ones.

Example 2:

47 – 16

40 – 10 = 30

7 – 6 = 1

30 + 1 = 31

Example 3:

38 – 21

30 – ___ = ___

8 – __ = __

___ + __ = ___

Example 4:

79 – 54

70 – ___ = ___

9 – __ = __

___ + __ = ___

Solve the subtractions using the split strategy.

● 59 – 24

50 – 20 = 30

9 – 4 = 5

30 + 5 = 35

a 89 – 32

___ – ___ = ___

__ – __ = __

___ + __ = ___

SELF CHECK Tick how you feel

Got it!	Need help...	I don't get it
☐	☐	☐

Check your answers

How many did you get correct? ☐

CATCH UP MATHS YEAR 4 BOOK A © PASCAL PRESS ISBN: 9781925726145

PRACTICE

1 Solve the subtractions using the split strategy.

● 73 − 22

70 − 20 = 50

3 − 2 = 1

50 + 1 = 51

a 27 − 13

___ − ___ = ___

___ − ___ = ___

___ + ___ = ___

b 59 − 20

___ − ___ = ___

___ − ___ = ___

___ + ___ = ___

c 78 − 32

___ − ___ = ___

___ − ___ = ___

___ + ___ = ___

d 66 − 32

___ − ___ = ___

___ − ___ = ___

___ + ___ = ___

e 92 − 20

___ − ___ = ___

___ − ___ = ___

___ + ___ = ___

f 83 − 12

___ − ___ = ___

___ − ___ = ___

___ + ___ = ___

g 49 − 18

___ − ___ = ___

___ − ___ = ___

___ + ___ = ___

h 67 − 26

___ − ___ = ___

___ − ___ = ___

___ + ___ = ___

i 59 − 42

___ − ___ = ___

___ − ___ = ___

___ + ___ = ___

SPLIT STRATEGY WITH THREE-DIGIT NUMBERS

The split strategy can also be used to solve subtraction with three-digit numbers.

Example 1:

Solve 783 – 42

700 – 0 = 700

80 – 40 = 40

3 – 2 = 1

700 + 40 + 1 = 741

Remember to add here.

Example 3:

Solve 465 – 24

400 – 0 = ____

60 – 20 = ____

5 – 4 = ____

____ + ____ + ____ = ____

SCAN to watch video

Example 2:

Solve 695 – 324

600 – 300 = 300

90 – 20 = 70

5 – 4 = 1

300 + 70 + 1 = 371

Example 4:

Solve 689 – 243

600 – ____ = ____

80 – ____ = ____

9 – ____ = ____

____ + ____ + ____ = ____

Check your answer on the video!

Your turn

Find the answers using the split strategy.

● 865 – 143

800 – 100 = 700

60 – 40 = 20

5 – 3 = 2

700 + 20 + 2 = 722

a 743 – 22

____ – ____ = ____

____ – ____ = ____

____ – ____ = ____

____ + ____ + ____ = ____

SELF CHECK Tick how you feel

Got it!	Need help...	I don't get it
 ☐	 ☐	 ☐

Check your answers

How many did you get correct? ☐

CATCH UP MATHS YEAR 4 BOOK A © PASCAL PRESS ISBN: 9781925726145

PRACTICE

1 Complete the subtractions using the split strategy.

● 782 – 20

700 – 0 = 700

80 – 20 = 60

2 – 0 = 2

700 + 60 + 2 = 762

a 379 – 63

____ – ____ = ____

___ – ___ = ___

__ – __ = __

____ + ___ + __ = ____

b 536 – 424

____ – ____ = ____

___ – ___ = ___

__ – __ = __

____ + ___ + __ = ____

c 849 – 535

____ – ____ = ____

___ – ___ = ___

__ – __ = __

____ + ___ + __ = ____

d 757 – 44

____ – ____ = ____

___ – ___ = ___

__ – __ = __

____ + ___ + __ = ____

e 465 – 53

____ – ____ = ____

___ – ___ = ___

__ – __ = __

____ + ___ + __ = ____

f 178 – 162

____ – ____ = ____

___ – ___ = ___

__ – __ = __

____ + ___ + __ = ____

g 682 – 10

____ – ____ = ____

___ – ___ = ___

__ – __ = __

____ + ___ + __ = ____

SPLIT STRATEGY WITH FOUR-DIGIT NUMBERS

Below are examples of how split strategy can be used to solve subtraction problems with four-digit numbers.

Example 1: 5947 – 325

5000 – 0 = 5000

900 – 300 = 600

40 – 20 = 20

7 – 5 = 2

5000 + 600 + 20 + 2 = 5622

Example 2: 7895 – 4263

7000 – 4000 = 3000

800 – 200 = 600

90 – 60 = 30

5 – 3 = 2

3000 + 600 + 30 + 2 = 3632

Example 3: 6582 – 461

6000 – ____ = ____

500 – ____ = ____

80 – ____ = ____

2 – ____ = ____

____ + ____ + ____ + ____ = ____

Example 4: 9786 – 1432

9000 – ____ = ____

700 – ____ = ____

80 – ____ = ____

6 – ____ = ____

____ + ____ + ____ + ____ = ____

Check your answer on the video!

Solve the subtractions using the split strategy.

● 4384 – 243

4000 – 0 = 4000

300 – 200 = 100

80 – 40 = 40

4 – 3 = 1

4000 + 100 + 40 + 1

= 4141

a 5836 – 1234

____ – ____ = ____

____ – ____ = ____

____ – ____ = ____

____ – ____ = ____

____ + ____ + ____ + ____

= ____

SELF CHECK Tick how you feel

Got it!	Need help...	I don't get it
 ☐	 ☐	 ☐

Check your answers

How many did you get correct? ☐

CATCH UP MATHS YEAR 4 BOOK A © PASCAL PRESS ISBN: 9781925726145

PRACTICE

1 Find the answers using the split strategy.

7594 – 1482

7000 – 1000 = 6000

500 – 400 = 100

90 – 80 = 10

4 – 2 = 2

6000 + 100 + 10 + 2

= 6112

a 3487 – 2275

____ – ____ = ____

____ – ____ = ____

___ – ___ = ___

__ – __ = __

____ + ____ + ___ + __

= ____

b 5378 – 167

____ – ____ = ____

____ – ____ = ____

___ – ___ = ___

__ – __ = __

____ + ____ + ___ + __

= ____

c 6265 – 3143

____ – ____ = ____

____ – ____ = ____

___ – ___ = ___

__ – __ = __

____ + ____ + ___ + __

= ____

d 4158 – 146

____ – ____ = ____

____ – ____ = ____

___ – ___ = ___

__ – __ = __

____ + ____ + ___ + __

= ____

e 2574 – 1352

____ – ____ = ____

____ – ____ = ____

___ – ___ = ___

__ – __ = __

____ + ____ + ___ + __

= ____

SUBTRACTION WITHOUT TRADING

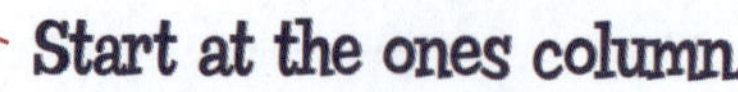

An algorithm is a way of setting out a maths question to work out the answer.

This is a subtraction algorithm:

Start at the ones column.

Always subtract downwards.

	5	8
−	3	2
	2	6

Example 1: Start here.

	Tens	Ones
	7	3
−	1	2
	6	1

Example 2:

	Tens	Ones
	8	5
−	3	3
	5	2

Example 3:

	Tens	Ones
	9	8
−	3	4
	6	4

Example 4:

	Tens	Ones
	6	4
−	3	2

Example 5:

	Tens	Ones
	7	7
−	4	6

Example 6:

	Tens	Ones
	8	3
−	5	0

Start at the ones column and subtract.

Complete these subtractions.

	Tens	Ones
	9	8
−	1	5
	8	3

a

	Tens	Ones
	6	9
−	4	2

b

	Tens	Ones
	5	7
−	3	5

c

	Tens	Ones
	4	7
−	2	4

Check your answers

How many did you get correct?

CATCH UP MATHS YEAR 4 BOOK A © PASCAL PRESS ISBN: 9781925726145

PRACTICE

1 Complete these subtractions.

Example:

Tens	Ones
5	8
− 2	4
3	4

a

Tens	Ones
6	9
− 3	7

b

Tens	Ones
5	7
− 4	1

c

Tens	Ones
3	8
− 1	7

d

Tens	Ones
4	6
− 2	4

e

Tens	Ones
4	3
− 1	1

f

Tens	Ones
3	9
− 2	0

g

Tens	Ones
2	6
− 1	5

2 Now complete these subtractions.

Example: 98 − 57 = 41

a 77 − 42 = ____

b 65 − 42 = ____

c 54 − 20 = ____

d 45 − 32 = ____

e 64 − 31 = ____

f 73 − 51 = ____

g 81 − 30 = ____

3 This is Mario's maths test. Work out the answers to the questions, mark his test and give him a score out of 8.

a 24 − 13 = 11 ✓

b 72 − 10 = 61

c 34 − 12 = 24

d 84 − 20 = 64

e 93 − 51 = 42

f 87 − 43 = 44

g 92 − 80 = 12

h 66 − 24 = 22

☐ out of 8

SUBTRACTION WITHOUT TRADING AND THREE-DIGIT NUMBERS

This is a subtraction problem using a three-digit number.

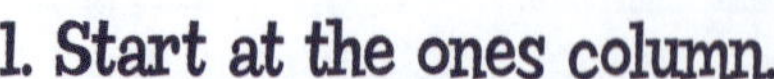

1. Start at the ones column.
2. Then go to the tens column.
3. Then the hundreds column.

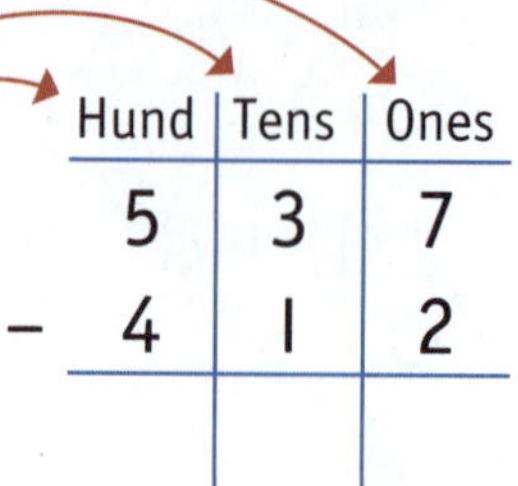

	Hund	Tens	Ones
	5	3	7
–	4	1	2

Always subtract downwards.

Example 1:

Start here.

	Hund	Tens	Ones
	4	9	6
–		2	4
	4	7	2

Example 2:

	Hund	Tens	Ones
	7	3	5
–		2	3
	7	1	2

Example 3:

	Hund	Tens	Ones
	4	9	8
–	1	5	4
	3	4	4

Example 4:

	Hund	Tens	Ones
	9	6	5
–		2	4
			1

Example 5:

	Hund	Tens	Ones
	4	9	6
–	1	0	3
			3

Example 6:

	Hund	Tens	Ones
	8	6	3
–	2	5	3
			0

Check your answer on the video!

First subtract the ones, then the tens, then the hundreds.

Complete the subtractions.

●

	H	T	O
	5	3	6
–	4	1	5
	1	2	1

a

	H	T	O
	7	3	9
–	3	2	4

b

	H	T	O
	8	6	8
–		1	2

SELF CHECK Tick how you feel

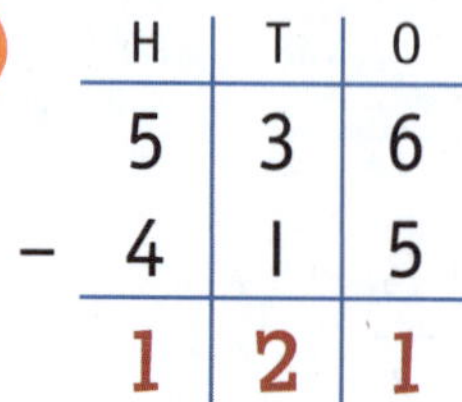

Got it! ☐ Need help... ☐ I don't get it ☐

Check your answers

How many did you get correct? ☐

CATCH UP MATHS YEAR 4 BOOK A © PASCAL PRESS ISBN: 9781925726145

PRACTICE

1 Complete these subtractions.

● 634 − 23 = 611

c 166 − 32 = ____

f 148 − 32 = ____

i 802 − 401 = ____

a 504 − 102 = ____

d 962 − 41 = ____

g 207 − 107 = ____

j 993 − 812 = ____

b 635 − 235 = ____

e 753 − 141 = ____

h 871 − 170 = ____

k 468 − 324 = ____

2 What is the difference between the numbers?

	Numbers	Working Out
●	536 and 115	536 − 115 = 421
a	657 and 241	
b	773 and 252	
c	895 and 50	

	Numbers	Working Out
d	897 and 881	
e	584 and 73	
f	779 and 779	
g	495 and 32	

SUBTRACTION WITHOUT TRADING AND FOUR-DIGIT NUMBERS

Subtraction of four-digit numbers extends into the thousands column.

Start at the ones column.

Always subtract downwards.

	Thou	Hund	Tens	Ones
	5	9	2	7
–	1	3	1	5
	4	6	1	2

Example 1:

Start here.

	Thou	Hund	Tens	Ones
	6	8	2	5
–	1	4	1	3
	5	4	1	2

Example 2:

	Thou	Hund	Tens	Ones
	9	3	7	7
–	4	2	1	3
	5	1	6	4

Example 3:

	Th	H	T	O
	3	7	8	1
–		3	4	1

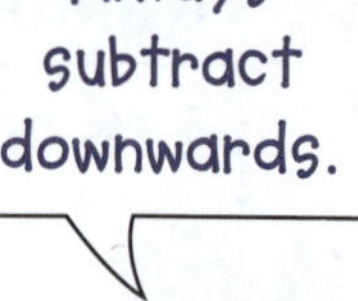

Example 4:

	Th	H	T	O
	6	3	0	3
–		2	0	2

Your turn

Solve these four-digit subtractions.

●

	Th	H	T	O
	5	8	6	3
–		3	2	1
	5	**5**	**4**	**2**

a

	Th	H	T	O
	7	2	6	7
–	1	1	3	1

SELF CHECK Tick how you feel

Got it!	Need help...	I don't get it
☐	☐	☐

Check your answers

How many did you get correct? ☐

CATCH UP MATHS YEAR 4 BOOK A © PASCAL PRESS ISBN: 9781925726145

PRACTICE

1 Complete the subtractions.

Example:
  7 4 2 5
− 1 3 2 1
  6 1 0 4

a
  8 5 9 3
− 4 1 3 1

b
  9 8 3 2
−   4 2 1

c
  5 7 4 3
− 3 3 0 2

d
  6 8 4 5
− 4 0 3 4

e
  5 2 4 3
−   1 4 0

f
  7 8 1 5
−   6 1 5

g
  3 0 9 6
− 1 0 3 1

h
  7 4 3 2
−   3 2 1

i
  6 5 6 0
−   4 4 0

j
  4 7 9 3
−   4 9 3

k
  4 3 0 9
− 1 2 0 5

2 Circle each correct answer.

Example: (1742) 1247
  5 8 9 3
− 4 1 5 1
  1 7 4 2

a 4422 4224
  7 3 4 9
− 3 1 2 5

b 7551 551
  7 8 9 9
−   3 4 8

c 9065 9075
  9 4 9 8
−   4 2 3

d 5117 5711
  6 5 3 8
− 1 4 2 1

e 1271 1111
  4 3 8 2
− 3 2 7 1

SUBTRACTION WITHOUT TRADING AND FIVE-DIGIT NUMBERS

Subtraction of five-digit numbers extends into the column for tens of thousands.

Start at the ones column.

Always subtract downwards.

	T Th	Thou	Hund	Tens	Ones
	5	9	8	3	2
–	1	2	4	2	1
	4	7	4	1	1

Example 1:

Start here.

	T Th	Thou	Hund	Tens	Ones
	8	7	5	6	3
–	1	4	2	1	2
	7	3	3	5	1

Example 3:

	T Th	Thou	Hund	Tens	Ones
	6	5	3	2	4
–		1	2	0	3

Example 2:

	T Th	Thou	Hund	Tens	Ones
	5	2	4	3	7
–			3	2	1
	5	2	1	1	6

Example 4:

	T Th	Thou	Hund	Tens	Ones
	7	2	6	7	7
–	5	1	5	3	4

Always subtract downwards.

Your turn

Solve these five-digit subtractions.

	T Th	Thou	Hund	Tens	Ones
	7	2	9	5	4
–			5	3	1
	7	2	4	2	3

a

	T Th	Thou	Hund	Tens	Ones
	6	8	4	7	1
–	5	1	3	6	0

SELF CHECK Tick how you feel

Check your answers
How many did you get correct?

CATCH UP MATHS YEAR 4 BOOK A © PASCAL PRESS ISBN: 9781925726145

PRACTICE

1 Complete the subtractions.

- e.g. 8 5 5 9 3 − 4 1 1 3 1 = ____
- a 1 5 3 7 9 − 4 2 4 3 = ____
- b 2 8 2 1 1 − 1 0 = ____
- c 2 2 4 8 1 − 1 3 4 0 = ____
- d 1 1 1 1 3 − 1 0 0 1 = ____
- e 1 0 5 4 8 − 1 0 3 4 2 = ____
- f 8 5 4 1 5 − 4 3 0 4 = ____
- g 6 3 4 7 9 − 2 4 6 3 = ____
- h 4 9 3 4 2 − 1 2 3 4 1 = ____
- i 5 4 9 8 3 − 4 3 5 7 2 = ____
- j 6 3 2 4 9 − 5 2 2 4 0 = ____
- k 7 2 7 6 7 − 1 0 3 4 5 = ____

2 Fill in the missing digits.

- e.g. 9 2 5 5 4 − 7 1 2 1 3 = 2 1 3 4 1
- a 7 4 _ 8 2 − 2 _ 1 _ 1 = _ 1 2 1 _
- b 8 _ _ 7 _ − _ 1 2 _ 2 = 1 8 2 2 1
- c 5 _ 3 _ 6 − _ 3 2 3 _ = 4 6 _ 1 1
- d 9 _ 8 7 _ − 5 6 _ 4 5 = _ 3 6 _ 4
- e 6 _ 4 8 7 − 1 4 _ 2 _ = _ 1 1 _ 4
- f 7 8 _ 9 3 − _ 2 3 _ 1 = 3 _ 2 2 _
- g 6 _ 4 3 _ − _ 4 _ 2 1 = 1 1 1 _ 1
- h 1 9 4 _ 2 − _ _ 0 1 _ = 0 6 _ 2 2

  ISBN: 9781925726145

SUBTRACTION WITH TRADING AND TWO-DIGIT NUMBERS

Here is a subtraction where we have to trade 1 ten for 10 ones to help us solve the question.

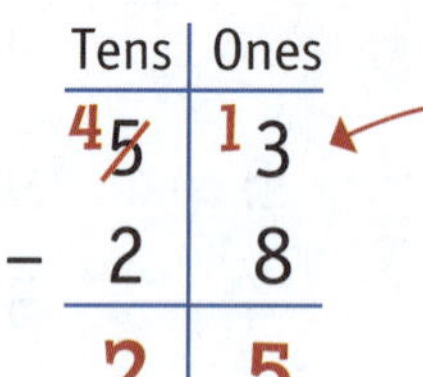

	Tens	Ones
	(4) ~~5~~	(1)3
−	2	8
	2	**5**

We cannot take 8 away from 3 so we trade 1 ten for 10 ones to make 13.

Example 1:

Start here.

	Tens	Ones
	(5) ~~6~~	(1)2
−	3	6
	2	6

Example 2:

	Tens	Ones
	(6) ~~7~~	(1)4
−	3	5
	3	9

Example 3:

	Tens	Ones
	(7) ~~8~~	(1)1
−	4	3

Example 4:

	Tens	Ones
	5	3
−	2	4

Trade 1 ten for 10 ones.

Use trading to complete these subtractions.

●

	Tens	Ones
	(2) ~~3~~	(1)2
−	1	5
	1	**7**

a

	Tens	Ones
	7	1
−	2	6

b

	Tens	Ones
	5	8
−	3	9

c

	Tens	Ones
	9	1
−	7	3

SELF CHECK Tick how you feel

Got it!	Need help...	I don't get it
☐	☐	☐

Check your answers

How many did you get correct? ☐

CATCH UP MATHS YEAR 4 BOOK A © PASCAL PRESS ISBN: 9781925726145

PRACTICE

1 Solve these subtractions.

●

	Tens	Ones
	4~~5~~	14
−	2	8
	2	6

a

	Tens	Ones
	2	5
−	1	9

b

	Tens	Ones
	3	1
−	2	5

c

	Tens	Ones
	4	3
−	2	4

d

	Tens	Ones
	7	2
−	5	3

e

	Tens	Ones
	8	1
−	4	8

f

	Tens	Ones
	6	2
−	4	3

g

	Tens	Ones
	9	0
−	3	5

h

	Tens	Ones
	6	8
−	5	9

i

	Tens	Ones
	7	4
−	5	6

j

	Tens	Ones
	8	1
−	4	3

k

	Tens	Ones
	6	1
−	3	2

2 Do these subtractions and then check your answers using addition.

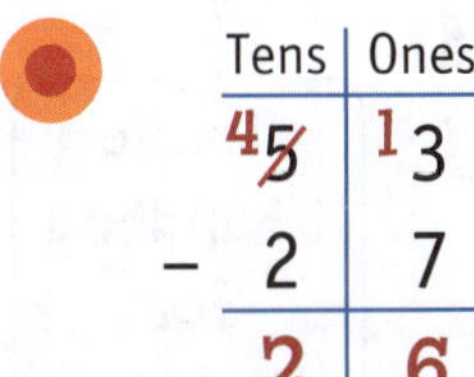

●

	Tens	Ones
	4~~5~~	13
−	2	7
	2	6

	Tens	Ones
	12	6
+	2	7
	5	3

a

	Tens	Ones
	7	1
−	3	3

	Tens	Ones
+	3	3

b

	Tens	Ones
	8	2
−	4	3

	Tens	Ones
+	4	3

c

	Tens	Ones
	9	3
−	6	4

	Tens	Ones
+	6	4

d

	Tens	Ones
	4	1
−	2	5

	Tens	Ones
+	2	5

e

	Tens	Ones
	6	0
−	3	4

	Tens	Ones
+	3	4

SUBTRACTION WITH TRADING AND THREE-DIGIT NUMBERS

When you cannot take a number away in a subtraction, you must trade.

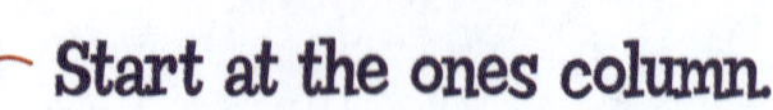

Start at the ones column.

Always subtract downwards.

	Hund	Tens	Ones
	6	4 ~~5~~	1 4
−	1	2	6
	5	2	8

Example 1:

Start here.

	Hund	Tens	Ones
	5	3 ~~4~~	1 8
−	2	3	9
	3	0	9

Example 3:

	Hund	Tens	Ones
	7	5	3
−	3	2	5

Example 2:

Traded twice.

	Hund	Tens	Ones
	3 ~~4~~	10 ~~1~~	1 5
−	1	3	6
	2	7	9

Example 4:

Traded twice.

	Hund	Tens	Ones
	8	1 ~~2~~	1 6
−	2	5	7

Check your answer on the video!

Use trading to complete these subtractions.

●

	Hund	Tens	Ones
	6 ~~7~~	10 ~~1~~	1 5
−	3	9	6
	3	1	9

a

	Hund	Tens	Ones
	4	2	3
−	1	2	6

b

	Hund	Tens	Ones
	5	2	5
−	2	3	9

SELF CHECK Tick how you feel

Got it!	Need help...	I don't get it
☐	☐	☐

Check your answers

How many did you get correct? ☐

CATCH UP MATHS YEAR 4 BOOK A © PASCAL PRESS ISBN: 9781925726145

PRACTICE

1 Solve these subtractions.

Example:

	Hund	Tens	Ones
	[2] ~~3~~	[11] ~~2~~	[1] 4
−	1	5	8
	1	6	6

a

	Hund	Tens	Ones
	5	6	3
−	1	4	7

b

	Hund	Tens	Ones
	6	2	5
−	2	3	4

c

	Hund	Tens	Ones
	4	2	6
−	2	0	7

d

	Hund	Tens	Ones
	9	3	0
−	2	4	6

e

	Hund	Tens	Ones
	5	0	6
−	2	4	3

f

	Hund	Tens	Ones
	7	3	1
−	2	4	5

g

	Hund	Tens	Ones
	8	1	5
−	1	2	7

h

	Hund	Tens	Ones
	7	2	2
−	4	3	5

i

	Hund	Tens	Ones
	8	1	0
−	6	3	9

j

	Hund	Tens	Ones
	6	3	2
−	1	2	5

k

	Hund	Tens	Ones
	9	1	5
−	2	3	6

2 Solve the subtractions and then use addition to check your answers.

Example:

	H	T	O
	[4] ~~5~~	[1] 1	9
−	1	3	3
	3	8	6

	H	T	O
	[1] 3	8	6
+	1	3	3
	5	1	9

a

	H	T	O
	6	2	4
−	2	0	6

	H	T	O
+	2	0	6

b

	H	T	O
	7	1	2
−	2	3	5

	H	T	O
+	2	3	5

c

	H	T	O
	9	1	3
−	4	2	5

	H	T	O
+	4	2	5

d

	H	T	O
	8	4	4
−	3	5	6

	H	T	O
+	3	5	6

e

	H	T	O
	4	0	9
−	1	2	3

	H	T	O
+	1	2	3

SUBTRACTION WITH TRADING AND FOUR-DIGIT NUMBERS

Four–digit subtraction can also use trading.

Start at the ones column.

	Thou	Hund	Tens	Ones
	3	3 ~~4~~	12 ~~3~~	1 2
–	1	3	5	7
	2	0	7	5

Always subtract downwards.

Example 1:

	Thou	Hund	Tens	Ones
	5	9	3 ~~4~~	1 2
–	1	4	2	3
	4	5	1	9

Example 2:

	Thou	Hund	Tens	Ones
	2	5	8 ~~9~~	1 3
–		2	8	7
	2	3	0	6

Example 3:

	Thou	Hund	Tens	Ones
	6	7 ~~8~~	1 7	8
–			9	3

Example 4:

	Thou	Hund	Tens	Ones
	9	0	3	0
–	5	8	2	7

Use trading to complete these subtractions.

●

	Thou	Hund	Tens	Ones
	1 ~~2~~	13 ~~4~~	15 ~~6~~	1 2
–		8	9	8
	1	5	6	4

a

	Thou	Hund	Tens	Ones
	3	8	2	4
–	1	3	1	7

b

	Thou	Hund	Tens	Ones
	4	0	6	9
–	3	8	7	8

c

	Thou	Hund	Tens	Ones
	5	2	5	6
–	4	7	2	9

SELF CHECK Tick how you feel

Got it!	Need help...	I don't get it
☐	☐	☐

Check your answers

How many did you get correct? ☐

CATCH UP MATHS YEAR 4 BOOK A © PASCAL PRESS ISBN: 9781925726145

PRACTICE

1 Solve these subtractions.

●

	Thou	Hund	Tens	Ones
	6 ~~7~~	11 ~~2~~	18 ~~9~~	1 0
−	3	8	9	7
	3	**3**	**9**	**3**

a

	Thou	Hund	Tens	Ones
	6	0	3	5
−		2	8	6

b

	Thou	Hund	Tens	Ones
	8	1	4	2
−	3	6	8	5

c

	Thou	Hund	Tens	Ones
	9	8	0	2
−		4	3	6

d

	Thou	Hund	Tens	Ones
	5	6	3	4
−	4	1	5	7

e

	Thou	Hund	Tens	Ones
	4	1	5	3
−	3	2	6	5

2 Solve these subtractions.

●

```
 ⁹1̶ 0̶ ¹3  6
 −  3  4  2
 ----------
    6  9  4
```

b

```
   5 0 3 5
 − 4 1 9 3
 ---------
```

d

```
   7 5 1 2
 −     5 5
 ---------
```

a

```
   7 1 3 2
 −     4 9
 ---------
```

c

```
   9 5 4 2
 −   3 5 4
 ---------
```

e

```
   3 2 1 7
 − 1 2 8 9
 ---------
```

3 Mark the following with ✓ if correct and ✗ if incorrect.

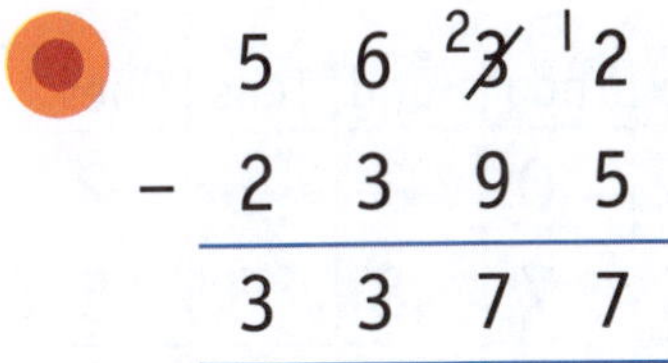

●

```
   5 6 ²3̸ ¹2
 − 2 3 9  5
 ----------
   3 3 7  7   ✗
```

Check:

```
   5 ⁵6̸ ¹²3̸ ¹2
 − 2  3   9  5
 -------------
   3  2   3  7
```

a

```
   8 ⁶7̸ ²3̸ ¹4
 − 4  3  8  5
 ------------
   4  3  6  9
```

Check:

```
   8 7 3 4
 − 4 3 8 5
 ---------
```

b

```
   ⁵6̸ ¹2 1 4
 − 3   8 0 3
 -----------
   2   4 1 1
```

Check:

```
   6 2 1 4
 − 3 8 0 3
 ---------
```

SUBTRACTING WITH TRADING AND FIVE-DIGIT NUMBERS

Trading can even be used to subtract large five-digit numbers.

Start at the ones column.

Always subtract downwards.

	T Th	Thou	Hund	Tens	Ones
	3	4 ~~5~~	1 6	3 ~~4~~	1 7
–	1	3	7	3	8
	2	1	9	0	9

Example 1:

	T Th	Thou	Hund	Tens	Ones
	8	6 ~~7~~	12 ~~3~~	1 0	4
–	3	4	9	8	3
	5	2	3	2	1

Example 2:

	T Th	Thou	Hund	Tens	Ones
	5 ~~6~~	1 2	6 ~~7~~	13 ~~4~~	1 3
–	1	5	3	5	8
	4	7	3	8	5

Example 3:

	T Th	Thou	Hund	Tens	Ones
	7	3 ~~4~~	1 1	2 ~~3~~	1 0
–		3	4	2	5

Example 4:

	T Th	Thou	Hund	Tens	Ones
	8	5	1	0	6
–			8	3	9

Use trading to complete these subtractions.

●

	T Th	Thou	Hund	Tens	Ones
	8	2 ~~3~~	15 ~~6~~	9 ~~0~~	1 1
–			7	2	5
	8	2	8	7	6

b

	T Th	Thou	Hund	Tens	Ones
	6	0	3	5	2
–	1	7	8	3	5

a

	T Th	Thou	Hund	Tens	Ones
	5	9	8	3	2
–		6	4	9	8

c

	T Th	Thou	Hund	Tens	Ones
	4	4	3	5	6
–			8	9	9

SELF CHECK Tick how you feel

Got it!	Need help...	I don't get it
☐	☐	☐

CATCH UP MATHS YEAR 4 BOOK A © PASCAL PRESS ISBN: 9781925726145

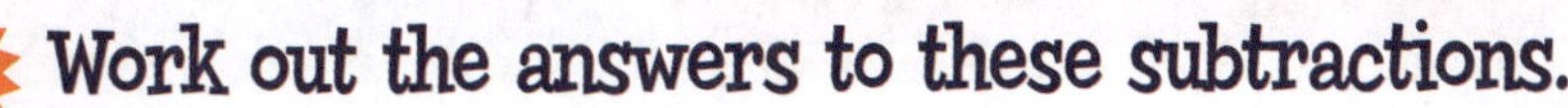

PRACTICE

1 Work out the answers to these subtractions.

	T Th	Thou	Hund	Tens	Ones
	6 ~~7~~	[1]4	8 ~~9~~	12 ~~3~~	[1]8
−	1	5	4	3	9
	5	9	4	9	9

a

	T Th	Thou	Hund	Tens	Ones
	5	9	3	4	1
−		1	7	5	0

b

	T Th	Thou	Hund	Tens	Ones
	6	8	2	1	5
−		4	7	9	1

c

	T Th	Thou	Hund	Tens	Ones
	4	2	5	8	4
−		3	7	4	8

2 Work out the answers to these subtractions.

a 89 003 − 1245 = ______

b 94 320 − 56 484 = ______

c 59 842 − 185 = ______

d 40 563 − 31 341 = ______

e 70 340 − 3435 = ______

f 40 531 − 245 = ______

3 Solve the subtractions, then check your answers by adding.

Example:

	2 ~~3~~	[1]4	6 ~~7~~	[1]3	5
−	1	5	3	4	3
	1	9	3	9	2

	[1]1	5	[1]3	4	3
+	1	9	3	9	2
	3	4	7	3	5

a 45 762 − 23 847 = ______

23 847 + ______ = ______

b 56 824 − 1949 = ______

1949 + ______ = ______

WHAT'S THE DIFFERENCE?

The difference is the amount one number is bigger or smaller than another number.

The difference between 8 and 11 is 3.

Example 1:

What is the difference between 21 and and 16?

21 − 16 = 5

The difference between 21 and 16 is 5.

Example 2:

What is the difference between 212 and 439?

439 − 212 = _____

The difference between 212 and 439 is _____.

Example 3:

Find the difference between 1536 and 1234.

 = 302

The difference between 1536 and 1234 is _____.

Your turn

What is the difference between the numbers?

● 763 and 421

763 − 421

= 342

b 742 and 132

____

= _______

a 4629 and 3113

____

= _______

c 5683 and 2767

____

= _______

SELF CHECK Tick how you feel

Got it!	Need help...	I don't get it
 ☐	 ☐	 ☐

Check your answers

How many did you get correct? ☐

CATCH UP MATHS YEAR 4 BOOK A © PASCAL PRESS ISBN: 9781925726145

PRACTICE

1 Find the difference between the numbers.

- 4932 and 1321
  4932 – 1321
  = 3611
- **b** 8374 and 2242
  ______
  = ______
- **d** 153 and 21
  ______
  = ______
- **a** 624 and 110
  ______
  = ______
- **c** 1526 and 12
  ______
  = ______
- **e** 98 and 13
  ______
  = ______

ute

sedan

four-wheel drive

hatchback

people-mover

motorbike

2 Find the difference in price between:

- **a** the people-mover and the ute
  ______
  = ______
- **b** the motorbike and the sedan
  ______
  = ______
- **c** the hatchback and the four-wheel drive
  ______
  = ______
- **d** the most expensive and least expensive vehicle
  ______
  = ______

3 If you have $10 000, how much would you have left if you bought the:

- **a** people-mover?
  ______
  = ______
- **b** ute?
  ______
  = ______
- **c** four-wheel drive?
  ______
  = ______

  ISBN: 9781925726145

ESTIMATING SUBTRACTION ANSWERS

Rounding can be used to get an answer close to the actual answer.

Example 1: Round off each number to the nearest 10 before finding the answer to the subtraction.

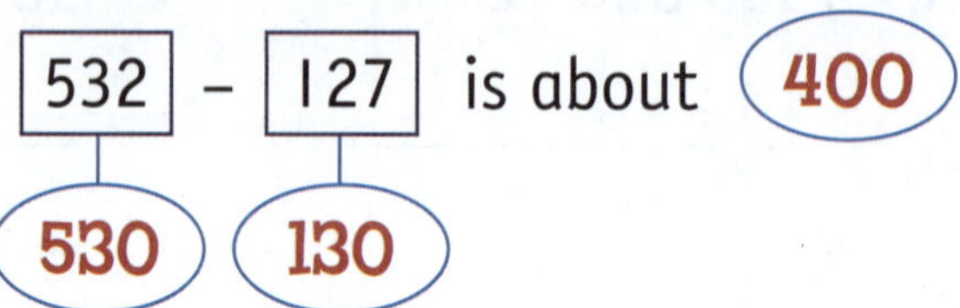

Example 2: Round off to the nearest 100 to estimate the answer.

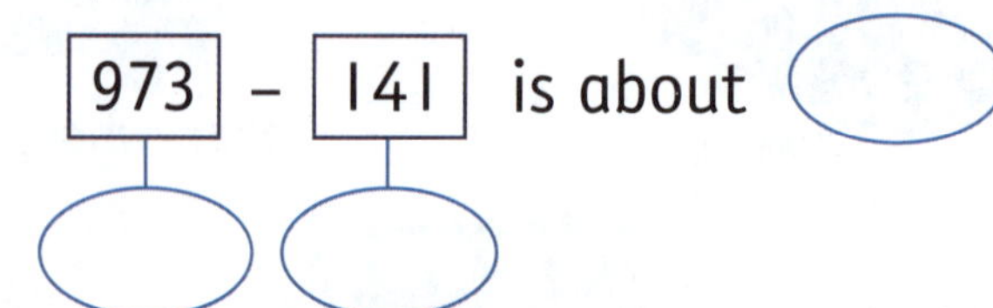

An estimate is near the answer, but it isn't perfectly accurate.

1 Round to the nearest 10 to estimate the answer.

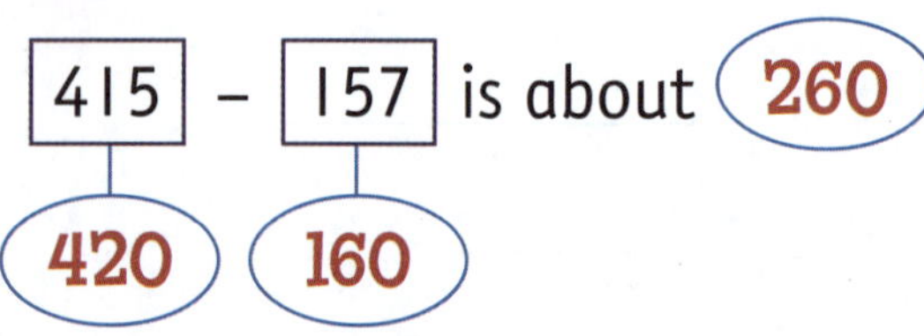

a

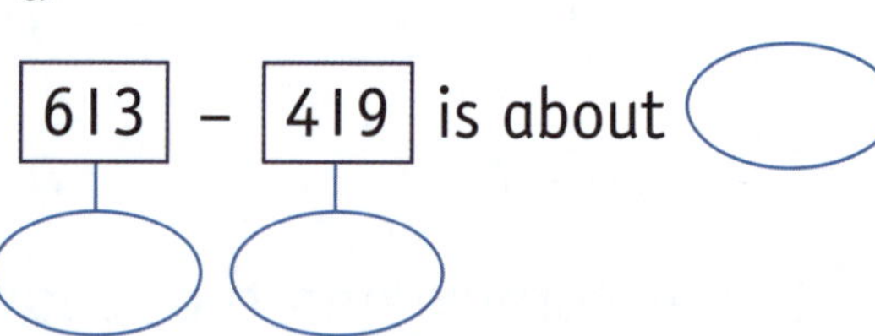

2 Round to the nearest 100 to estimate the answer.

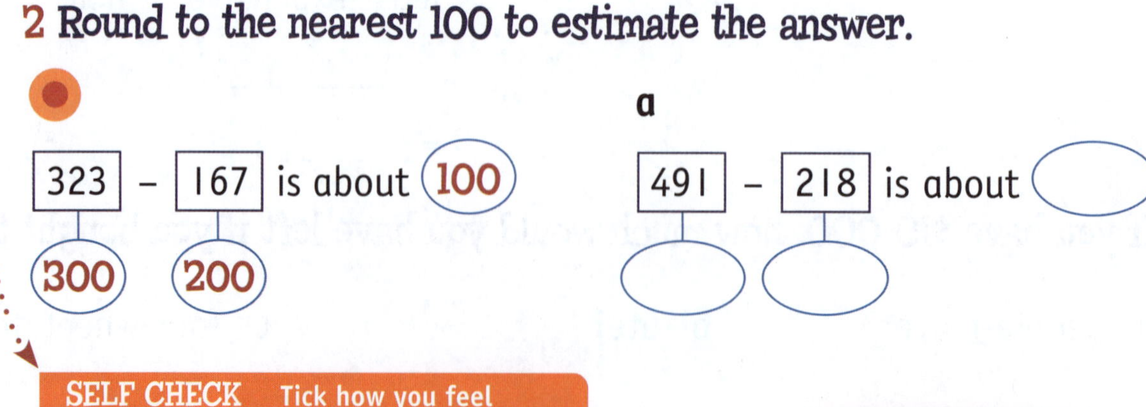

Check your answers

How many did you get correct?

CATCH UP MATHS YEAR 4 BOOK A © PASCAL PRESS ISBN: 9781925726145

PRACTICE

1 Round off to the nearest 10 to estimate the answer.

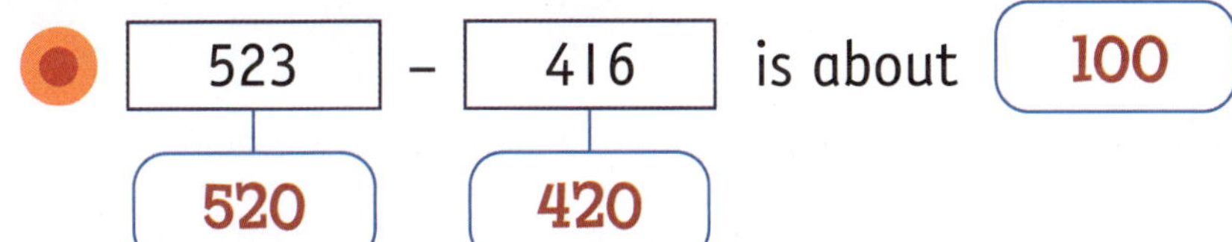

- 523 – 416 is about 100 (520, 420)

a 1352 – 383 is about ___ (___ , ___)

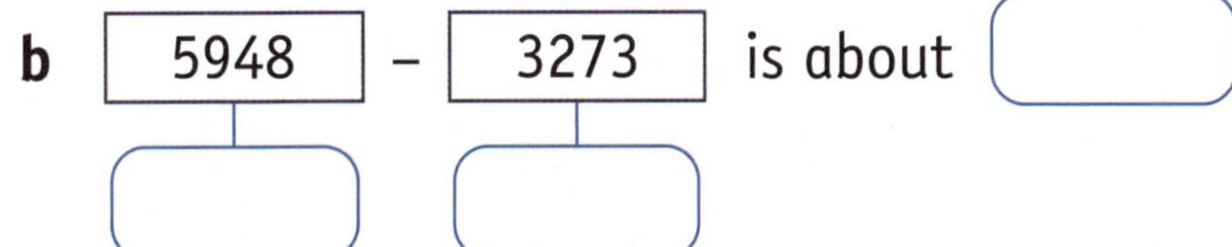

b 5948 – 3273 is about ___ (___ , ___)

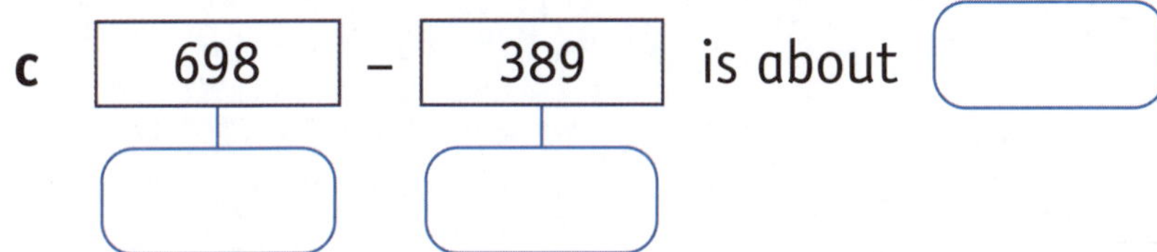

c 698 – 389 is about ___ (___ , ___)

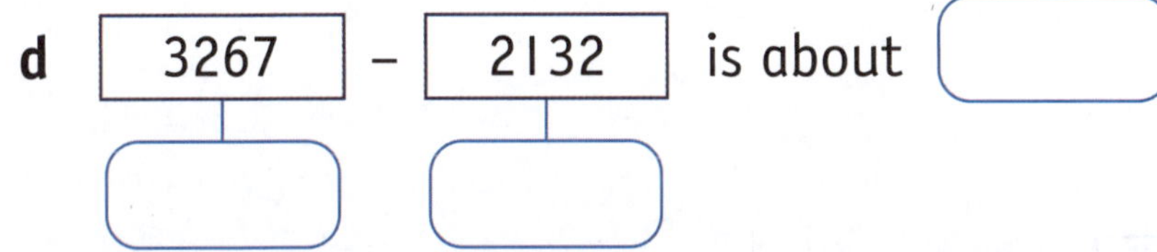

d 3267 – 2132 is about ___ (___ , ___)

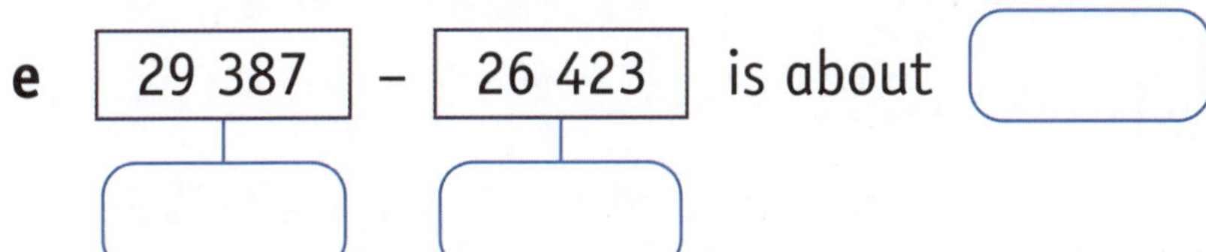

e 29 387 – 26 423 is about ___ (___ , ___)

f 542 – 169 is about ___ (___ , ___)

g 63 525 – 62 349 is about ___ (___ , ___)

Round off to the nearest 100 to estimate the answer.

●	58 462	–	35 498	is about	23 000
	58 500		35 500		
a	7389	–	4214	is about	
b	51 752	–	34 635	is about	
c	569	–	317	is about	
d	4382	–	1216	is about	
e	24 362	–	19 431	is about	
f	893	–	179	is about	
g	4373	–	2536	is about	
h	6136	–	5988	is about	

CATCH UP MATHS YEAR 4 BOOK A © PASCAL PRESS ISBN: 9781925726145

SUBTRACTION REVIEW

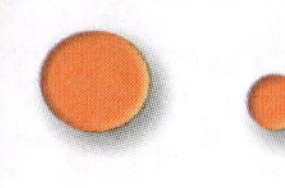

1 Solve using the jump strategy and the number line.

a 83 – 24 = ___

b 59 – 32 = ___

c 342 – 63 = _____

d 598 – 51 = _____

e 765 – 613 = _____

f 8419 – 502 = ______

g 7353 – 415 = ______

h 9823 – 2401 = ______

  ISBN: 9781925726145

REVIEW

2 Solve using the split strategy.

a 54 – 23

___ – ___ = ___

__ – __ = __

___ + __ = ___

b 85 – 42

___ – ___ = ___

__ – __ = __

___ + __ = ___

c 564 – 32

____ – ____ = ____

___ – ___ = ___

__ – __ = __

____ + ___ + __ = ____

d 754 – 41

____ – ____ = ____

___ – ___ = ___

__ – __ = __

____ + ___ + __ = ____

e 537 – 123

____ – ____ = ____

___ – ___ = ___

__ – __ = __

____ + ___ + __ = ____

f 3495 – 173

_____ – _____ = _____

____ – ____ = ____

___ – ___ = ___

__ – __ = __

_____ + ____ + ___ + __

= _____

g 7467 – 5231

_____ – _____ = _____

____ – ____ = ____

___ – ___ = ___

__ – __ = __

_____ + ____ + ___ + __

= _____

h 8342 – 5241

_____ – _____ = _____

____ – ____ = ____

___ – ___ = ___

__ – __ = __

_____ + ____ + ___ + __

= _____

  ISBN: 9781925726145

3 Complete these subtractions.

a
```
   5 3
 - 2 1
 -----
```

d
```
   9 5
 - 4 3
 -----
```

g
```
   5 0
 - 3 6
 -----
```

j
```
   4 3 4
 - 1 2 3
 -------
```

b
```
   7 5
 - 1 3
 -----
```

e
```
   7 5
 - 3 8
 -----
```

h
```
   7 1
 - 5 4
 -----
```

k
```
   5 3 7
 - 2 4 8
 -------
```

c
```
   8 4
 - 2 2
 -----
```

f
```
   8 8
 - 6 9
 -----
```

i
```
   1 2 9
 -   1 8
 -------
```

l
```
   6 2 4
 -   3 5
 -------
```

4 Complete these subtractions.

a
```
   4 5 9 7
 - 2 1 3 2
 ---------
```

e
```
   5 0 3 5
 -   2 5 6
 ---------
```

i
```
   9 2 0 0 7
 -   3 2 6 5
 -----------
```

b
```
   5 9 3 2
 -   4 5 1
 ---------
```

f
```
   6 8 4 1
 -   2 3 0
 ---------
```

j
```
   8 3 2 7 2
 - 1 4 5 8 5
 -----------
```

c
```
   6 5 3 7
 - 4 2 1 5
 ---------
```

g
```
   5 2 4 7 3
 - 2 4 5 1 8
 -----------
```

k
```
   7 2 9 5 2
 - 1 5 3 4 5
 -----------
```

d
```
   7 5 0 3
 - 1 2 8 5
 ---------
```

h
```
   6 5 9 3 7
 -   5 3 9 7
 -----------
```

l
```
   4 9 8 3 1
 - 3 9 3 5 1
 -----------
```

5 Solve these and check your answers with addition.

a

```
  5 8        
- 3 2      + 3 2
-----      -----

-----      -----
```

b

```
  7 4
- 2 5      + 2 5
-----      -----

-----      -----
```

c

```
  5 4 7
- 1 2 3      + 1 2 3
-------      -------

-------      -------
```

d

```
  6 1 4
- 2 5 3      + 2 5 3
-------      -------

-------      -------
```

6 Solve these and check your answers with addition.

a

```
  4 3 5 7
-   1 4 3      +   1 4 3
---------      ---------

---------      ---------
```

b

```
  8 7 3 4
- 2 3 0 5      + 2 3 0 5
---------      ---------

---------      ---------
```

c

```
  2 9 8 5 6
-   3 2 4 3      +   3 2 4 3
-----------      -----------

-----------      -----------
```

d

```
  5 9 8 1 0
-   1 2 4 5      +   1 2 4 5
-----------      -----------

-----------      -----------
```

e

```
  8 9 5 3 6
-   6 9 8 9      +   6 9 8 9
-----------      -----------

-----------      -----------
```


CATCH UP MATHS YEAR 4 BOOK A © PASCAL PRESS ISBN: 9781925726145

7 Complete these subtractions.

a
$$\begin{array}{r} 15379 \\ -\ 4243 \\ \hline \\ \hline \end{array}$$

d
$$\begin{array}{r} 11113 \\ -\ 1001 \\ \hline \\ \hline \end{array}$$

g
$$\begin{array}{r} 63479 \\ -\ 2463 \\ \hline \\ \hline \end{array}$$

b
$$\begin{array}{r} 28211 \\ -\ 10 \\ \hline \\ \hline \end{array}$$

e
$$\begin{array}{r} 10548 \\ -\ 10342 \\ \hline \\ \hline \end{array}$$

h
$$\begin{array}{r} 49342 \\ -\ 12341 \\ \hline \\ \hline \end{array}$$

c
$$\begin{array}{r} 22481 \\ -\ 1340 \\ \hline \\ \hline \end{array}$$

f
$$\begin{array}{r} 85415 \\ -\ 4304 \\ \hline \\ \hline \end{array}$$

i
$$\begin{array}{r} 75031 \\ -\ 6329 \\ \hline \\ \hline \end{array}$$

8 Hassan's test paper is shown below. Work out if his answers are correct. Tick the correct answers and put a cross next to the wrong ones. Give Hassan a final score out of 10.

a
$$\begin{array}{r} 53 \\ -\ 14 \\ \hline 41 \\ \hline \end{array}$$

b
$$\begin{array}{r} 72 \\ -\ 51 \\ \hline 21 \\ \hline \end{array}$$

c
$$\begin{array}{r} 157 \\ -\ 48 \\ \hline 111 \\ \hline \end{array}$$

d
$$\begin{array}{r} {}^{5}\not{6}\ {}^{1}0\ 3 \\ -\ 1\ 4\ 2 \\ \hline 4\ 6\ 1 \\ \hline \end{array}$$

e
$$\begin{array}{r} 1385 \\ -\ 735 \\ \hline 1450 \\ \hline \end{array}$$

f
$$\begin{array}{r} 4\ {}^{7}\not{8}\ {}^{1}5\ 2 \\ -\ 1\ 0\ 7\ 1 \\ \hline 4\ 7\ 8\ 1 \\ \hline \end{array}$$

g
$$\begin{array}{r} 5982 \\ -\ 2431 \\ \hline 3551 \\ \hline \end{array}$$

h
$$\begin{array}{r} 2\ {}^{7}\not{8}\ {}^{13}\not{4}\ {}^{1}6\ 0 \\ -\ 1\ 3\ 5\ 7\ 9 \\ \hline 1\ 4\ 8\ 9\ 9 \\ \hline \end{array}$$

i
$$\begin{array}{r} 73972 \\ -\ 12241 \\ \hline 61732 \\ \hline \end{array}$$

j
$$\begin{array}{r} {}^{5}\not{6}\ {}^{1}4\ {}^{1}\not{2}\ {}^{1}0\ 5 \\ -\ 5\ 0\ 3\ 0 \\ \hline 5\ 9\ 1\ 7\ 5 \\ \hline \end{array}$$

☐ out of 10

REVIEW

9 What is the difference?

a 31 – 15 = ______

b 49 – 31 = ______

c 57 – 32 = ______

d 143 – 48 = ______

e 531 – 42 = ______

f 241 – 18 = ______

g 5243 – 137 = ______

h 6245 – 314 = ______

i 8473 – 1852 = ______

j 36 241 – 2342 = ______

k 58 245 – 37 841 = ______

l 73 249 – 62 109 = ______

10 Round to the nearest 10, then estimate the answer.

a 529 – 137 is about ______

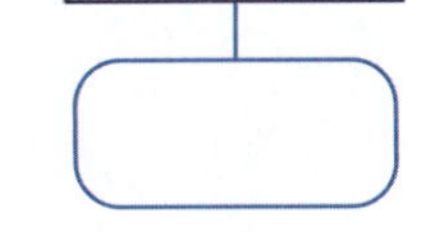

b 324 – 213 is about ______

c 3521 – 2989 is about ______

d 7436 – 4321 is about ______

e 8265 – 1653 is about ______

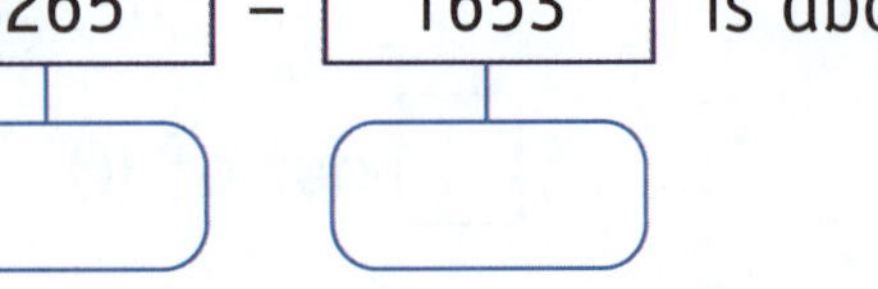

CATCH UP MATHS YEAR 4 BOOK A © PASCAL PRESS ISBN: 9781925726145

11 Round to the nearest 100, then estimate the answer.

a 637 – 241 is about ☐

☐ ☐

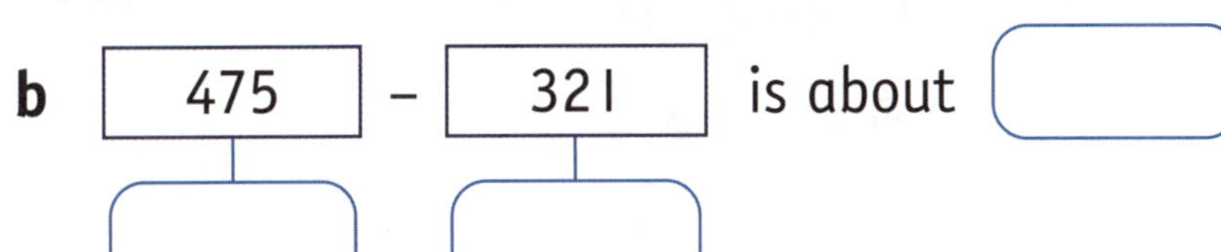

b 475 – 321 is about ☐

☐ ☐

c 1831 – 428 is about ☐

☐ ☐

d 2895 – 342 is about ☐

☐ ☐

e 24 357 – 15 659 is about ☐

☐ ☐

f 53 273 – 45 284 is about ☐

☐ ☐

g 54 321 – 51 839 is about ☐

☐ ☐

h 73 246 – 41 288 is about ☐

☐ ☐

GROUPS AND ROWS

Objects can be put in groups or rows to help solve multiplication.

Example 1:

2 rows of 5

These circles are in rows.

There are 2 rows of circles and 5 circles in each row.

Example 2:

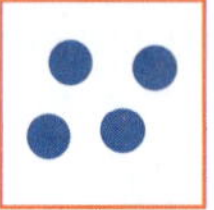 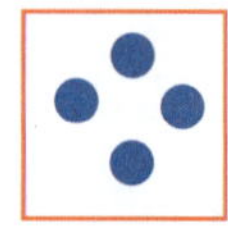 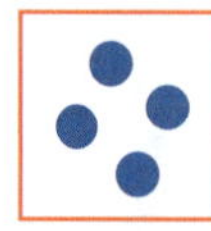

3 groups of 4

These circles are in groups.

There are 3 groups of circles and 4 circles in each group.

Example 3:

__ row of 6

These circles are in a row.

There is __ row of circles and __ circles in the row.

Example 4:

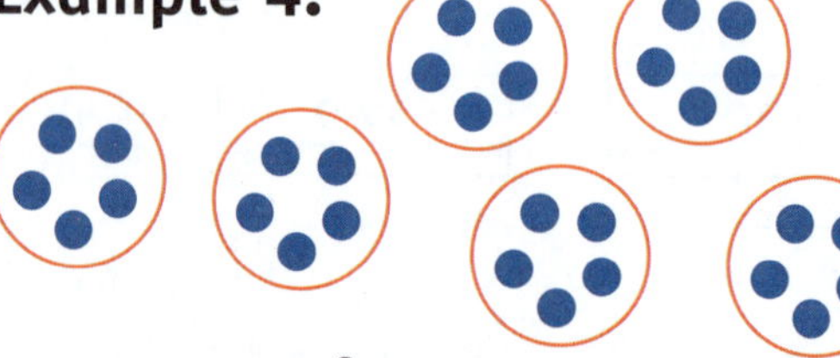

__ groups of 5

These circles are in groups.

There are __ groups of circles and __ circles in each group.

SCAN to watch video

1 How many circles are in each row?

3 in each row

a __ in each row

b __ in each row

2 How many circles are in each group?

4 in each group

a 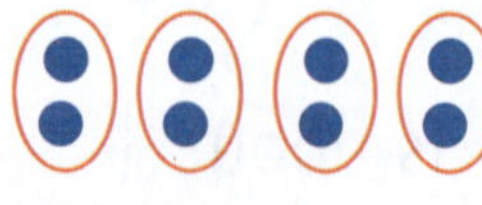

__ in each group

b 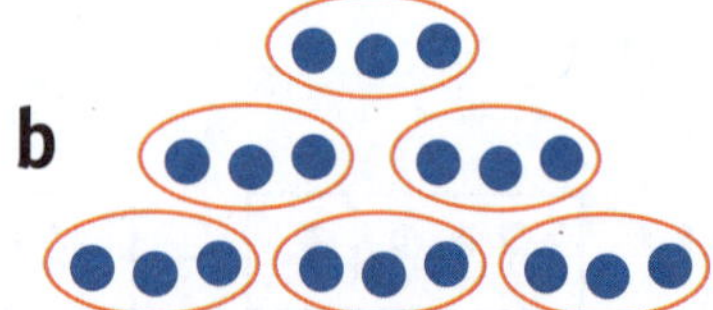

__ in each group

SELF CHECK Tick how you feel

Got it!	Need help...	I don't get it
 ☐	 ☐	  ☐

Check your answers

How many did you get correct? ☐

CATCH UP MATHS YEAR 4 BOOK A © PASCAL PRESS ISBN: 9781925726145

PRACTICE

1 Draw circles to make each group equal.

3 groups of 4

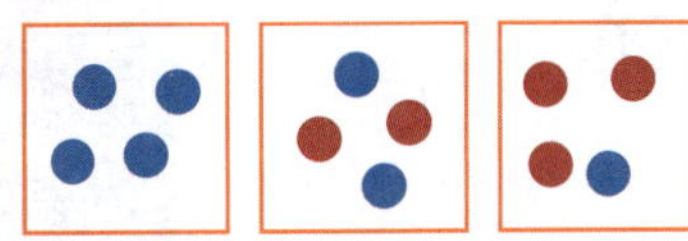

a 4 groups of 3

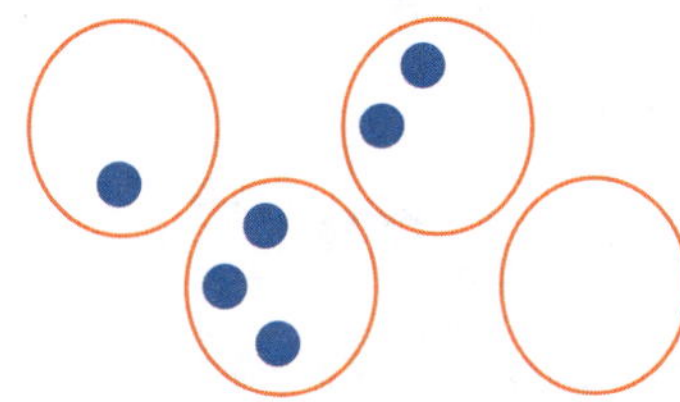

b 3 groups of 8

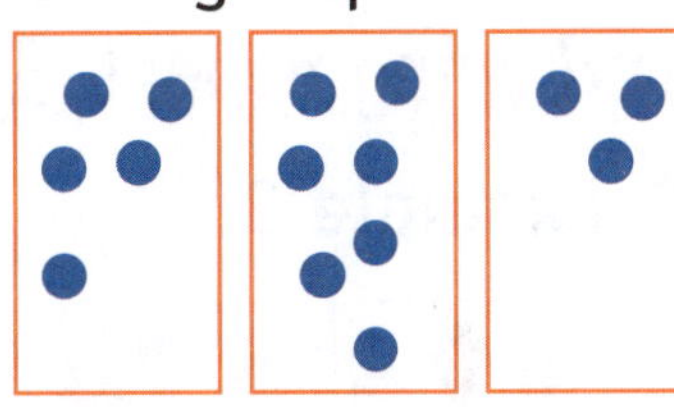

c 5 groups of 2

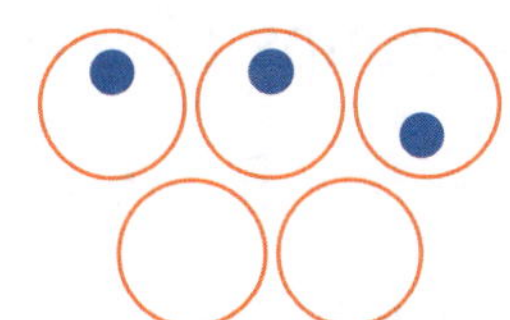

d 4 groups of 5

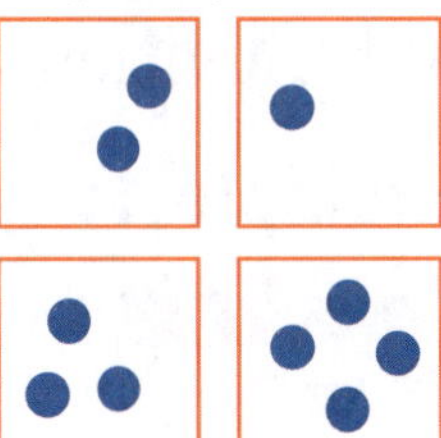

e 6 groups of 5

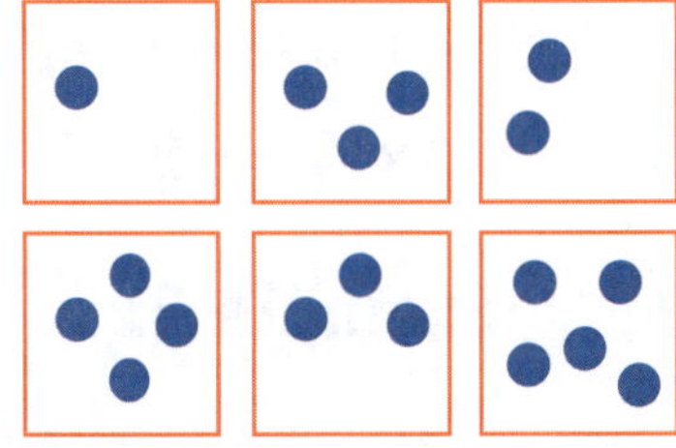

2 Make these rows equal.

4 rows of 3

a 5 rows of 4

b 3 rows of 6

c 2 rows of 7

3 Fill in the missing numbers.

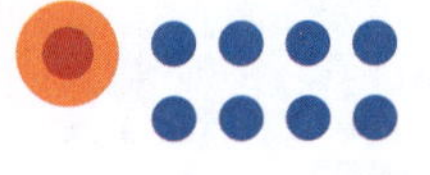

2 rows of 4

a

__ rows of __

b

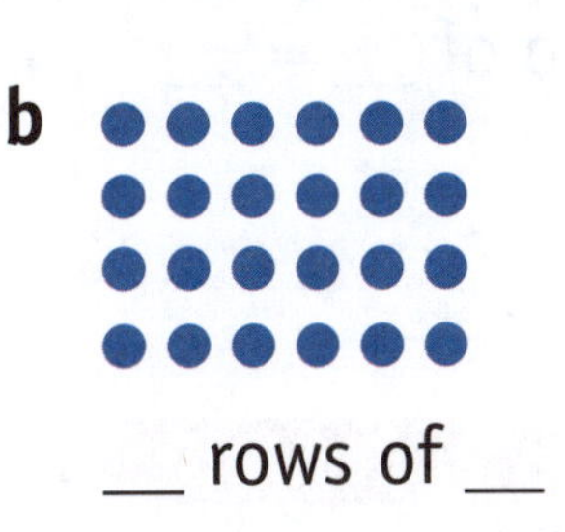

__ rows of __

c

__ rows of __

d

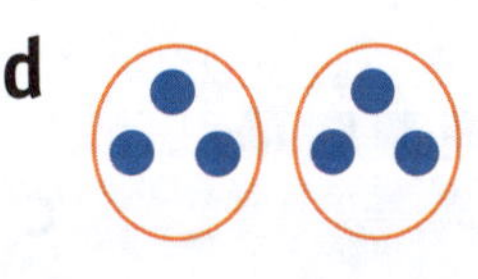

__ groups of __

e

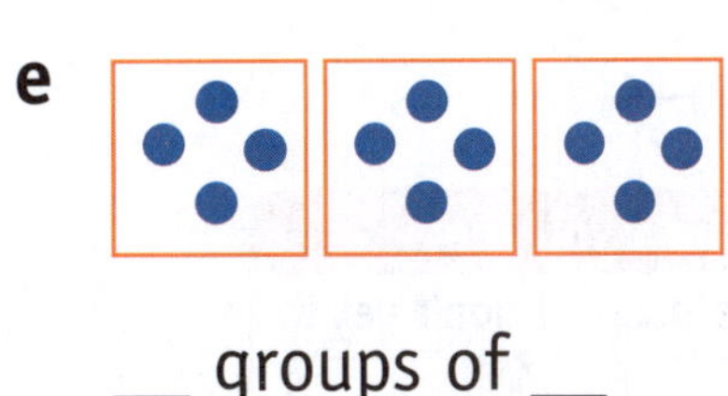

__ groups of __

f

__ groups of __

g

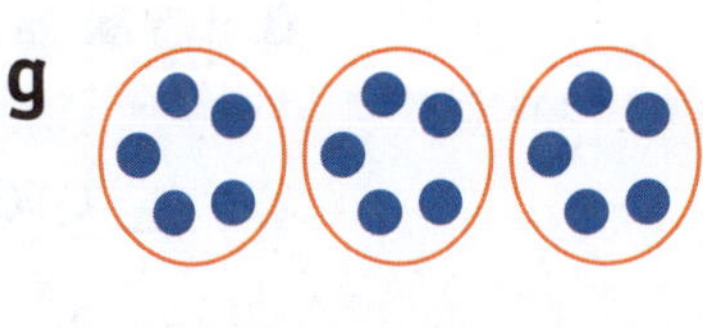

__ groups of __

h

__ rows of __

  ISBN: 9781925726145

REPEATED ADDITION TO SOLVE MULTIPLICATION

Multiplication can be solved using repeated addition.

Example 1:

3 rows of 4 = 12

4 + 4 + 4 = 12

3 × 4 = 12

Example 2:

3 groups of 6 = 18

6 + 6 + 6 = 18

3 × 6 = 18

Example 3:

4 groups of 5 = ___

5 + 5 + 5 + 5 = ___

4 × 5 = ___

SCAN to watch video

Example 4:

___ rows of 5 = 30

5 + 5 + 5 + 5 + 5 + 5 = ___

___ × 5 = ___

Your turn

Fill in the missing numbers.

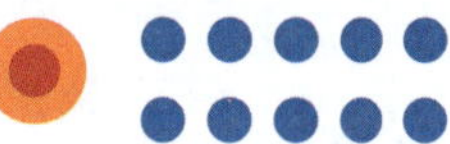

2 rows of 5 = 10

5 + 5 = 10

2 × 5 = 10

a 

___ groups of ___ = ___

___ + ___ + ___ = ___

___ × ___ = ___

b  

___ rows of ___ = ___

___ + ___ + ___ = ___

___ × ___ = ___

c  

___ group of ___ = ___

___ = ___

___ × ___ = ___

SELF CHECK Tick how you feel

Got it!	Need help...	I don't get it

Check your answers

How many did you get correct?

CATCH UP MATHS YEAR 4 BOOK A © PASCAL PRESS ISBN: 9781925726145

PRACTICE

1 Complete.

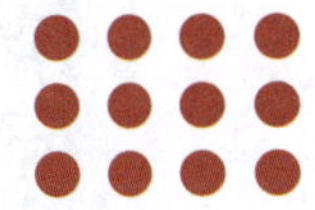

3 rows of 4

3 rows of 4 = 12

4 + 4 + 4 = 12

3 × 4 = 12

b 3 rows of 3

3 rows of 3 = __

__ + __ + __ = __

__ × __ = __

a 5 groups of 1

5 groups of 1 = __

__ + __ + __ + __ + __ = __

__ × __ = ___

c 8 groups of 2

8 groups of 2 = __

__ + __ + __ + __ + __ + __

+ __ + __ = __

__ × __ = __

2 Fill in the table.

	8 + 8 + 8	=	8 × 3	=	24
a	6 + 6	=		=	
b	2 + 2 + 2 + 2	=		=	
c	9 + 9 + 9	=		=	
d	1 + 1 + 1 + 1 + 1	=		=	
e	4 + 4 + 4 + 4 + 4	=		=	
f	5 + 5 + 5 + 5 + 5	=		=	
g	10 + 10 + 10 + 10	=		=	
h	4 + 4 + 4 + 4 + 4 + 4 + 4	=		=	
i	7 + 7 + 7 + 7 + 7	=		=	
j	6 + 6 + 6 + 6 + 6 + 6	=		=	

COMMUTATIVE PROPERTY

The commutative property of addition and multiplication means that the order of the numbers being added or multiplied can be changed and the answer is still the same.

Example 1:

4 + 3 = 7 | 3 + 4 = 7

Example 2:

2 × 8 = 16 | 8 × 2 = 16

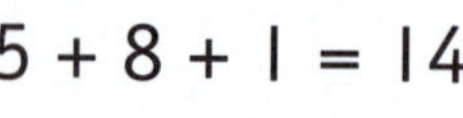

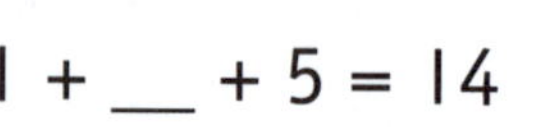

Example 3:

5 + 8 + 1 = 14

1 + ___ + 5 = 14

8 + ___ + 1 = 14

Example 4:

4 × 3 = ___

___ × ___ = ___

Check your answer on the video!

Your turn

Show that:

3 × 4 = 4 × 3

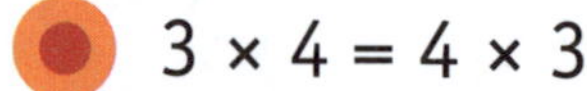

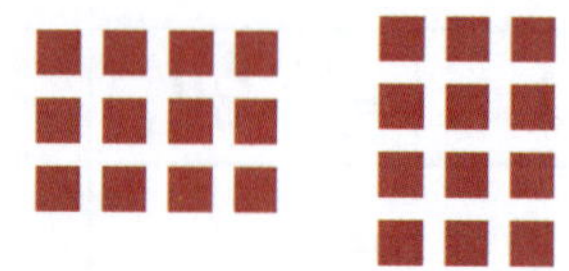

a 6 + 3 = 3 + 6

b 2 + 4 + 8 = 8 + 2 + 4

c 5 × 2 = 2 × 5

d 3 × 7 = 7 × 3

e 5 + 3 + 2 = 2 + 3 + 5

SELF CHECK Tick how you feel

Got it!	Need help...	I don't get it
 ☐	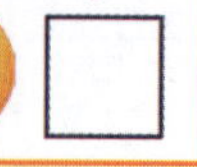 ☐	 ☐

Check your answers
How many did you get correct? ☐

CATCH UP MATHS YEAR 4 BOOK A © PASCAL PRESS ISBN: 9781925726145

PRACTICE

1 **Complete the tables.**

●	2 × 4 = 4 × 2
a	9 × 3 =
b	4 + 5 =
c	7 × 1 =
d	8 + 2 =
e	9 + 3 =
f	8 × 9 =
g	9 × 10 =
h	1 × 11 =
i	2 × 6 =

j	6 + 4 =
k	3 × 8 =
l	1 + 9 + 3 =
m	7 + 4 + 2 =
n	3 × 4 =
o	5 × 1 =
p	8 + 8 =
q	9 × 4 =
r	5 × 9 =
s	6 × 7 =

2 **Show that:**

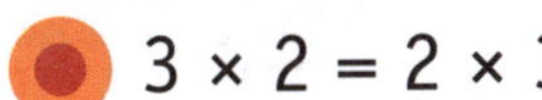

● 3 × 2 = 2 × 3

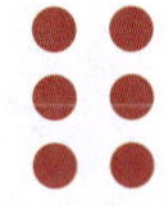 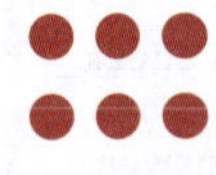

Both equal 6

b 3 + 2 = 2 + 3

Both equal ___

d 5 + 4 = 4 + 5

Both equal ___

a 7 × 4 = 4 × 7

Both equal ___

c 9 × 3 = 3 × 9

Both equal ___

e 6 × 3 = 3 × 6

Both equal ___

INVERSE OPERATIONS OF MULTIPLICATION AND DIVISION

Inverse means opposite. In Maths, × is the opposite of ÷.
Multiplication is the inverse operation of division.

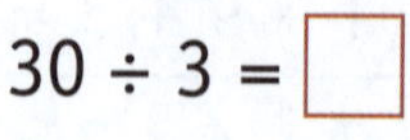

30 ÷ 3 = ☐ ☐ = 10

3 × ☐ = 30

Example 1:

The inverse of 15 ÷ 5 = 3 is 5 × 3 = 15.

Example 2:

The inverse of 4 × 3 = 12 is 12 ÷ 4 = 3.

Example 3:

The inverse of 35 ÷ 7 = 5 is 7 × __ = 35.

Example 4:

The inverse of 6 × 10 = 60 is ___ ÷ 6 = ___.

Multiplication is the inverse of division, and division is the inverse of multiplication.

Check your answer on the video!

Fill in the missing numbers.

● 20 ÷ 4 = 5
4 × 5 = 20

a 24 ÷ 6 = __
6 × __ = 24

b 32 ÷ 8 = __
8 × __ = 32

c 12 ÷ 3 = __
3 × __ = 12

d 28 ÷ 7 = __
7 × __ = 28

e 9 ÷ 3 = __
3 × __ = 9

SELF CHECK Tick how you feel		
Got it! ☐	Need help... ☐	I don't get it ☐

Check your answers
How many did you get correct? ☐

CATCH UP MATHS YEAR 4 BOOK A © PASCAL PRESS ISBN: 9781925726145

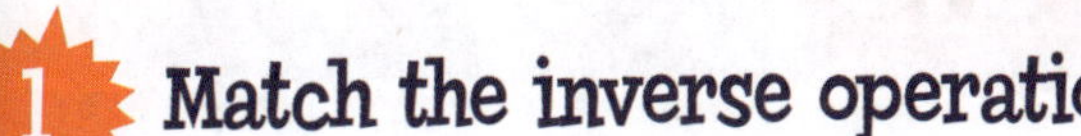

PRACTICE

1 Match the inverse operations.

● 2 × 4 = 8	21 ÷ 3 = 7
a 5 × 5 = 25	48 ÷ 4 = 12
b 7 × 3 = 21	9 ÷ 1 = 9
c 10 × 4 = 40	36 ÷ 6 = 6
d 6 × 6 = 36	40 ÷ 4 = 10
e 9 × 1 = 9	25 ÷ 5 = 5
f 12 × 4 = 48	8 ÷ 4 = 2
g 3 × 8 = 24	24 ÷ 8 = 3

2 Complete the table.

	×	÷	×	÷
●	3 × 4 = 12	12 ÷ 4 = 12	4 × 3 = 12	12 ÷ 3 = 4
a	6 × 2 =		2 × 6 =	
b		10 ÷ 5 = 2		
c			5 × 8 =	
d	1 × 7 =			
e				24 ÷ 2 =
f			5 × 4 =	
g	7 × 4 =			
h		32 ÷ 4 =		
i				33 ÷ 3 =
j	6 × 7 =			
k		90 ÷ 10 =		
l				18 ÷ 9 =

SKIP COUNTING

When counting by ones, we count every number:

1, 2, 3, 4, 5 ...

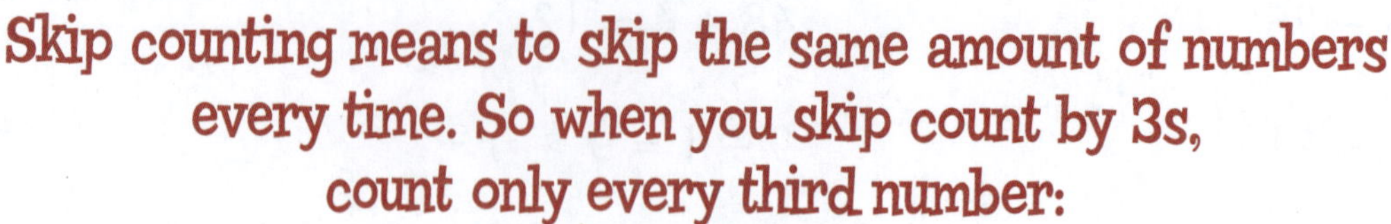

Skip counting means to skip the same amount of numbers every time. So when you skip count by 3s, count only every third number:

3, 6, 9, 12, 15 ...

You can skip count forwards or backwards.

SCAN to watch video

Example 1:

Skip count forwards by 5, from 5 to 50.

5, 10, 15, 20, 25, 30, 35, 40, 45, 50

When you practise counting in equal groups, it helps you learn to multiply.

Example 2:

Skip count backwards by 7, starting at 42.

42, 35, 28, 21, 14, 7

Example 3:

Skip count forwards by 4, starting at 12.

12, 16, 20, 24, ___, ___, ___, ___, ___

Example 4:

Skip count backwards by 9, starting at 99.

99, 90, 81, 72, ___, ___, ___, ___, ___

Your turn

1 Skip count forwards by:

- 8: 8, 16, 24, 32, 40
- a 6: 6, ___, ___, ___, ___
- b 4: 4, ___, ___, ___, ___

2 Skip count backwards by:

- 5: 40, 35, 30, 25, 20
- a 2: 24, ___, ___, ___, ___
- b 4: 24, ___, ___, ___, ___

SELF CHECK Tick how you feel		
Got it! ☐	Need help... ☐	I don't get it ☐

Check your answers
How many did you get correct? ☐

CATCH UP MATHS YEAR 4 BOOK A © PASCAL PRESS ISBN: 9781925726145

PRACTICE

1 Fill in the circles.

● Count by 2s

10, 12, 14, 16, 18, 20, 22, 24

b Count by 6s

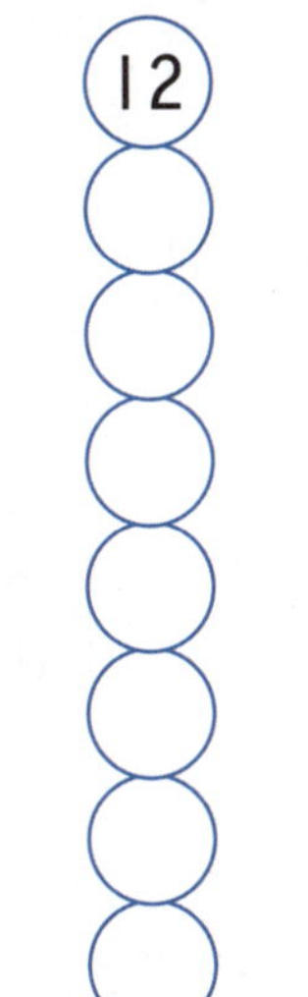

12

d Count back by 3s

36

f Count back by 9s

81

a Count by 4s

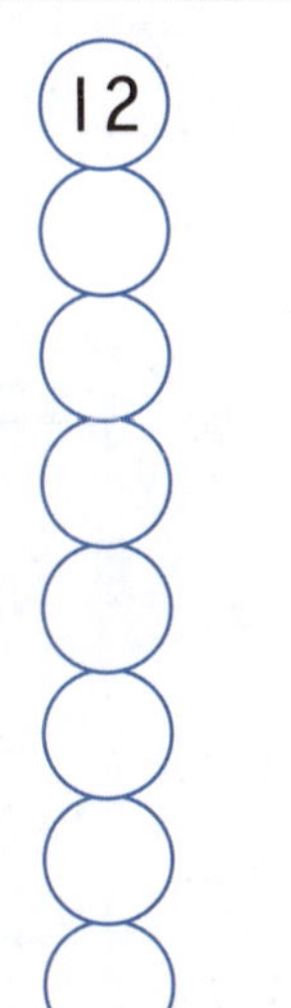

12

c Count by 7s

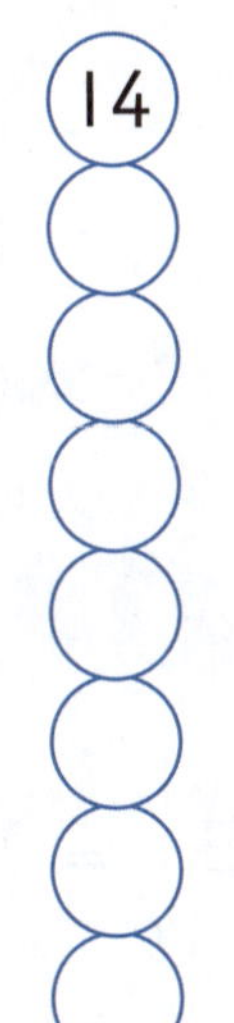

14

e Count back by 8s

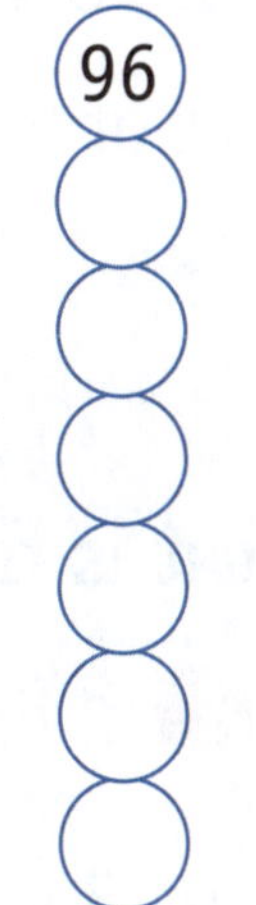

96

g Count back by 5s

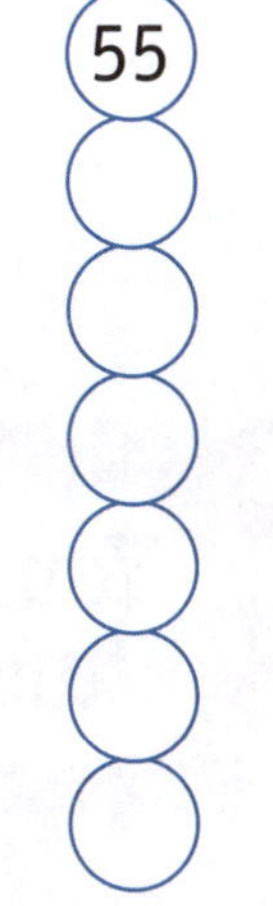

55

2 Fill in the missing numbers.

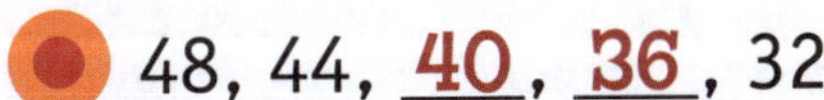

● 48, 44, 40, 36, 32

a 9, 18, ____, ____, 45

b 14, ____, 28, ____, ____

c 12, 18, ____, ____, 36

d 88, 80, ____, 64, ____

e 1, 2, ____, ____, 5

f 24, 32, ____, ____, 56

g 12, ____, ____, 21, 24

h 60, 55, ____, ____, 40

i 36, 33, ____, 27, ____

PRODUCT OF NUMBERS

When numbers are multiplied, the answer is called the product.

Example 1: 3 × 8 = 24

The **product** of 3 and 8 is 24.

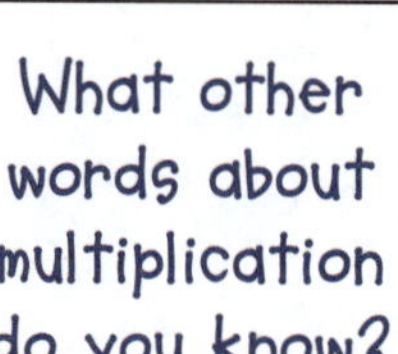

Example 2: 7 × 4 = 28

The **product** of 7 and 4 is 28.

Example 3: 6 × 2 = 12

The  of 6 and __ is 12.

Example 4: 8 × 8 = 64

The **product** of 8 and __ is ___.

What is the product?

- 2 and 5   2 × 5 = 10
- **a** 3 and 6 ________ = ___
- **b** 1 and 9 ________ = ___
- **c** 5 and 5 ________ = ___
- **d** 2 and 8 ________ = ___
- **e** 7 and 7 ________ = ___

SELF CHECK Tick how you feel		
Got it! ☐	Need help... ☐	I don't get it ☐

Check your answers
How many did you get correct? ☐

CATCH UP MATHS YEAR 4 BOOK A © PASCAL PRESS ISBN: 9781925726145

PRACTICE

1 Complete the statements.

- The product of 5 and 2 is 10.
- a The product of 6 and 7 is ___.
- b The product of 8 and 6 is ___.
- c The product of 3 and 9 is ___.
- d The product of 4 and 4 is ___.
- e The product of 7 and 1 is ___.
- f The product of 0 and 4 is ___.
- g The product of 6 and 6 is ___.
- h The product of 10 and 12 is ___.
- i The product of 9 and 8 is ___.

2 Match the number sentence with its product.

●	4 × 8	24
a	5 × 7	40
b	2 × 9	32
c	7 × 9	63
d	3 × 10	7
e	8 × 5	35
f	1 × 7	30
g	6 × 4	18

3 Circle the correct product.

- The product of 6 and 2 is 12 21 3.
- a The product of 5 and 9 is 9 45 54.
- b The product of 8 and 7 is 65 56 42.
- c The product of 2 and 12 is 14 6 24.
- d The product of 8 and 10 is 88 80 8.

FACTORS AND MULTIPLES

Factors
A factor is a number that can multiply with another to give a multiple.

Multiples
A multiple is the answer you get when you multiply two numbers together.

3 and 4 are factors of 12. 12 is a multiple of 3 and 4.

3 × 4 = 12 ← multiple

factors

Example 1: What are the factors of 15?

1 × 15 = 15
3 × 5 = 15

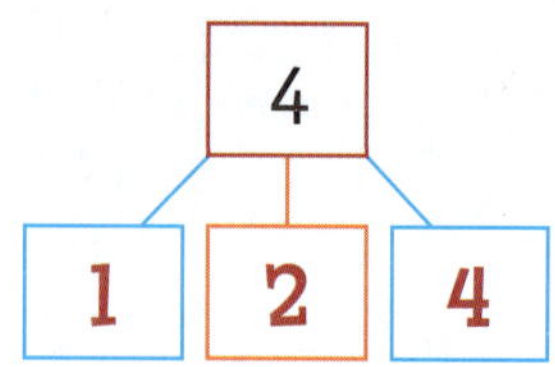

1, 3, 5 and 15 are the factors of 15.

15 is a multiple of 1, 3, 5 and 15.

Example 2: What are the factors of 4?

4 → 1, 2, 4

1, 2 and 4 are the factors of 4.

4 is a multiple of 1, 2 and 4.

Example 3: What are the factors of 8?

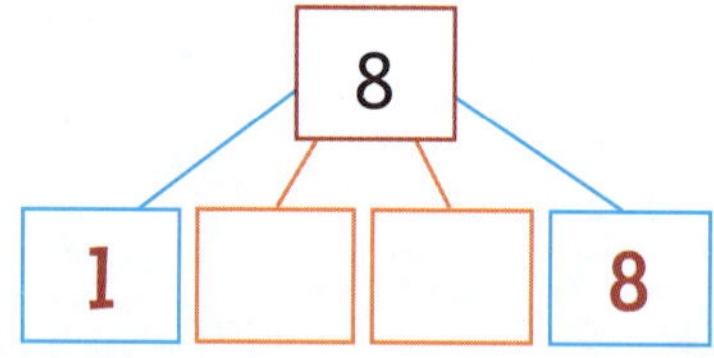

Your turn

Write T for true or F for false.

- 4 is a factor of 20. T
- **a** A factor of 10 is 3. ___
- **b** 50 is a multiple of 25. ___
- **c** 2 is a factor of 4. ___
- **d** A multiple of 10 is 25. ___
- **e** A factor of 50 is 5. ___

Check your answers
How many did you get correct?

CATCH UP MATHS YEAR 4 BOOK A © PASCAL PRESS ISBN: 9781925726145

PRACTICE

1 Write the factors.

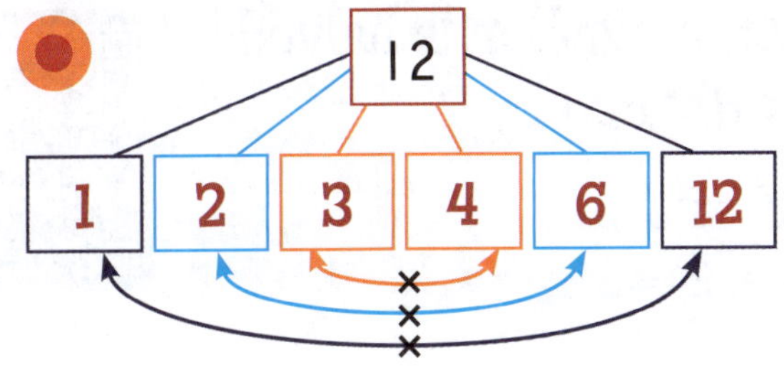

a

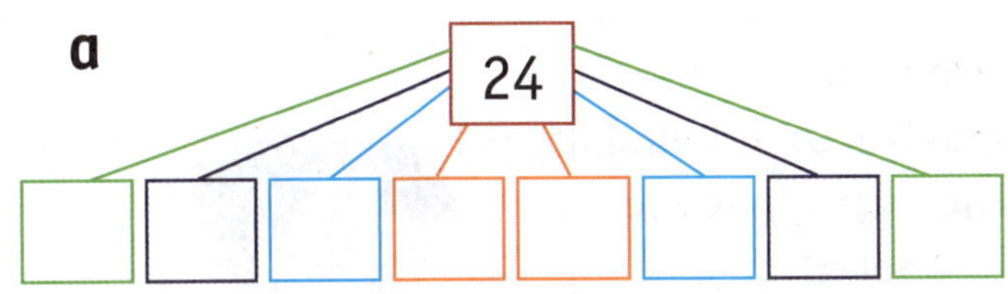

b

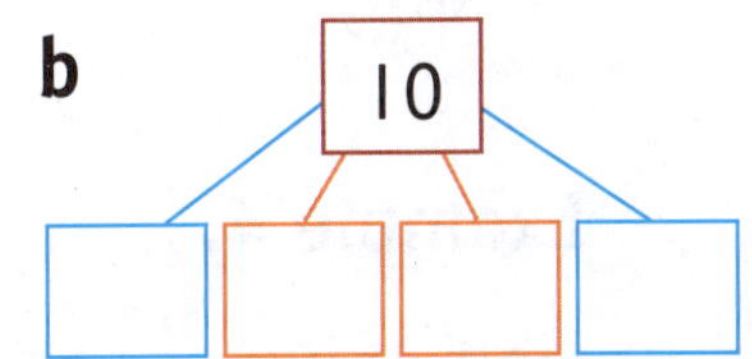

c

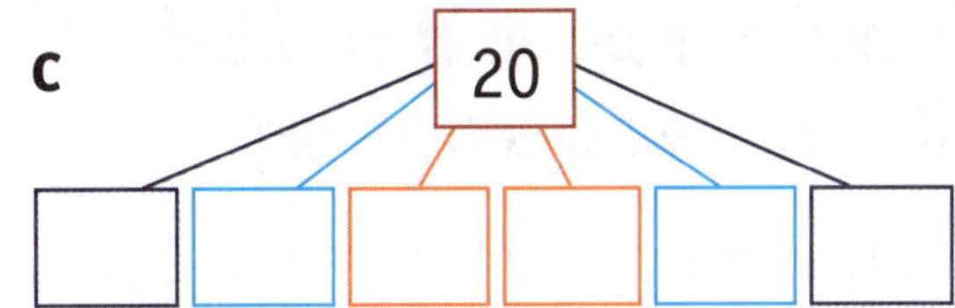

d

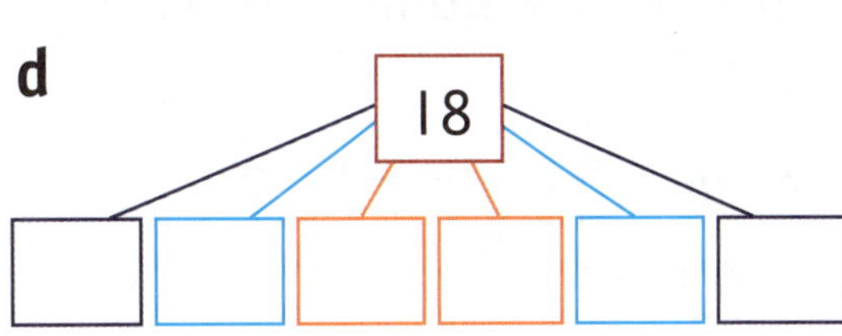

e

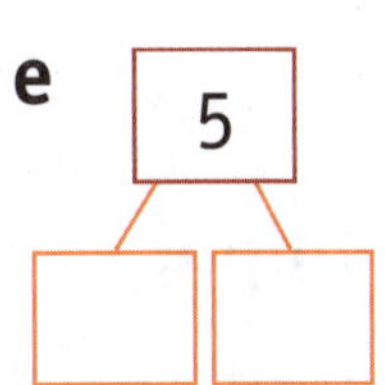

2 Write the first five multiples.

- 8: 8, 16, 24, 32, 40
- a 4: ___, ___, ___, ___, ___
- b 7: ___, ___, ___, ___, ___
- c 6: ___, ___, ___, ___, ___
- d 5: ___, ___, ___, ___, ___
- e 10: ___, ___, ___, ___, ___

3 Write in the missing factors.

- 3 × 4 = 12
- a 4 × __ = 20
- b __ × 3 = 18
- c 4 × __ = 36
- d 5 × __ = 45
- e 9 × __ = 63
- f __ × 4 = 16
- g 6 × __ = 30
- h 7 × __ = 63
- i __ × 8 = 64
- j __ × 9 = 72
- k 8 × __ = 56

4 Complete the tables.

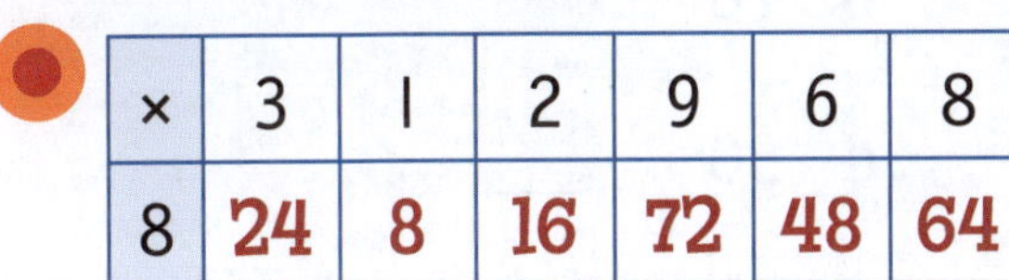

×	3	1	2	9	6	8
8	24	8	16	72	48	64

a

×	2	5	7	8	3	9
7						

b

×	3	5	9	8	7	4
6						

c

×	1	9	10	12	6	8
9						

DOUBLING AND HALVING

When a number is doubled it becomes twice as big.

You can double a number by:

- adding the number to itself
  3 + 3 = 6
- or multiplying it by 2
  3 × 2 = 6

When a number is halved it is divided by 2.

Half of 6
= 6 ÷ 2 = 3

Doubling is the opposite of halving. They are inverse operations.

Example 1:

Double 9

9 + 9 = 18

9 × 2 = 18

9	9
18	

Example 2:

Half of 16

16 ÷ 2 = 8

8	16
8	

Example 3:

Double 11

11 + 11 = ___

11 × ___ = ___

11	

Example 4:

Half of 14

14 ÷ 2 = ___

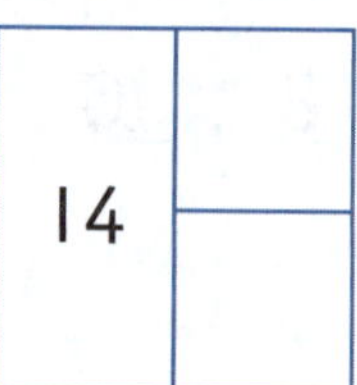

Check your answer on the video!

Your turn

1 What is half?

a 20 ___

b 2 ___

c 100 ___

2 Double each number.

a 6 ___

b 10 ___

c 50 ___

SELF CHECK Tick how you feel		
Got it! ☐	Need help... ☐	I don't get it ☐

Check your answers
How many did you get correct? ☐

CATCH UP MATHS YEAR 4 BOOK A © PASCAL PRESS ISBN: 9781925726145

PRACTICE

1 Double or halve the numbers.

Example		
7	7	14

Item	Parts	Whole
a	2, 2	
b	,	24
c	16, 16	
d	,	30
e	,	144
f	,	130
g	200, 200	
h	,	120
i	45, 45	
j	,	250
k	,	350

2 Complete the tables.

	Half	Number	Double
Example	5	10	20
a		20	
b		38	
c		90	
d		160	

	Half	Number	Double
e		270	
f		480	
g		500	
h		50	
i		110	

3 Complete the tables.

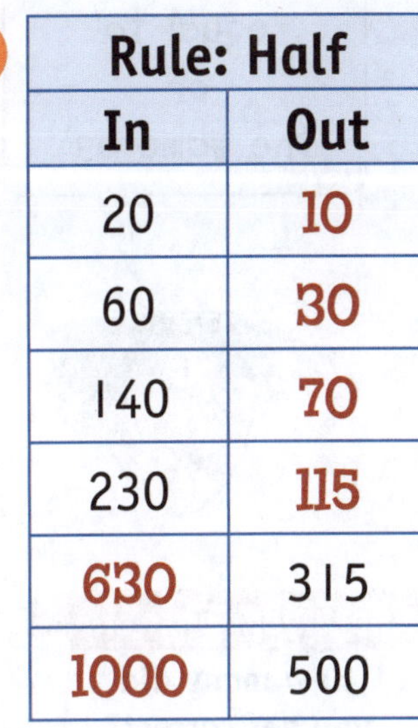

Rule: Half	
In	**Out**
20	10
60	30
140	70
230	115
630	315
1000	500

a

Rule: Double	
In	**Out**
	40
30	
	80
	300
125	
	800

b

Rule: Half	
In	**Out**
16	
28	
	58
	106
44	
	380

EQUIVALENT NUMBER RELATIONSHIPS

Equivalent means to have the same value.

SCAN to watch video

Example 1:

4 × 2 is equivalent to 1 × 8.

equal to or the same as

The answer to both is 8.

4 × 2 = 1 × 8

Example 2:

3 × 4 is equivalent to 2 × 6.

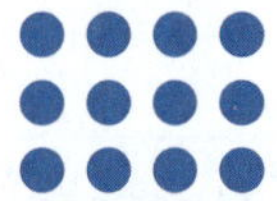

The answer to both is 12.

3 × 4 = 2 × 6

Example 3:

4 × 4 = 2 × __

The answer to both is ___.

Example 4:

3 × __ = 1 × 9

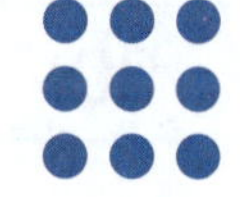

The answer to both is __.

Check your answer on the video!

Your turn

Join the equivalent multiplications.

	4 × 4	2 × 14
a	6 × 3	8 × 2
b	7 × 4	3 × 12
c	6 × 6	2 × 9
d	1 × 6	3 × 2
e	12 × 2	4 × 6

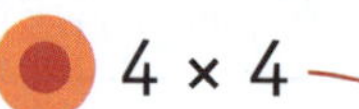

'Equivalent to' means 'equal to' or 'the same as'.

SELF CHECK Tick how you feel

Got it!	Need help...	I don't get it
 ☐	 ☐	☐

Check your answers

How many did you get correct?

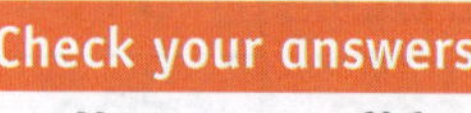

CATCH UP MATHS YEAR 4 BOOK A © PASCAL PRESS ISBN: 9781925726145

PRACTICE

1 Write an equivalent multiplication.

- 2 × 8 = 3 × 6
- **a** 1 × 9 = ______
- **b** 2 × 12 = ______
- **c** 1 × 10 = ______
- **d** 4 × 3 = ______
- **e** 6 × 5 = ______
- **f** 8 × 6 = ______
- **g** 15 × 1 = ______
- **h** 4 × 5 = ______

2 Cross out the multiplication that is NOT equivalent.

●	5 × 2	2 × 5	~~5 × 3~~	1 × 10
a	2 × 12	6 × 4	8 × 3	9 × 2
b	6 × 5	15 × 2	30 × 1	1 × 12
c	4 × 11	10 × 4	5 × 8	20 × 2
d	9 × 2	6 × 3	18 × 1	3 × 7
e	2 × 10	5 × 3	4 × 5	10 × 2
f	8 × 1	2 × 4	6 × 2	1 × 8
g	2 × 6	12 × 1	3 × 4	12 × 11

3 Write the missing number.

- 3 × 8 = 12 × 2
- **a** 4 × 10 = 5 × ___
- **b** 12 × ___ = 9 × 4
- **c** 6 × ___ = 12 × 12
- **d** 4 × 0 = 0 × ___
- **e** 10 × 1 = ___ × 2
- **f** 15 × 1 = 5 × ___
- **g** 15 × 2 = ___ × 6
- **h** ___ × 6 = 2 × 9
- **i** 4 × 4 = 8 × ___
- **j** 12 × ___ = 8 × 6
- **k** 21 × 2 = ___ × 6

MULTIPLYING THREE SINGLE-DIGIT NUMBERS

Three numbers can be multiplied together in two steps.

SCAN to watch video

First multiply two of the numbers together. Two numbers that multiply to 10, 20 or 30 will make the next step easier.

Then multiply the product of that multiplication by the third number.

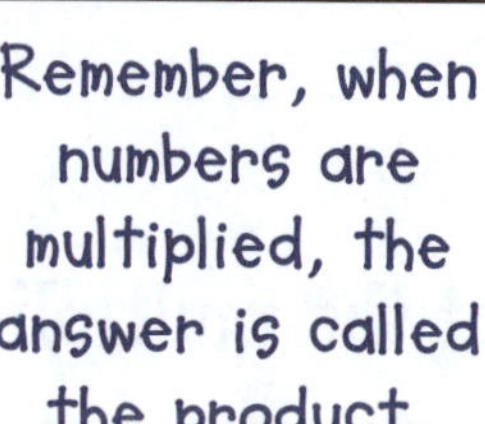

Example 1:

2 × 3 × 5

= 2 × 5 × 3

= 10 × 3

= 30

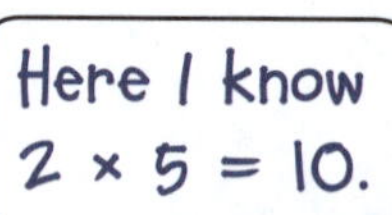

Example 2:

5 × 4 × 2

= 5 × 4 × 2

= 20 × 2

= 40

Here I know 5 × 4 = 20.

Example 3:

8 × 5 × 2

= 5 × ___ × 8

= 10 × 8

= ___

Example 4:

4 × 10 × 3

= 4 × ___ × 3

= 40 × ___

= ___

Solve the multiplications.

	a	b
5 × 4 × 9	6 × 4 × 5	5 × 6 × 2
= 5 × 4 × 9	= ___ × ___ × ___	= ___ × ___ × ___
= 20 × 9	= ___ × ___	= ___ × ___
= 180	= ___	= ___

SELF CHECK Tick how you feel

Got it!	Need help...	I don't get it

Check your answers

How many did you get correct?

CATCH UP MATHS YEAR 4 BOOK A © PASCAL PRESS ISBN: 9781925726145

PRACTICE

1 Look for numbers that multiply to 10, 20 or 30 to help you solve the multiplications.

● 5 × 9 × 6
= 5 × 6 × 9
= 30 × 9
= 270

a 5 × 7 × 2
= __ × __ × __
= ___ × __
= _____

b 5 × 4 × 8
= __ × __ × __
= ___ × __
= _____

c 4 × 5 × 4
= __ × __ × __
= ___ × __
= _____

d 6 × 7 × 5
= __ × __ × __
= ___ × __
= _____

e 9 × 5 × 4
= __ × __ × __
= ___ × __
= _____

f 8 × 2 × 5
= __ × __ × __
= ___ × __
= _____

g 3 × 4 × 5
= __ × __ × __
= ___ × __
= _____

h 6 × 4 × 5
= __ × __ × __
= ___ × __
= _____

i 2 × 5 × 8
= __ × __ × __
= ___ × __
= _____

j 2 × 2 × 5
= __ × __ × __
= ___ × __
= _____

k 5 × 5 × 2
= __ × __ × __
= ___ × __
= _____

l 8 × 3 × 10
= __ × __ × __
= ___ × __
= _____

m 4 × 7 × 5
= __ × __ × __
= ___ × __
= _____

n 2 × 6 × 5
= __ × __ × __
= ___ × __
= _____

MULTIPLYING TWO-DIGIT NUMBERS BY SINGLE-DIGIT NUMBERS

Here are three different ways to multiply two-digit numbers by single-digit numbers.

Example 1: 13 × 9

Using known facts

13 × 9

10 × 9 = 90

90 + 9 + 9 + 9
(3 lots of 9)

= 117

Multiplying the tens and then the units

13 × 9

= 9 tens + 9 threes

= 90 + 27

= 117

Using an area model

13 × 9

	10	3
9	90	27

= 90 + 27

= 117

Example 2: 15 × 8

Using known facts

15 × 8

10 × __ = 80

80 + __ + __ + __ + __ + __
(5 lots of 8)

= ____

Multiplying the tens and then the units

15 × 8

= __ tens + __ fives

= 80 + ___

= ____

Using an area model

15 × 8

	10	5
8	80	

= 80 + ___

= ____

Check your answer on the video!

Your turn

Solve using the three different methods.

42 × 7

40 × 7 = ________

____

____

42 × 7

4 tens + ________

____

____

42 × 7

	40	2
__		

____

____

SELF CHECK Tick how you feel

Got it!	Need help...	I don't get it
 ☐	 ☐	 ☐

Check your answers

How many did you get correct? ☐

CATCH UP MATHS YEAR 4 BOOK A © PASCAL PRESS ISBN: 9781925726145

PRACTICE

1 Solve using known facts.

27 × 4

20 × 4 = 80

80 + 4 + 4 + 4 + 4 + 4 + 4 + 4

= 108

a 52 × 5

______

______

= ______

b 19 × 6

______

______

= ______

c 28 × 3

______

______

= ______

d 97 × 4

______

______

= ______

e 88 × 8

______

______

= ______

f 34 × 7

______

______

= ______

g 45 × 8

______

______

= ______

2 Solve by multiplying the tens and then the units.

23 × 9

= 9 × 2 tens + 9 threes

= 180 + 27

= 207

a 58 × 4

= ______

= ______

= ______

b 46 × 3

= ____________________

= ____________

= ______

c 39 × 7

= ____________________

= ____________

= ______

d 71 × 8

= ____________________

= ____________

= ______

e 62 × 5

= ____________________

= ____________

= ______

3 Solve using area models.

45 × 9

	40	5
9	360	45

= 360 + 45

= 405

a 63 × 7

	__	__
__		

= ____ + ____

= ____

b 92 × 6

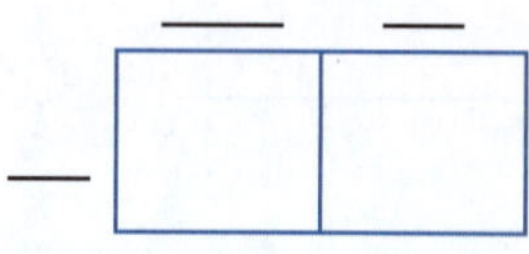

= ____ + ____

= ____

c 32 × 8

	__	__
__		

= ____ + ____

= ____

d 19 × 4

	__	__
__		

= ____ + ____

= ____

e 86 × 3

	__	__
__		

= ____ + ____

= ____

CATCH UP MATHS YEAR 4 BOOK A © PASCAL PRESS ISBN: 9781925726145

MULTIPLICATION REVIEW

1 Fill in the missing numbers.

a

__ rows of __

b

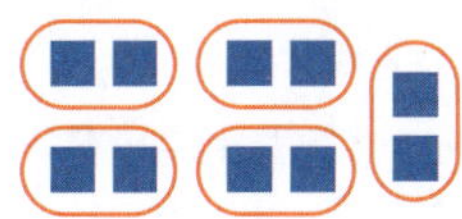

__ groups of __

c

__ rows of __

2 Complete.

a

__ rows of __

__ + __ + __ + __ + __ + __ = __

__ × __ = __

b

__ groups of __

__ + __ + __ + __ = __

__ × __ = __

c

__ rows of __

__ + __ + __ = ___

__ × __ = ___

d

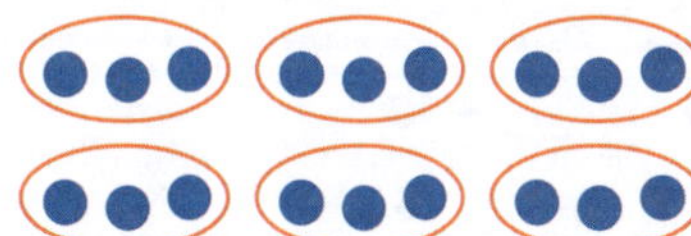

__ groups of __

__ + __ + __ + __ = __

__ × __ = __

3 Fill in the missing nmbers

a 8 + 8 + 8 + 8 = __ × __ = ___

b 5 + 5 + 5 + 5 + 5 = __ × __ = ___

c 6 + 6 + 6 + 6 = __ × __ = ___

d 2 + 2 + 2 + 2 + 2 + 2 = __ × __ = ___

e 1 + 1 + 1 + 1 = __ × __ = ___

f 9 + 9 + 9 + 9 + 9 + 9 = __ × __ = ___

REVIEW

4 Show each commutative property.

a 5 × 9 = __ × __

b 7 × 4 = __ × __

c 2 + 3 + 4 = __ + __ + __

d 5 + 2 = __ + __

e 6 × 8 = __ × __

f 5 + 7 + 8 = __ + __ + __

5 Complete these inverse operations.

a 4 × 9 = ___
__ ÷ 9 = ___

b 24 ÷ ___ = 8
__ × 8 = 24

c 5 × 6 = ___
__ ÷ 6 = ___

d 35 ÷ __ = 5
__ × 5 = 35

e 42 ÷ __ = 6
__ × 6 = 42

f 90 ÷ ___ = 9
___ × 9 = 90

g 8 × __ = 32
__ ÷ 8 = 4

h 4 × ___ = 44
44 ÷ ___ = 4

6 Skip count forwards by:

a 9: ___, ___, ___, ___, ___

b 5: ___, ___, ___, ___, ___

c 8: ___, ___, ___, ___, ___

d 3: ___, ___, ___, ___, ___

e 4: ___, ___, ___, ___, ___

f 6: ___, ___, ___, ___, ___

g 10: ___, ___, ___, ___, ___

h 12: ___, ___, ___, ___, ___

7 Count backwards by:

a 2: 22, ___, ___, ___, ___, ___

b 4: 44, ___, ___, ___, ___, ___

c 6: 48, ___, ___, ___, ___, ___

d 7: 77, ___, ___, ___, ___, ___

e 3: 39, ___, ___, ___, ___, ___

f 5: 55, ___, ___, ___, ___, ___

g 9: 81, ___, ___, ___, ___, ___

h 11: 99, ___, ___, ___, ___, ___

CATCH UP MATHS YEAR 4 BOOK A © PASCAL PRESS ISBN: 9781925726145

8 What is the product?

a 3 and 9 ___

b 9 and 8 ___

c 7 and 7 ___

d 1 and 11 ___

e 4 and 0 ___

f 10 and 11 ___

9 Write the factors.

a 12 ______________________

b 20 ______________________

c 36 ______________________

d 40 ______________________

10 Write T for True or F for False.

a 4 is a factor of 16. ___

b 20 is a multiple of 2. ___

c A factor of 30 is 6. ___

d 9 is a multiple of 28. ___

e 5 is a factor of 21. ___

f 60 is a multiple of 1. ___

g 20 is a factor of 40. ___

h A factor of 50 is 7. ___

11 Write the first 5 multiples.

a 9: ___, ___, ___, ___, ___

b 3: ___, ___, ___, ___, ___

c 10: ___, ___, ___, ___, ___

d 2: ___, ___, ___, ___, ___

e 1: ___, ___, ___, ___, ___

f 5: ___, ___, ___, ___, ___

12 Complete the tables.

a

×	1	6	3	8	11	4
5						
4						
7						
2						
9						

b

×	5	2	12	10	0	7
8						
3						
10						
1						
6						

REVIEW

Complete by halving or doubling.

a

	12
	12

b

18	18

c

	48

d

60	

e

75	
75	

f

290	

g

150	

h

	170
	170

Complete the table.

	Half	Number	Double
a			20
b		40	
c	8		
d			100
e		80	
f	10		
g	4		
h	6		

	Half	Number	Double
i		16	
j	9		
k	22		
l		48	
m			128
n		136	
o			700
p			1200

15 Write an equivalent multiplication.

a 2 × 8 = ___ × ___

b 10 × 1 = ___ × ___

c 5 × 6 = ___ × ___

d 4 × 8 = ___ × ___

e 6 × 3 = ___ × ___

f 8 × 7 = ___ × ___

g 6 × 2 = ___ × ___

h 4 × 2 = ___ × ___

i 5 × 8 = ___ × ___

j 7 × 9 = ___ × ___

16 **Multiply.**

a 6 × 5 × 2

= __ × __ × __

= ___ × __

= _____

b 4 × 8 × 5

= __ × __ × __

= ___ × __

= _____

c 9 × 2 × 5

= __ × __ × __

= ___ × __

= _____

17 **Solve.**

a Use known facts

16 × 3

___ × ___ = ___

__ + __ + __ + __ + __ + __

= _____

Multiply tens and then units

16 × 3

= 3 × __ tens + __ sixes

= ___ + ___

= ____________

Use an area model

16 × 3

	10	6
3		

= _____ + _____

= _____

b Use known facts

42 × 6

___ × ___ = ___

_____ + _____

= _____

Multiply tens and then units

42 × 6

= 6 × __ tens + __________

= ___ + ___

= ____________

Use an area model

42 × 6

= _____ + _____

= _____

GROUPING

Grouping is sharing (or dividing) objects into groups of the same size.

Example 1: Share 12 balls between 6 children.

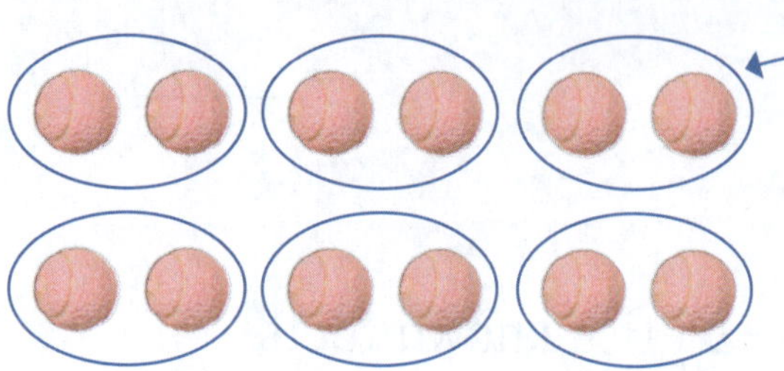

There will be 6 groups with 2 balls each.

$12 \div 6 = 2$ ← How many in each group

The number to be shared — The number of groups

Example 2: Share 16 balls between 4 children.

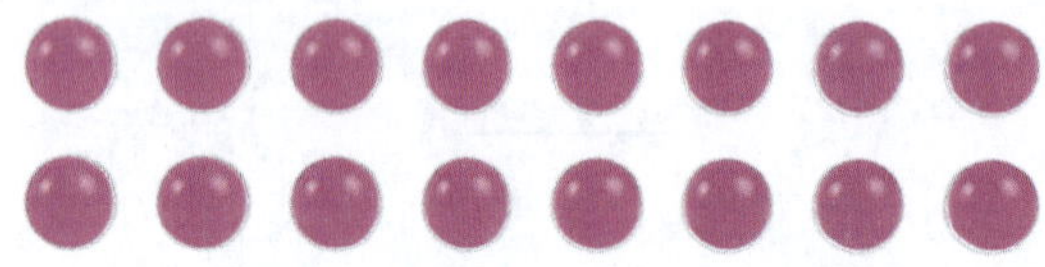

There will be 4 groups with 4 balls each.

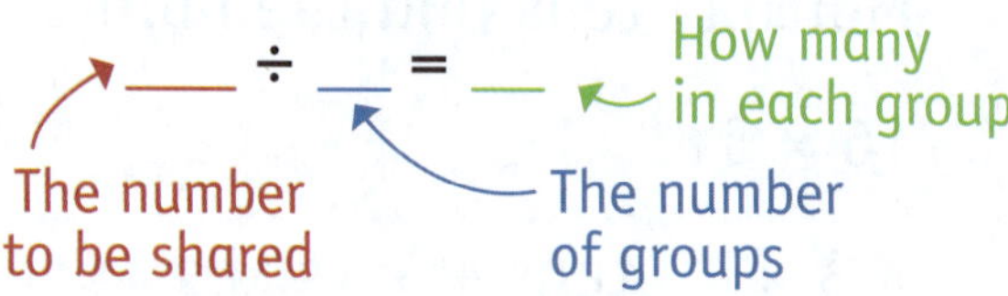

___ ÷ ___ = ___ ← How many in each group

The number to be shared — The number of groups

Draw the ■ then complete the sentence.

● Share 15 ■ between 3 children.
Each child will get ___ ■.

a Share 16 ■ between 8 children.
Each child will get ___ ■.

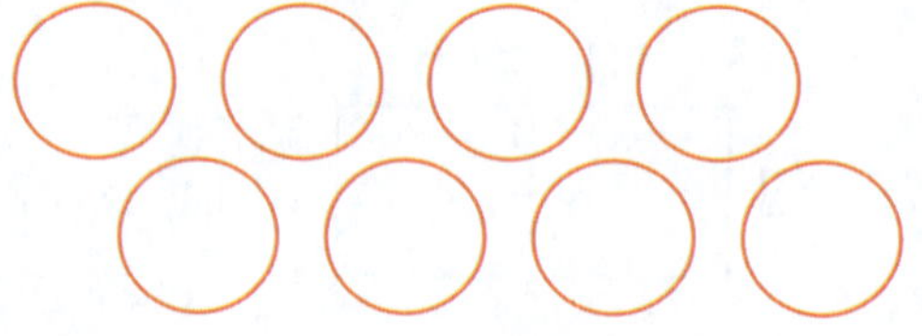

b Share 12 ■ between 4 children.
Each child will get ___ ■.

SELF CHECK Tick how you feel

Got it!	Need help...	I don't get it
☐	☐	☐

Check your answers
How many did you get correct? ☐

CATCH UP MATHS YEAR 4 BOOK A © PASCAL PRESS ISBN: 9781925726145

PRACTICE

1 Complete the sentences.

● There are 2 equal groups with 3 in each group.

a 
There are __ equal groups with __ in each group.

b 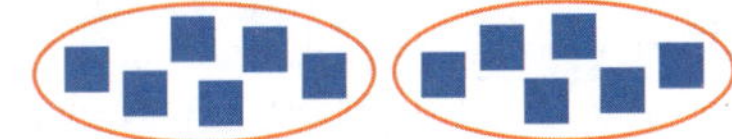
There are __ equal groups with __ in each group.

c 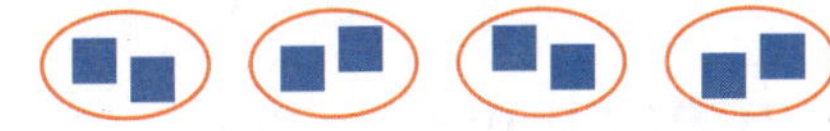
There are __ equal groups with __ in each group.

d 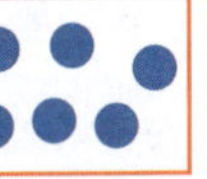  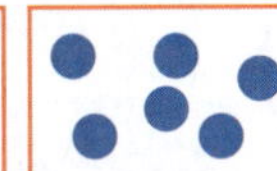
There are __ equal groups with __ in each group.

2 Make equal groups then complete the number sentence.

● Groups of 2	a Groups of 4	b Groups of 3
 There are 5 groups of 2. $10 \div \underline{5} = 2$	 There are __ groups of 4. $20 \div$ __ $= 4$	 There are __ groups of 3. $12 \div$ __ $= 3$

3 Circle the groups and complete the sentence.

● Groups of 5	a Groups of 6	b Groups of 7
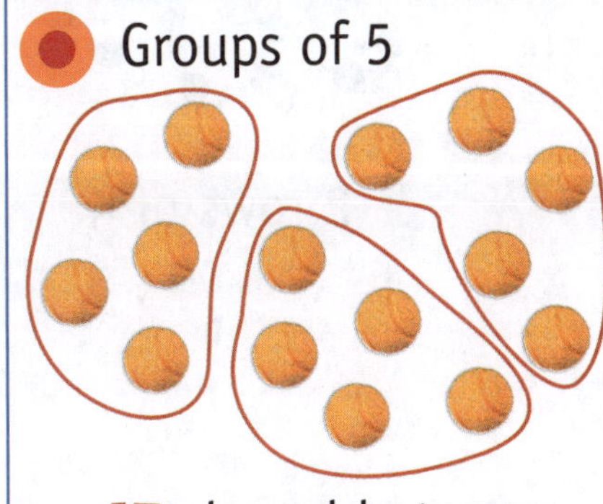 15 shared between 3 equals 5.	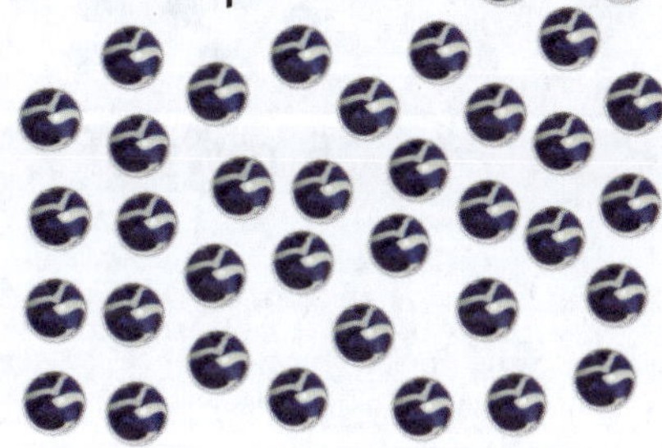 __ shared between __ equals 6.	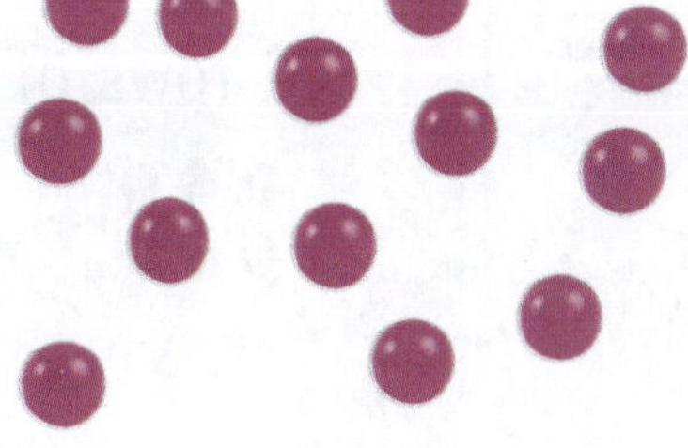 __ shared between __ equals 7.

EQUAL ROWS

An equal row is when the number in each row is the same.

Example 1:

Here are 20 counters.

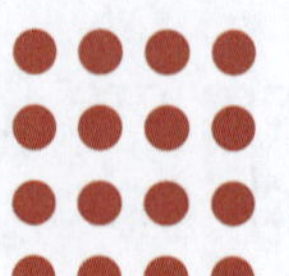

Now the 20 counters have been arranged in rows.

5 rows of 4

Example 2:

Here are 8 counters.

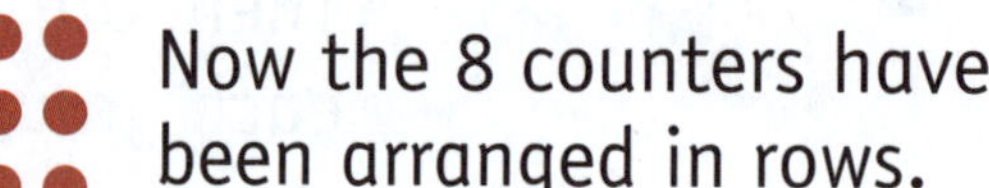

Now the 8 counters have been arranged in rows.

4 rows of 2

Example 3:

Here are 15 counters.

Now the 15 counters have been arranged in __ rows of __.

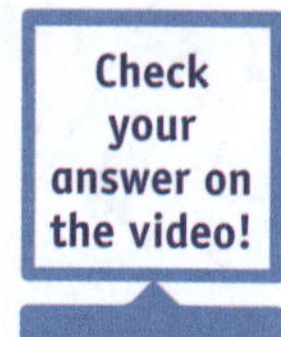

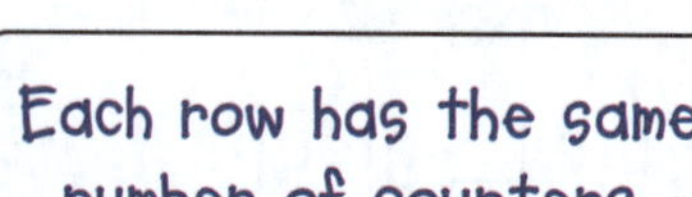

Arrange the counters in rows.

● 6 counters

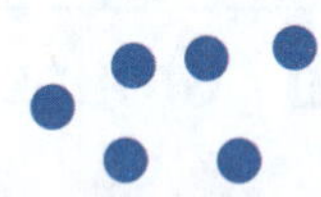

2 rows of 3

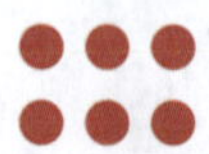

a 10 counters

5 rows of 2

b 8 counters

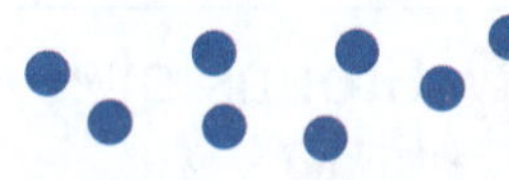

2 rows of 4

SELF CHECK Tick how you feel		
Got it!	Need help...	I don't get it

Check your answers

How many did you get correct?

CATCH UP MATHS YEAR 4 BOOK A © PASCAL PRESS ISBN: 9781925726145

PRACTICE

1 **Draw ● in each box below to show:**

● 24 in 3 equal rows

3 rows of 8 = 24

24 ÷ 3 = 8

a 20 in 4 equal rows

4 rows of __ = 20

20 ÷ 4 = __

b 12 in 1 row

1 row of __ = 12

12 ÷ 1 = __

c 18 in 9 equal rows

9 rows of __ = 18

18 ÷ 9 = __

d 6 in 6 equal rows

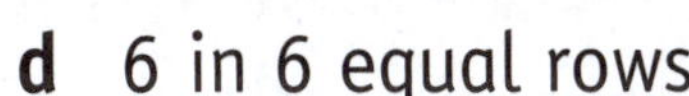

6 rows of __ = 6

6 ÷ 6 = __

e 15 in 5 equal rows

5 rows of __ = 15

15 ÷ 5 = __

2 **Match the description to the right picture.**

● 

a 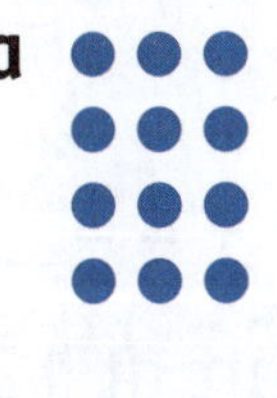

b 

c 

d 

8 rows of 5	2 rows of 9	4 rows of 3	6 rows of 3	3 rows of 4

REPEATED SUBTRACTION TO SOLVE DIVISION

One way division can be solved is by using repeated subtraction.

Example 1: Solve 24 ÷ 6.

Start at 24

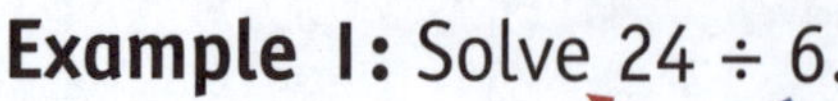

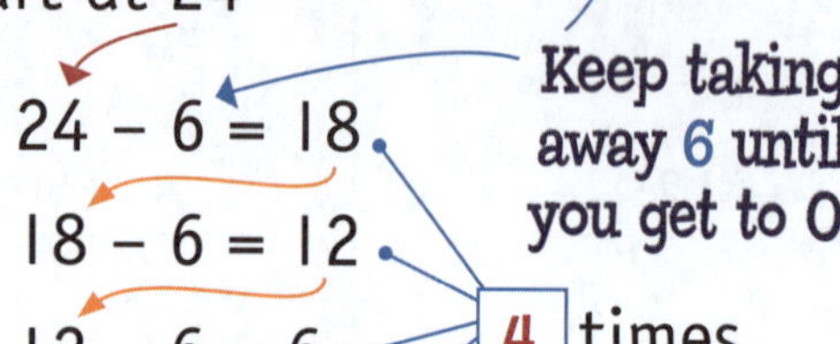

24 – 6 = 18

18 – 6 = 12

12 – 6 = 6

6 – 6 = 0

4 times

so 24 ÷ 6 = 4

Example 2: Solve 32 ÷ 8.

Start at 32.

32 – 8 = 24

24 – 8 = 16

16 – 8 = 8

8 – 8 = 0

4 times

so 32 ÷ 8 = 4

Example 3: Solve 18 ÷ 3.

Start at ___.

___ – 3 = ___

___ – 3 = ___

___ – 3 = ___

___ – 3 = ___

___ – 3 = ___

___ – 3 = ___

☐ times

so 18 ÷ 3 = ___

Your turn

Complete these division problems.

10 ÷ 5 = **2**

Start at 10

10 – 5 = **5**

5 – 5 = 0

2

a 9 ÷ 3 = ___

Start at 9

9 – 3 = ___

___ – 3 = ___

___ – 3 = ___

☐

b 14 ÷ 7 = ___

Start at 14

14 – 7 = ___

___ – 7 = ___

☐

c 12 ÷ 4 = ___

Start at 12

12 – 4 = ___

___ – 4 = ___

___ – 4 = ___

☐

SELF CHECK Tick how you feel

Got it!	Need help...	I don't get it
 ☐	 ☐	 ☐

Check your answers

How many did you get correct? ☐

  ISBN: 9781925726145

PRACTICE

1 Solve these division problems using repeated subtraction.

21 ÷ 3 = 7

Start at 21

21 − 3 = 18

18 − 3 = 15

15 − 3 = 12

12 − 3 = 9

9 − 3 = 6

6 − 3 = 3

3 − 3 = 0

[7] times

a 45 ÷ 5 = __

Start at 45

[ ] times

b 24 ÷ 8 = __

Start at 24

[ ] times

c 30 ÷ 6 = __

Start at 30

[ ] times

d 35 ÷ 7 = __

Start at 35

[ ] times

e 12 ÷ 2 = __

Start at 12

[ ] times

INVERSE OPERATIONS OF MULTIPLICATION AND DIVISION

Multiplication and division are inverse operations.
This means they are the opposite to each other.

The inverse (opposite) of × is ÷, and the inverse of ÷ is ×.

$6 \times 4 = 24$ $24 \div 4 = 6$

Example 1:

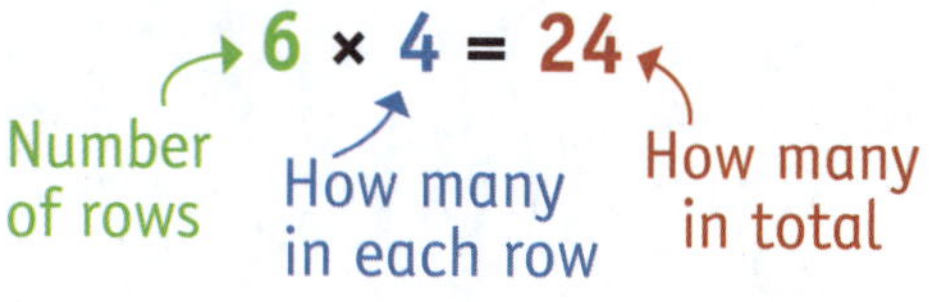

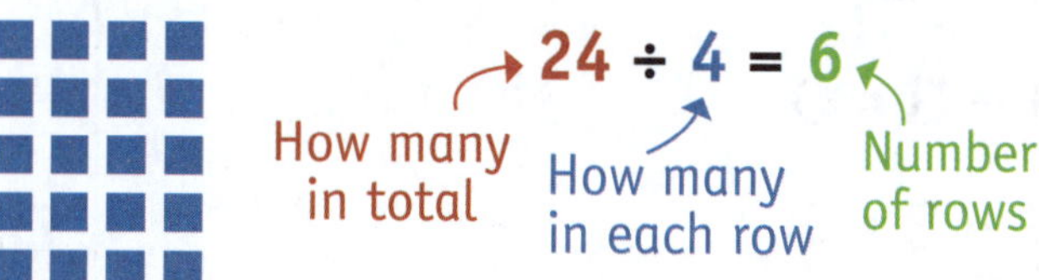

Example 2:

$9 \times 6 = 54$ $54 \div 6 = 9$

Example 3:

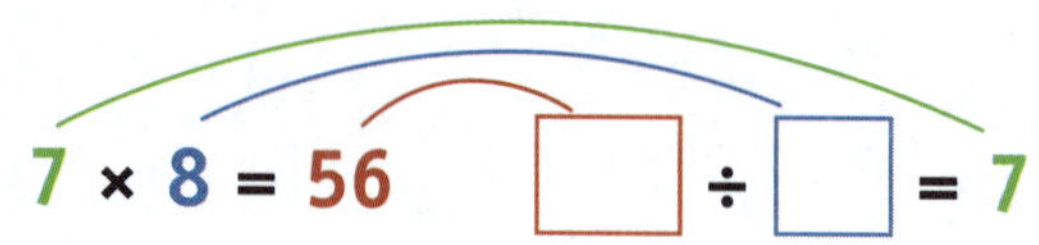

Match each with their inverse operation.

●	5 × 6 = 30	28 ÷ 7 = 4
a	4 × 7 = 28	42 ÷ 6 = 7
b	2 × 9 = 18	45 ÷ 9 = 5
c	7 × 6 = 42	30 ÷ 6 = 5
d	5 × 9 = 45	18 ÷ 9 = 2
e	3 × 8 = 24	24 ÷ 8 = 3

Multiplication 'undoes' division and division 'undoes' multiplication.

SELF CHECK Tick how you feel

Got it!	Need help...	I don't get it

Check your answers

How many did you get correct?

CATCH UP MATHS YEAR 4 BOOK A © PASCAL PRESS ISBN: 9781925726145

PRACTICE

1 Write ÷ and × facts for each set of numbers.

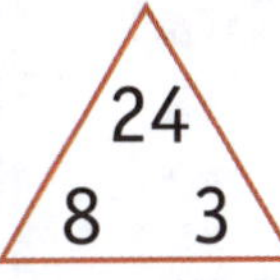

8 × 3 = 24
3 × 8 = 24
24 ÷ 3 = 8
24 ÷ 8 = 3

b

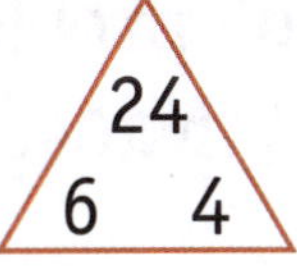

___ × ___ = ___
___ × ___ = ___
___ ÷ ___ = ___
___ ÷ ___ = ___

d

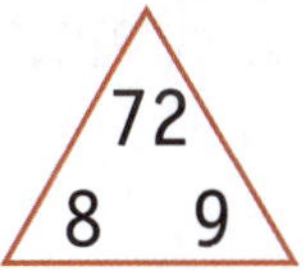

___ × ___ = ___
___ × ___ = ___
___ ÷ ___ = ___
___ ÷ ___ = ___

a

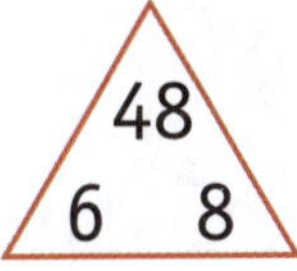

___ × ___ = ___
___ × ___ = ___
___ ÷ ___ = ___
___ ÷ ___ = ___

c

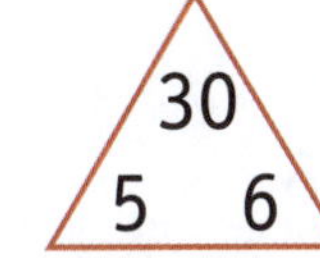

___ × ___ = ___
___ × ___ = ___
___ ÷ ___ = ___
___ ÷ ___ = ___

e

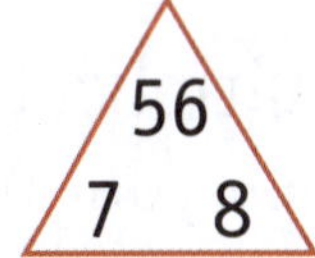

___ × ___ = ___
___ × ___ = ___
___ ÷ ___ = ___
___ ÷ ___ = ___

2 Write the ÷ sign and = sign in the correct boxes.

●	48 [÷] 8 [=] 6	**c**	16 [ ] 4 [ ] 4	**f**	72 [ ] 9 [ ] 8
a	14 [ ] 7 [ ] 2	**d**	66 [ ] 11 [ ] 6	**g**	42 [ ] 6 [ ] 7
b	21 [ ] 3 [ ] 7	**e**	48 [ ] 12 [ ] 4	**h**	45 [ ] 9 [ ] 5

3 Fill in the boxes below.

●	[20] ÷ 5 = 4	**c**	[ ] ÷ 3 = 4	**f**	[ ] ÷ 7 = 10
a	[ ] ÷ 2 = 7	**d**	[ ] ÷ 10 = 11	**g**	[ ] ÷ 8 = 8
b	[ ] ÷ 3 = 6	**e**	[ ] ÷ 12 = 5	**h**	[ ] ÷ 1 = 9

FORMAL DIVISION

Formal division is where division problems are written using the $\overline{)\quad}$ symbol instead of ÷.

Example 1:

$$\begin{array}{r} 9 \\ 2\overline{)18} \end{array}$$

9 ← Answer (Quotient)

2 → Number dividing by

18 ← Number being divided

Example 2:

$$\begin{array}{r} 12 \\ 3\overline{)36} \end{array}$$

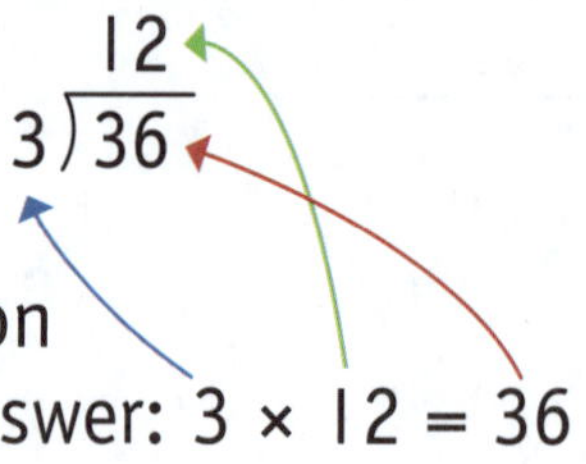

Use multiplication to check your answer: 3 × 12 = 36

Example 3:

$$8\overline{)24}$$

Check using multiplication: 8 × __ = ___

Your turn

Do the division, then check your answer with multiplication.

●	$\begin{array}{r} 8 \\ 2\overline{)16} \end{array}$	Check: 2 × 8 = 16
a	$6\overline{)36}$	Check: 6 × __ = 36
b	$7\overline{)42}$	Check: 7 × __ = 42
c	$9\overline{)81}$	Check: 9 × __ = 81
d	$4\overline{)48}$	Check: 4 × __ = 48

Check your answers
How many did you get correct?

CATCH UP MATHS YEAR 4 BOOK A © PASCAL PRESS ISBN: 9781925726145

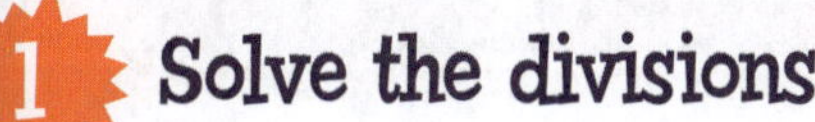

PRACTICE

1 **Solve the divisions.**

$\begin{array}{r}4\\5\overline{)20}\end{array}$

a $10\overline{)40}$

b $3\overline{)33}$

c $7\overline{)56}$

d $9\overline{)90}$

e $4\overline{)36}$

f $7\overline{)63}$

g $8\overline{)56}$

h $8\overline{)64}$

i $4\overline{)20}$

j $3\overline{)21}$

k $5\overline{)30}$

2 **Fill in the missing numbers below, then check your answer with multiplication.**

$\begin{array}{r}5\\6\overline{)30}\end{array}$ Check: 6 × 5 = 30

a $\begin{array}{r}7\\4\overline{)}\end{array}$ Check: 4 × 7 = ___

b $7\overline{)56}$ Check: 7 × ___ = 56

c $\begin{array}{r}10\\\overline{)60}\end{array}$ Check: ___ × 10 = 60

d $3\overline{)15}$ Check: 3 × ___ = 15

e $9\overline{)45}$ Check: 9 × ___ = 45

f $\begin{array}{r}11\\2\overline{)}\end{array}$ Check: 2 × 11 = ___

g $\begin{array}{r}12\\5\overline{)}\end{array}$ Check: 5 × 12 = ___

3 **Write True or False.**

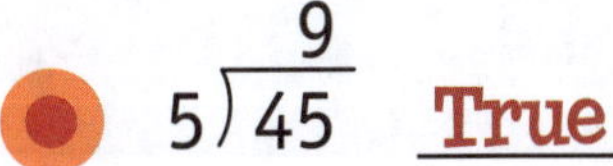

$\begin{array}{r}9\\5\overline{)45}\end{array}$ True

a $\begin{array}{r}11\\4\overline{)40}\end{array}$ ________

b $\begin{array}{r}12\\5\overline{)60}\end{array}$ ________

c $\begin{array}{r}10\\4\overline{)40}\end{array}$ ________

d $\begin{array}{r}4\\2\overline{)6}\end{array}$ ________

e $\begin{array}{r}6\\2\overline{)8}\end{array}$ ________

f $\begin{array}{r}6\\5\overline{)30}\end{array}$ ________

g $\begin{array}{r}2\\10\overline{)12}\end{array}$ ________

h $\begin{array}{r}12\\3\overline{)36}\end{array}$ ________

i $\begin{array}{r}8\\6\overline{)72}\end{array}$ ________

j $\begin{array}{r}7\\7\overline{)49}\end{array}$ ________

k $\begin{array}{r}11\\8\overline{)96}\end{array}$ ________

DIVISION WITH REMAINDERS

When a number cannot be divided exactly, the leftover is called the remainder.

Example 1:

I have 15 balls and share them equally between 4 people.

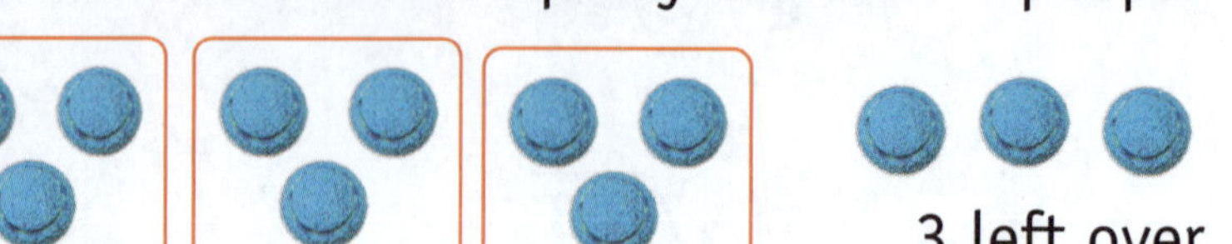

Each person gets 3 balls.

There are 3 balls left over and this is called the remainder.

So 15 ÷ 4 = 3 remainder 3.

Example 2:

I have ___ strawberries and share them equally between __ people.

1 left over

Each person gets __ strawberries and there is __ left over.

The remainder is __ strawberry.

So ___ ÷ __ = __ remainder __.

Share the objects equally.

- Share 10 balls between 4 people.

Each person gets 2 balls and the remainder is 2.

10 ÷ 4 = 2 remainder 2

a Share 12 balls between 5 people.

Each person gets __ balls and there are __ remainder.

12 ÷ 5 = __ remainder __

SELF CHECK Tick how you feel

Got it!	Need help...	I don't get it
☐	☐	☐

Check your answers

How many did you get correct? ☐

CATCH UP MATHS YEAR 4 BOOK A © PASCAL PRESS ISBN: 9781925726145

PRACTICE

1 Write the number sentence and the answer.

13 apples shared equally between 5 people

13 ÷ 5 = 2 remainder 3

a 15 jelly beans shared equally between 2 people

___ ÷ ___ = ___ remainder ___

b 18 balls shared equally between 4 people

___ ÷ ___ = ___ remainder ___

c 48 bones shared equally between 8 dogs

___ ÷ ___ = ___ remainder ___

d 25 fish shared equally between 3 seals

___ ÷ ___ = ___ remainder ___

e 21 oranges shared equally between 7 people

___ ÷ ___ = ___ remainder ___

f 29 chocolates shared equally between 4 children

___ ÷ ___ = ___ remainder ___

g 83 balls shared equally between 9 people

___ ÷ ___ = ___ remainder ___

2 Solve the divisions.

14 ÷ 3 = 4 remainder 2

a 22 ÷ 5 = ___ remainder ___

b 29 ÷ 3 = ___ remainder ___

c 43 ÷ 10 = ___ remainder ___

d 62 ÷ 12 = ___ remainder ___

e 83 ÷ 9 = ___ remainder ___

f 101 ÷ 10 = ___ remainder ___

g 146 ÷ 12 = ___ remainder ___

3 Match the problem with its remainder.

87 ÷ 9 = 9	remainder 2
a 93 ÷ 10 = 9	remainder 1
b 27 ÷ 5 = 5	remainder 5
c 17 ÷ 2 = 8	remainder 6
d 64 ÷ 5 = 12	remainder 4
e 47 ÷ 7 = 6	remainder 3

DIVISION REVIEW

1 Complete the statements.

a 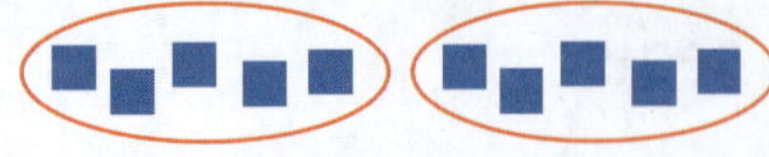

There are ___ equal groups with ___ in each group.

b 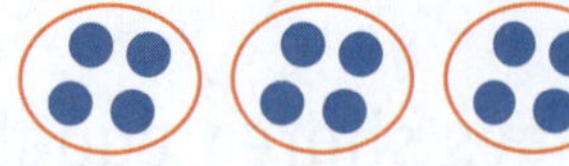

There are ___ equal groups with ___ in each group.

c 

There are __ equal groups with ___ in each group.

2 Circle equal groups, then complete the number sentences.

a Groups of 3	b Groups of 5	c Groups of 6

There are __ groups of 3.	There are __ groups of 5.	There are __ groups of 6.
15 ÷ __ = 3	30 ÷ __ = 5	42 ÷ __ = 6

3 Draw ● in each box below to show.

a 20 in 5 equal rows

5 rows of __ = 20

20 ÷ 5 = __

b 12 in 3 equal rows

3 rows of __ = 12

12 ÷ 3 = __

c 32 in 4 equal rows

4 rows of __ = 32

32 ÷ 4 = __

4 Draw counters in:

a 3 rows of 6

b 5 rows of 2

c 8 rows of 9

5 Use repeated subtraction to solve the divisions.

a 32 ÷ 4 = ___

Start at 32

b 48 ÷ 12 = ___

Start at 48

c 24 ÷ 6 = ___

Start at 24

d 18 ÷ 9 = ___

Start at 18

e 6 ÷ 6 = ___

Start at 6

REVIEW

Write the inverse operations for the multiplications below.

a 6 × 5 = 30

____

or ________________

b 4 × 9 = 36

____

or ________________

c 2 × 12 = 24

____

or ________________

d 3 × 11 = 33

____

or ________________

e 8 × 9 = 72

____

or ________________

f 8 × 4 = 32

____

or ________________

g 12 × 4 = 48

____

or ________________

h 10 × 2 = 20

____

or ________________

i 6 × 6 = 36

____

j 4 × 4 = 16

____

k 9 × 3 = 27

____

or ________________

Write × and ÷ facts for each set of numbers.

a

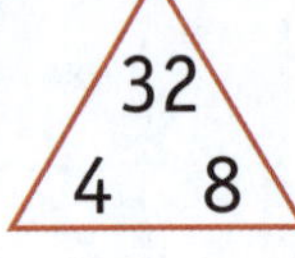

____ × ____ = ____

____ × ____ = ____

____ ÷ ____ = ____

____ ÷ ____ = ____

b

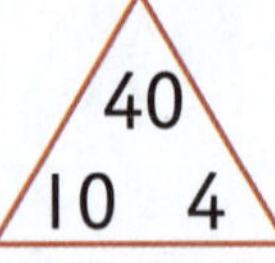

____ × ____ = ____

____ × ____ = ____

____ ÷ ____ = ____

____ ÷ ____ = ____

CATCH UP MATHS YEAR 4 BOOK A © PASCAL PRESS ISBN: 9781925726145

c

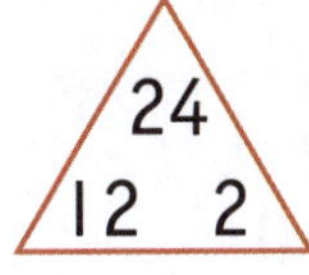

____ × ____ = ____

____ × ____ = ____

____ ÷ ____ = ____

____ ÷ ____ = ____

d

49
7 7

____ × ____ = ____

____ × ____ = ____

____ ÷ ____ = ____

____ ÷ ____ = ____

8 Write one multiplication fact about each division.

a 30 ÷ 3 __________

b 72 ÷ 9 __________

c 80 ÷ 10 __________

d 40 ÷ 5 __________

e 24 ÷ 4 __________

f 21 ÷ 7 __________

g 64 ÷ 8 __________

h 20 ÷ 5 __________

i 12 ÷ 6 __________

j 36 ÷ 12 __________

9 Fill in the boxes below.

a ☐ ÷ 2 = 6

b ☐ ÷ 1 = 8

c ☐ ÷ 9 = 2

d ☐ ÷ 3 = 9

e ☐ ÷ 4 = 7

f ☐ ÷ 7 = 11

g ☐ ÷ 5 = 10

h ☐ ÷ 6 = 1

i ☐ ÷ 8 = 12

j ☐ ÷ 2 = 7

REVIEW

10 **Place the ÷ and = signs in the correct spaces.**

a 42 □ 6 □ 7

b 33 □ 3 □ 11

c 24 □ 8 □ 3

d 64 □ 8 □ 8

e 72 □ 9 □ 8

f 100 □ 10 □ 10

g 121 □ 11 □ 11

h 132 □ 12 □ 11

i 99 □ 9 □ 11

j 60 □ 6 □ 10

11 **Solve these formal division questions.**

a $3\overline{)36}$

b $5\overline{)20}$

c $4\overline{)32}$

d $6\overline{)42}$

e $6\overline{)48}$

f $2\overline{)24}$

g $3\overline{)30}$

h $9\overline{)54}$

i $7\overline{)56}$

12 **Fill in the missing numbers below, then check your answer with multiplication.**

a $6\overline{)24}$ Check: 6 × ___ = 24

b $\overset{12}{\overline{)60}}$ Check: ___ × 12 = 60

c $8\overline{)56}$ Check: 8 × ___ = 56

d $4\overset{10}{\overline{)}}$ Check: 4 × 10 = ___

e $\overset{6}{\overline{)42}}$ Check: ___ × 6 = 42

f $7\overline{)84}$ Check: 7 × ___ = 84

g $8\overline{)64}$ Check: 8 × ___ = 64

h $7\overline{)21}$ Check: 7 × ___ = 21

i $9\overset{11}{\overline{)}}$ Check: 9 × 11 = ___

j $\overset{12}{\overline{)72}}$ Check: ___ × 12 = 72

CATCH UP MATHS YEAR 4 BOOK A © PASCAL PRESS ISBN: 9781925726145

13 **Complete the number sentence and the answer.**

a 23 balls shared equally between 4 people

___ ÷ ___ = ___ remainder ___

b 37 fish shared equally between 12 people

___ ÷ ___ = ___

c 43 balls shared equally between 5 people

___ ÷ ___ = ___ remainder ___

d 92 lollies shared equally between 10 people

___ ÷ ___ = ___ remainder ___

e 14 jelly beans shared equally between 7 people

___ ÷ ___ = ___ remainder ___

f 28 pens shared equally between 9 people

___ ÷ ___ = ___ remainder ___

g 63 pencils shared equally between 6 people

___ ÷ ___ = ___ remainder ___

h 48 flowers shared equally between 10 people

___ ÷ ___ = ___ remainder ___

i 25 jelly beans shared equally between 5 people

___ ÷ ___ = ___ remainder ___

j 59 marbles shared equally between 12 people

___ ÷ ___ = ___ remainder ___

14 **Solve the division.**

a 23 ÷ 4 = ___ remainder ___

b 28 ÷ 5 = ___ remainder ___

c 45 ÷ 6 = ___ remainder ___

d 62 ÷ 7 = ___ remainder ___

e 57 ÷ 11 = ___ remainder ___

f 133 ÷ 12 = ___ remainder ___

g 106 ÷ 10 = ___ remainder ___

h 93 ÷ 9 = ___ remainder ___

i 78 ÷ 9 = ___ remainder ___

j 34 ÷ 3 = ___ remainder ___

k 69 ÷ 8 = ___ remainder ___

l 19 ÷ 3 = ___ remainder ___

m 21 ÷ 2 = ___ remainder ___

n 38 ÷ 6 = ___ remainder ___

NUMERATORS AND DENOMINATORS

A fraction has three parts: the numerator, the vinculum (the line) and the denominator.

vinculum (the line) → $\frac{1}{4}$

numerator
The top number in a fraction is the number of parts in this fraction.

denominator
The bottom number in a fraction is the total number of parts.

SCAN to watch video

Example 1:
two-fifths = $\frac{2}{5}$

Example 2:
one-eighth = $\frac{1}{8}$

Example 3:
five-eighths = $\frac{5}{8}$

Example 4:
one-quarter = $\frac{}{4}$

Example 5:
three-tenths = $\frac{3}{}$

Example 6:
________-half = $\frac{1}{2}$

Colour the numerator red, the vinculum green and the denominator blue.

	$\frac{2}{5}$	d	$\frac{1}{8}$	h	$\frac{5}{8}$
a	$\frac{1}{3}$	e	$\frac{1}{2}$	i	$\frac{2}{4}$
b	$\frac{2}{8}$	f	$\frac{7}{8}$	j	$\frac{6}{8}$
c	$\frac{3}{4}$	g	$\frac{3}{5}$	k	$\frac{2}{3}$

SELF CHECK Tick how you feel

Got it!	Need help...	I don't get it
 ☐	 ☐	 ☐

Check your answers
How many did you get correct?

CATCH UP MATHS YEAR 4 BOOK A © PASCAL PRESS ISBN: 9781925726145

PRACTICE

1 What is the numerator in these fractions?

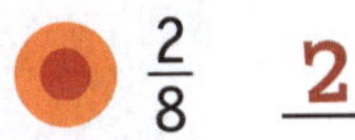

- ● $\frac{2}{8}$ <u>2</u>
- a $\frac{3}{4}$ ___
- b $\frac{1}{5}$ ___
- c $\frac{7}{8}$ ___
- d $\frac{2}{2}$ ___
- e $\frac{4}{5}$ ___
- f $\frac{4}{8}$ ___
- g $\frac{3}{5}$ ___

2 What is the denominator in these fractions?

- ● $\frac{2}{5}$ <u>5</u>
- a $\frac{3}{8}$ ___
- b $\frac{2}{4}$ ___
- c $\frac{3}{5}$ ___
- d $\frac{7}{8}$ ___
- e $\frac{1}{2}$ ___
- f $\frac{1}{4}$ ___
- g $\frac{1}{5}$ ___

3 Write the name of the fraction.

- ● $\frac{2}{5}$ is <u>two-fifths</u>
- a $\frac{3}{8}$ is ______________________
- b $\frac{1}{5}$ is ______________________
- c $\frac{7}{8}$ is ______________________
- d $\frac{3}{5}$ is ______________________
- e $\frac{3}{4}$ is ______________________
- f $\frac{2}{4}$ is ______________________
- g $\frac{1}{8}$ is ______________________
- h $\frac{5}{8}$ is ______________________
- i $\frac{1}{3}$ is ______________________

4 Cross out the fraction with the different denominator.

- ● $\frac{2}{5}, \frac{1}{5}, \frac{3}{5}, \cancel{\frac{4}{10}}, \frac{4}{5}$
- a $\frac{1}{8}, \frac{3}{8}, \frac{7}{8}, \frac{4}{8}, \frac{3}{4}$
- b $\frac{1}{4}, \frac{2}{4}, \frac{4}{5}, \frac{3}{4}, \frac{4}{4}$
- c $\frac{1}{3}, \frac{2}{3}, \frac{3}{3}, \frac{3}{8}$
- d $\frac{3}{5}, \frac{4}{10}, \frac{6}{10}, \frac{3}{10}, \frac{8}{10}$
- e $\frac{2}{8}, \frac{3}{3}, \frac{6}{8}, \frac{5}{8}, \frac{8}{8}$

5 Write the name of each fraction you crossed out in Question 4.

- ● <u>four-tenths</u>
- a ______________________
- b ______________________
- c ______________________
- d ______________________
- e ______________________

FRACTIONS – HALVES

Numbers that are parts of a whole are called fractions. When there are two equal parts, each part is called one-half ($\frac{1}{2}$).

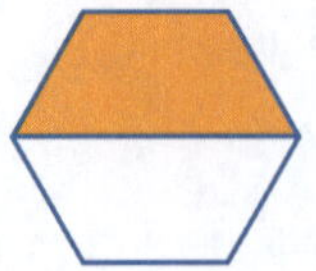 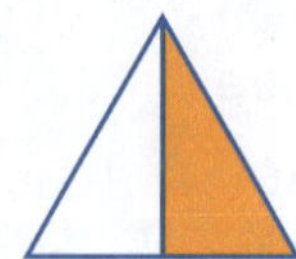 

These objects have been divided into two equal parts.

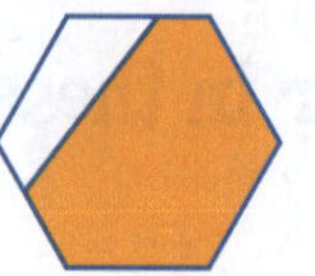 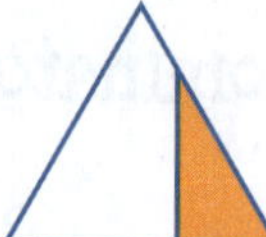 

These objects have NOT been divided into two equal parts.

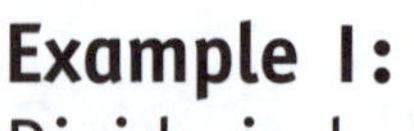

Example 1:
Divide in half.

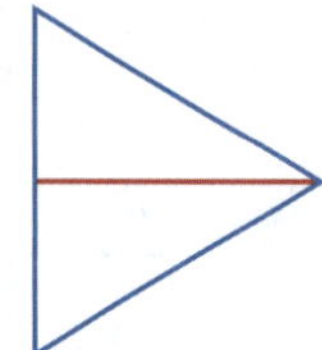

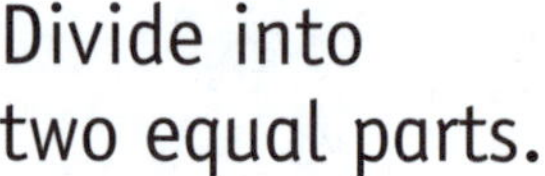

Example 3:
Divide into two equal parts.

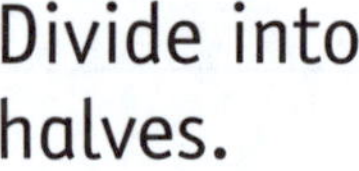

Example 2:
Divide into halves.

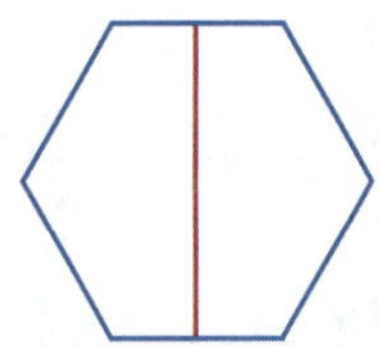

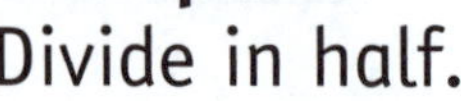

Example 4:
Divide in half.

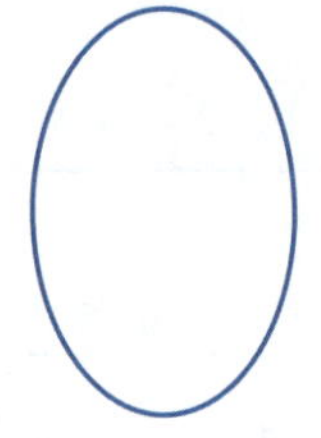

Your turn

Circle the shapes that have been divided into two equal parts.

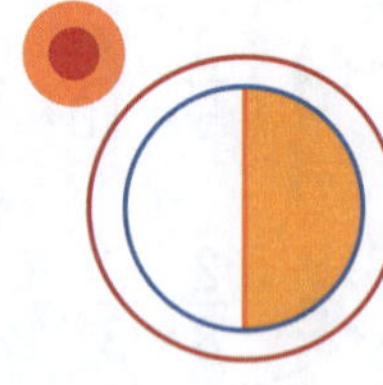

b 

d 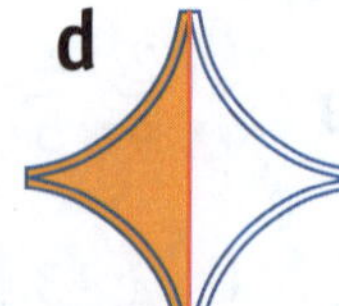

f 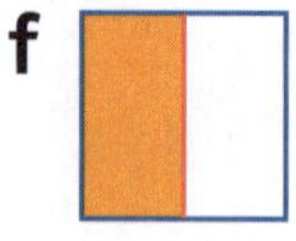

h 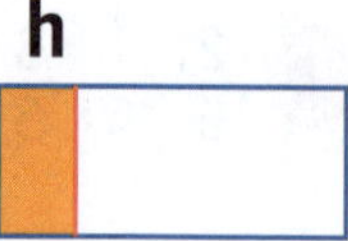

a 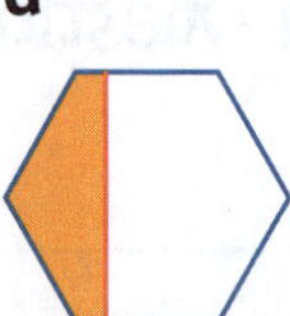

c

e

g

i 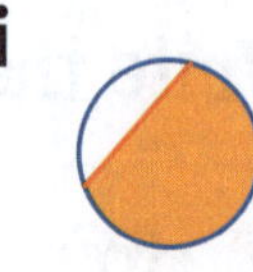

SELF CHECK Tick how you feel

Got it!	Need help...	I don't get it
 ☐	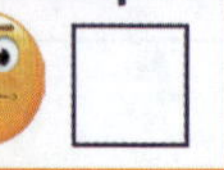 ☐	 ☐

Check your answers
How many did you get correct? ☐

CATCH UP MATHS YEAR 4 BOOK A © PASCAL PRESS ISBN: 9781925726145

PRACTICE

1 Tick the shapes that have been divided into two equal shares.

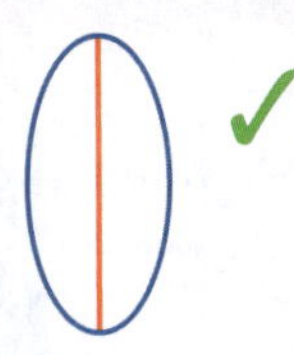

b 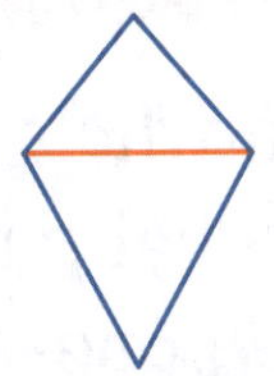

d 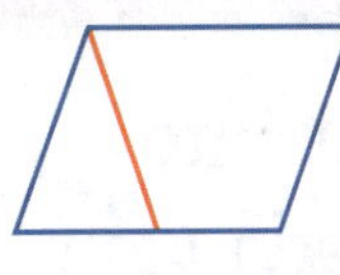

f 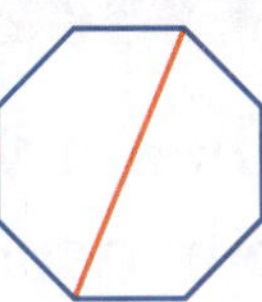

a 

c 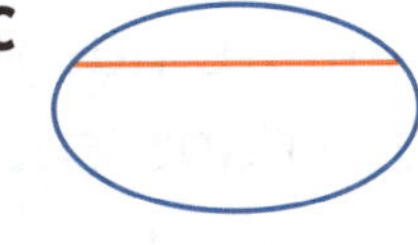

e 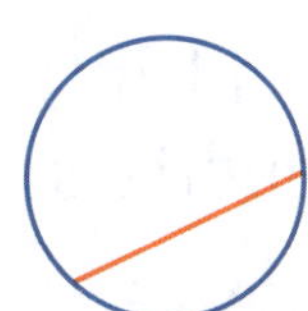

g 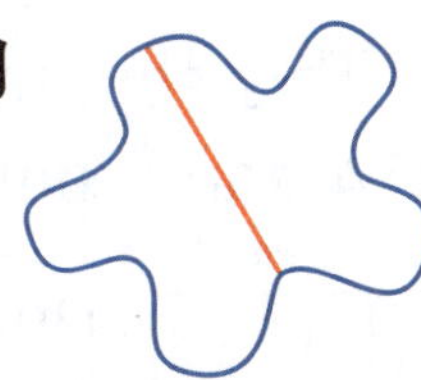

2 Cut the following shapes in half.

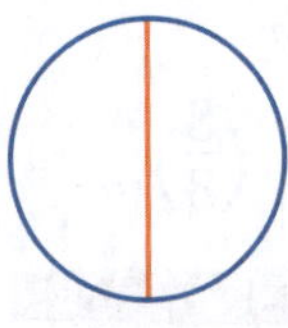

a 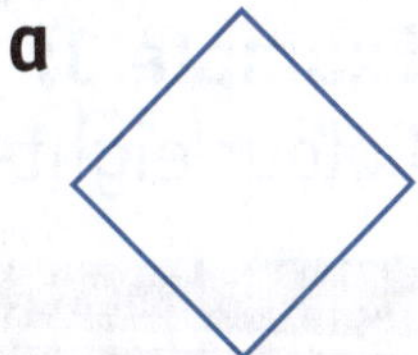

b 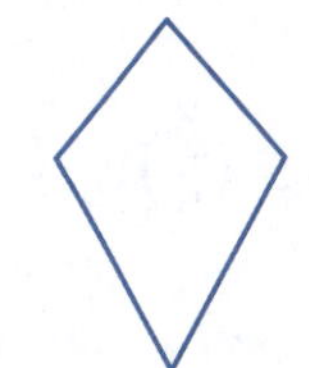

c 

3 Label the shapes that have been cut in half ($\frac{1}{2}$). Cross out the shapes that have not been cut in half.

$\frac{1}{2}$ $\frac{1}{2}$

b 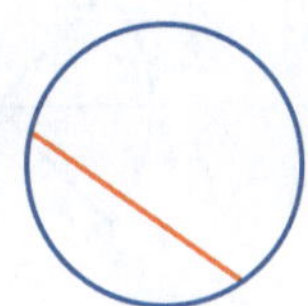

e 

h 

c 

f 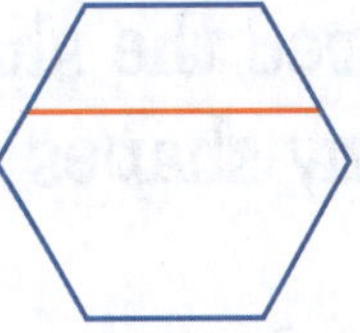

i 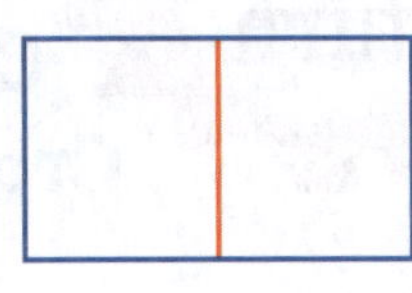

a 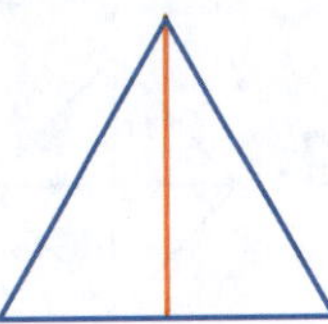

d 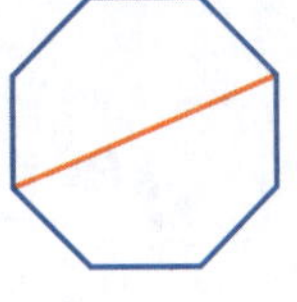

g 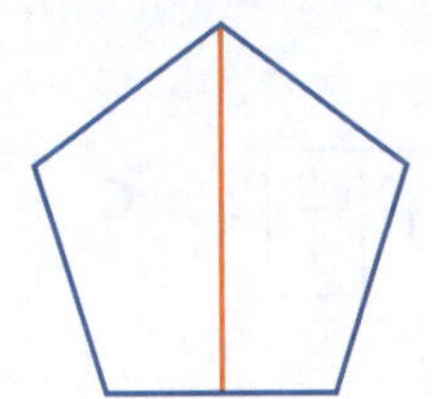

j 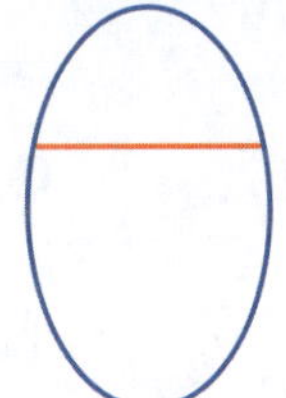

FRACTIONS – QUARTERS AND EIGHTHS

When a whole is cut into four equal parts, each part is called one-quarter ($\frac{1}{4}$).

This square has been cut into 4 equal parts called quarters.

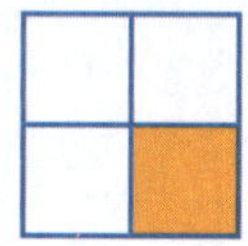

One-quarter ($\frac{1}{4}$) of the square has been coloured orange.

Example 1:
Colour three-quarters ($\frac{3}{4}$).

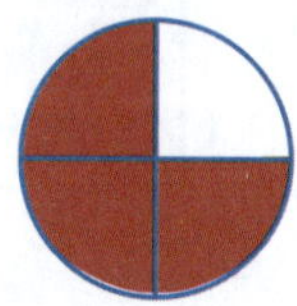

Example 2:
Colour two-quarters ($\frac{2}{4}$).

When a whole is cut into eight equal parts, each part is called one-eighth ($\frac{1}{8}$).

SCAN to watch video

This square has been cut into 8 equal parts called eighths.

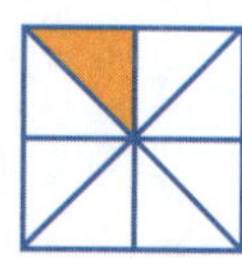

One-eighth ($\frac{1}{8}$) of the square has been coloured orange.

Example 3:
Colour eight-eighths ($\frac{8}{8}$).

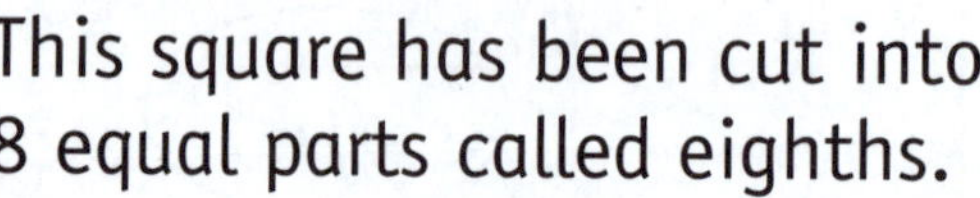

Example 4:
Colour five-eighths ($\frac{5}{8}$).

Your turn

Circle in blue the shapes that are cut into quarters.
Circle in red the shapes that are cut into eighths.
Cross out any shapes that are not cut into quarters or eighths.

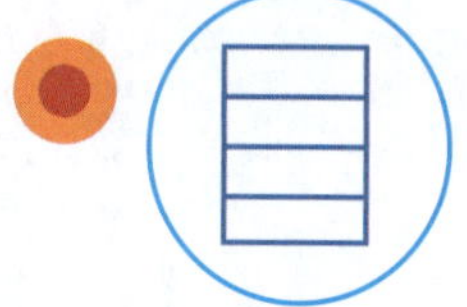

b

d

f

a

c

e

g

SELF CHECK Tick how you feel

Got it!	Need help...	I don't get it
☐	☐	☐

Check your answers

How many did you get correct? ☐

CATCH UP MATHS YEAR 4 BOOK A © PASCAL PRESS ISBN: 9781925726145

PRACTICE

1 Complete the table.

	Fraction name	Fraction	Picture
●	one-eighth	$\frac{1}{8}$	
a		$\frac{7}{8}$	
b	three-quarters		
c			
d	two-quarters		
e			
f		$\frac{4}{4}$	

Cut these shapes into quarters.

a 

b 

c 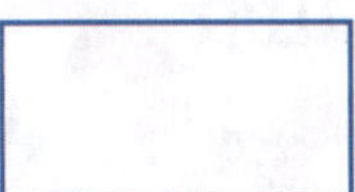

Cut these shapes into eighths.

a 

b 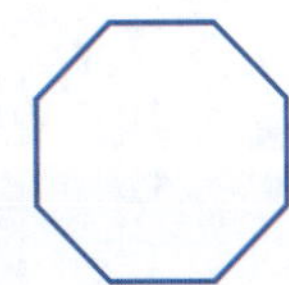

c 

4 Complete each fraction. Then number the boxes to order the fractions from smallest (1) to largest (6).

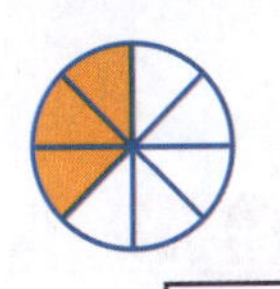 $\frac{\quad}{8}$

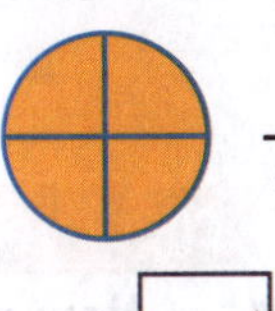 $\frac{\quad}{4}$

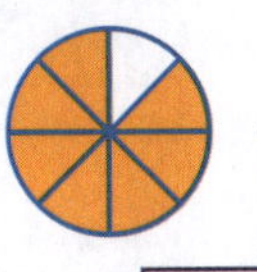 $\frac{\quad}{8}$

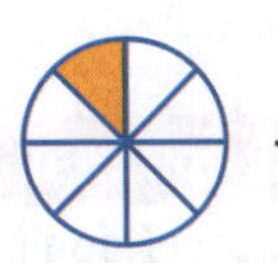 $\frac{\quad}{8}$

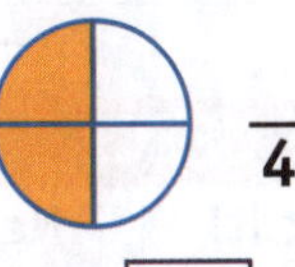 $\frac{\quad}{4}$

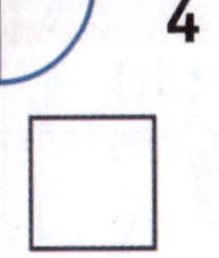

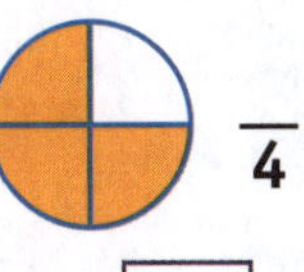 $\frac{\quad}{4}$

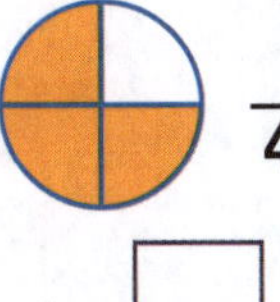

FRACTIONS – THIRDS AND FIFTHS

When a whole is divided into three equal parts, each part is called one-third ($\frac{1}{3}$).

This rectangle has been cut into three equal parts.

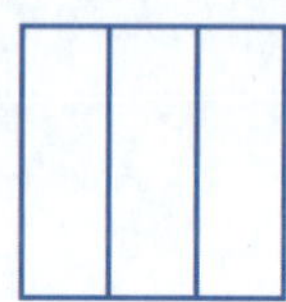

Example 1:
Colour two-thirds ($\frac{2}{3}$).

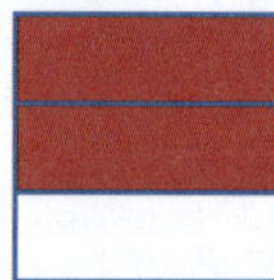

Example 2:
Colour one-third ($\frac{1}{3}$).

When a whole is divided into five equal parts, each part is called one-fifth ($\frac{1}{5}$).

This rectangle has been cut into five equal parts.

Example 3:
Colour two-fifths ($\frac{2}{5}$).

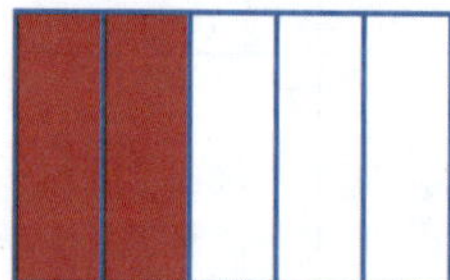

Example 4:
Colour one-fifth ($\frac{1}{5}$).

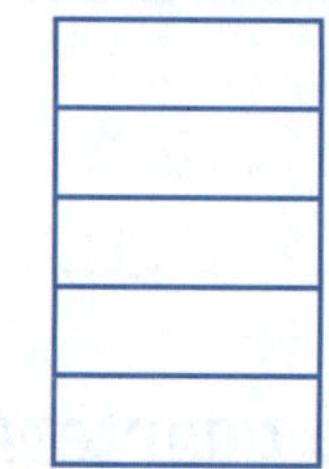

Your turn

Colour with purple the shapes that have been cut into thirds. Colour with green the shapes that have been cut into fifths. Cross out the shapes that have not been cut into thirds or fifths.

 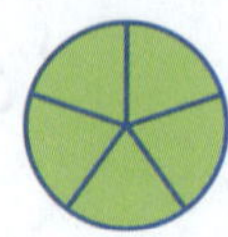

b 

d 

f 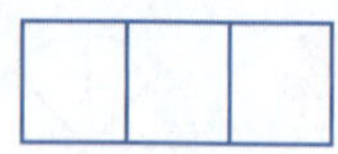

a 

c 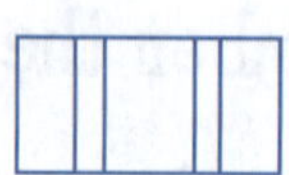

e 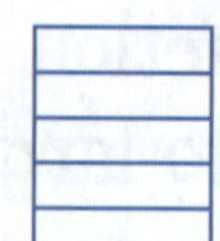

g

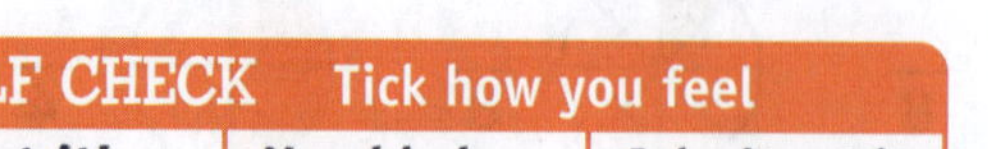

Check your answers
How many did you get correct?

CATCH UP MATHS YEAR 4 BOOK A © PASCAL PRESS ISBN: 9781925726145

PRACTICE

1 Complete the table.

	Fraction name	Fraction	Picture
●	one-third	$\frac{1}{3}$	
a	two-fifths		
b		$\frac{2}{3}$	
c			
d		$\frac{4}{5}$	
e	five-fifths		

2 Complete each fraction. Then number the boxes to order the fractions from smallest (1) to largest (6).

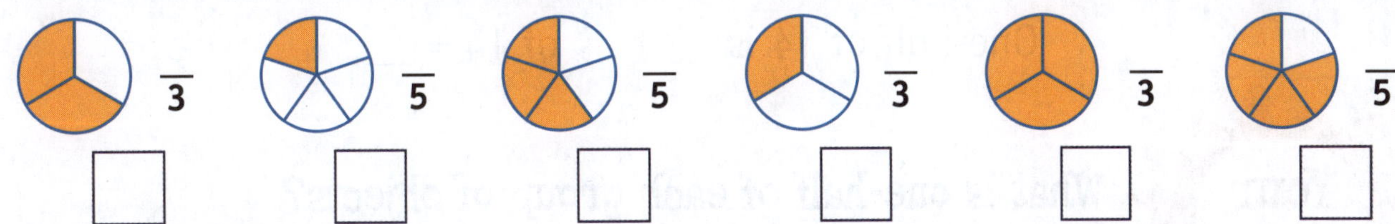

3 What fraction is coloured?

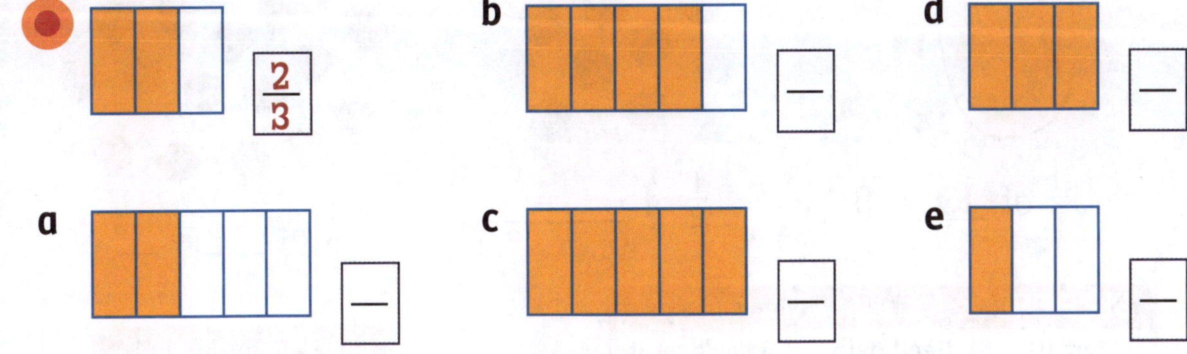

  ISBN: 9781925726145

HALVES OF A COLLECTION

If a group of objects is divided into two equal parts, each part is called one-half ($\frac{1}{2}$).

SCAN to watch video

Example 1: These 8 marbles are divided into halves. There are two equal parts.

If the 8 marbles are shared into two equal parts, each equal part has 4 marbles.

One-half of 8 is 4 $\quad \frac{1}{2}$ of 8 = 4

Example 2: These 14 jelly beans are divided into halves. There are two equal parts.

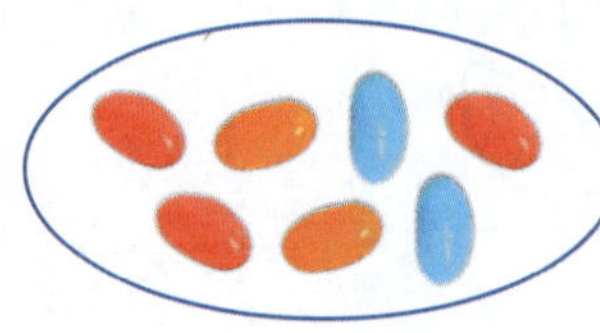
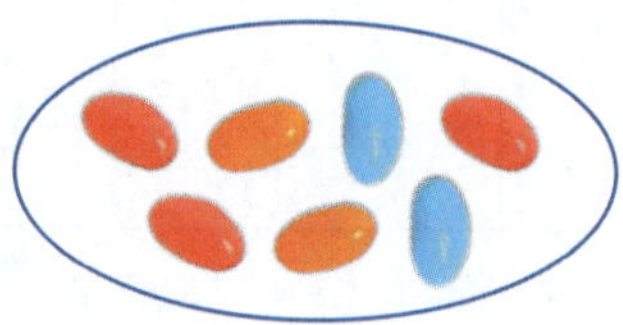

If the 14 jelly beans are shared into two equal parts, each equal part has 7 jelly beans.

One-half of 14 is __ $\quad \frac{1}{2}$ of 14 = __

What is one-half of each group of objects? Circle one-half ($\frac{1}{2}$), then fill in the missing numbers.

$\frac{1}{2}$ of 10 = 5

a

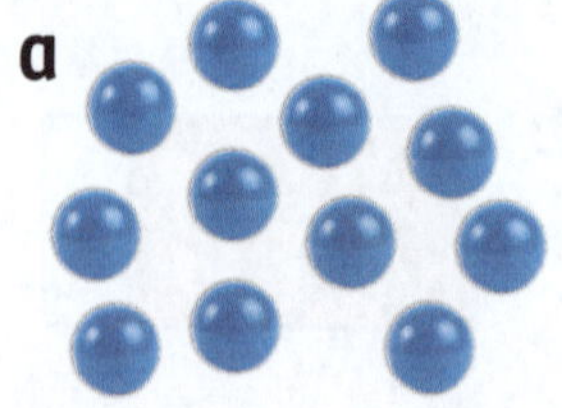

$\frac{1}{2}$ of ___ = __

b

$\frac{1}{2}$ of ___ = __

SELF CHECK Tick how you feel

Got it!	Need help...	I don't get it
 ☐	 ☐	 ☐

Check your answers
How many did you get correct? ☐

CATCH UP MATHS YEAR 4 BOOK A © PASCAL PRESS ISBN: 9781925726145

1 Colour in one-half of each group.

●	●●●●○○○○	$\frac{1}{2}$ = 4
a	○○○○	$\frac{1}{2}$ =
b	○○	$\frac{1}{2}$ =
c	○○○○○○○○○○	$\frac{1}{2}$ =
d	○○○○○○○○○○○○○○○○○○○○	$\frac{1}{2}$ =
e	○○○○○○	$\frac{1}{2}$ =
f	○○○○○○○○○○○○○○○○○○○○○○○○	$\frac{1}{2}$ =
g	○○○○○○○○○○○○	$\frac{1}{2}$ =

2 Circle $\frac{1}{2}$ of each collection and fill in the missing numbers.

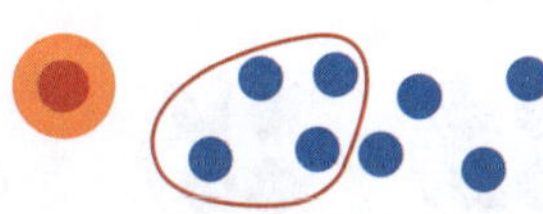

Half of 8 = 4

a 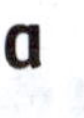 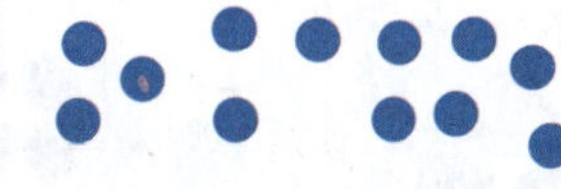

Half of __ = __

b 

Half of __ = __

3 These groups have been divided into halves. Fill in the missing numbers.

Each group has 5 of the 10 marbles.

b  

Each group has __ of the __ marbles.

a 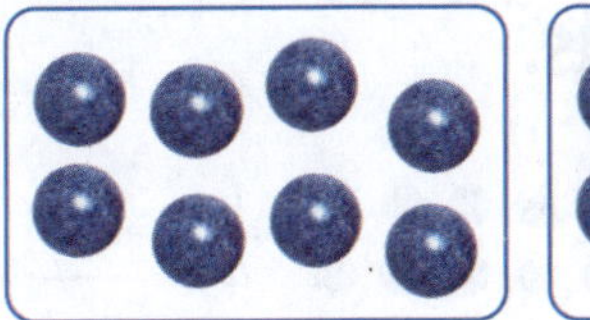

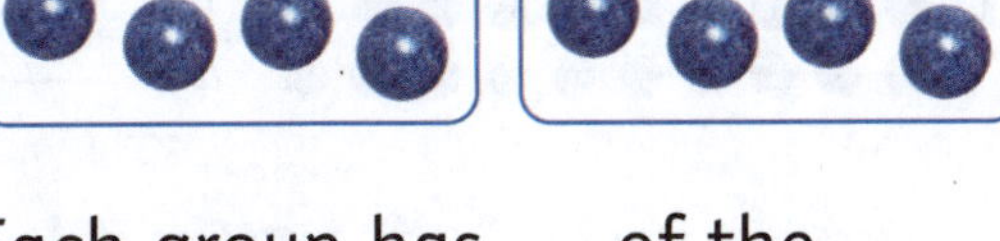

Each group has __ of the __ marbles.

c  

Each group has __ of the __ marbles.

QUARTERS AND EIGHTHS OF A COLLECTION

When a group of objects is divided into eight equal parts, each part is called one-eighth ($\frac{1}{8}$).

The 8 marbles are divided into eighths. There are 8 equal parts.

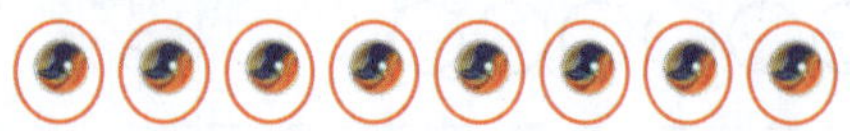

When 8 marbles are shared into 8 equal parts, each equal part has 1 marble.

One-eighth of 8 is 1. $\frac{1}{8}$ of 8 = 1

Example 1:
24 marbles are shared into 8 equal parts.

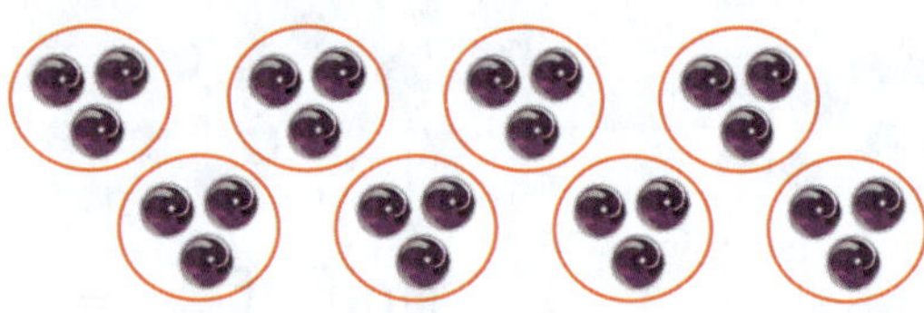

Each part has __ marbles.

One-eighth of 24 is __

$\frac{1}{8}$ of 24 = __

When a group of objects is divided into four equal parts, each part is called one-quarter ($\frac{1}{4}$).

The 8 marbles are divided into quarters. There are 4 equal parts.

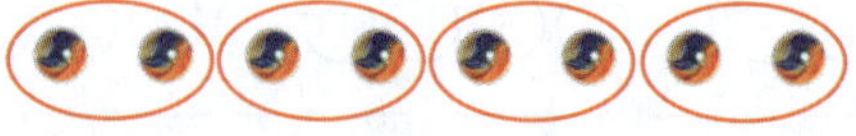

When 8 marbles are shared into 4 equal parts, each equal part has 2 marbles.

One-quarter of 8 is 2. $\frac{1}{4}$ of 8 = 2

Example 2:
16 marbles are shared into 4 equal parts.

Each part has __ marbles.

One-quarter of 16 is __

$\frac{1}{4}$ of 16 = __

Your turn

1 Circle $\frac{1}{8}$ then write how many that is.

 $\frac{1}{8}$ = 1

a  $\frac{1}{8}$ = __

2 Circle $\frac{1}{4}$ then write how many that is.

 $\frac{1}{4}$ = 2

a  $\frac{1}{4}$ = __

SELF CHECK Tick how you feel

Got it!	Need help...	I don't get it
 ☐	 ☐	 ☐

Check your answers
How many did you get correct? ☐

CATCH UP MATHS YEAR 4 BOOK A © PASCAL PRESS ISBN: 9781925726145

1 Colour one-eighth of each group and write how many that is.

●	●○○○○○○○	$\frac{1}{8}$ = 1
a	△△△△△△△△△△△△△△△△	$\frac{1}{8}$ =
b	⬡⬡	$\frac{1}{8}$ =
c	▽▽▽▽▽▽▽▽▽▽▽▽▽▽▽▽▽▽▽▽▽▽▽▽▽▽▽▽▽▽▽▽	$\frac{1}{8}$ =
d	◇◇	$\frac{1}{8}$ =

2 Colour one-quarter of each group and write how many that is.

●	▲▲△△△△△△	$\frac{1}{4}$ = 2
a	○○○○○○○○○○○○	$\frac{1}{4}$ =
b	□□□□□□□□□□□□□□□□□□□□□□□□	$\frac{1}{4}$ =
c	⬡⬡⬡⬡⬡⬡⬡⬡⬡⬡⬡⬡⬡⬡⬡⬡⬡⬡⬡⬡	$\frac{1}{4}$ =
d	□□□□	$\frac{1}{4}$ =
e	0000000000000000000000000000	$\frac{1}{4}$ =

3 What is one-eighth of each number? Hint: divide by 8.

● 16 2   b 56 ___   d 40 ___   f 24 ___

a 32 ___   c 96 ___   e 64 ___   g 48 ___

4 What is one-quarter of each number? Hint: divide by 4.

● 20 5   b 4 ___   d 48 ___   f 12 ___

a 44 ___   c 36 ___   e 24 ___   g 28 ___

SIMPLIFYING FRACTIONS

When you simplify a fraction, you make the numerator and denominator as small as possible.

When a fraction is written in its simplest form, the top and bottom numbers can no longer be divided by the same whole number exactly or evenly.

Example 1: Simplify $\frac{8}{12}$.

$\frac{8}{12} = \frac{2}{3}$

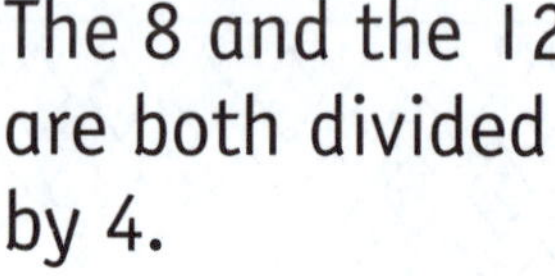

The 8 and the 12 are both divided by 4.

Example 3: Simplify $\frac{10}{24}$.

$\frac{10}{24} = \frac{5}{12}$

The 10 and the 24 are both divided by 2.

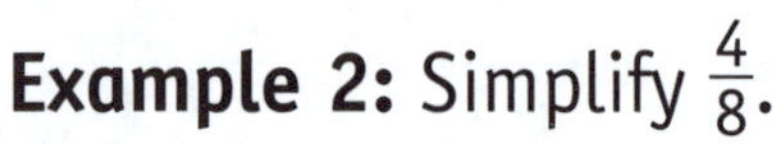

Example 2: Simplify $\frac{4}{8}$.

$\frac{4}{8} = \frac{2}{4} = \frac{1}{2}$

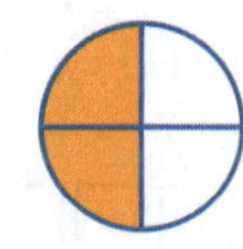

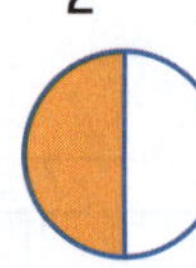

The 4 and 8 are divided by 2 to give $\frac{2}{4}$ and then the 2 and 4 are divided again by 2.

Or you could just divide the 4 and 8 by 4 to get $\frac{1}{2}$!

Example 4: Simplify $\frac{6}{10}$.

$\frac{6}{10} = \frac{}{5}$

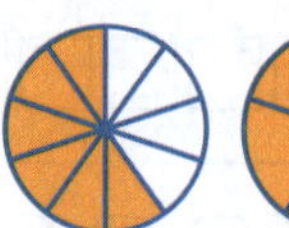

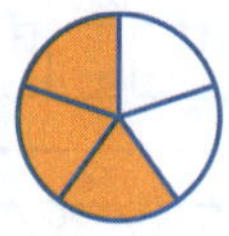

The 6 and 10 are both divided by ___.

Check your answer on the video!

Your turn

Simplify the fractions.

● $\frac{6}{8} = \frac{3}{4}$ (÷ 2, ÷ 2)

a $\frac{8}{10} = \frac{4}{}$ (÷ 2, ÷ 2)

b $\frac{30}{100} = \frac{}{10}$ (÷ 10, ÷ 10)

c $\frac{10}{35} = \frac{}{}$ (÷ 5, ÷ 5)

SELF CHECK Tick how you feel		
Got it! ☐	Need help... ☐	I don't get it ☐

Check your answers

How many did you get correct? ☐

CATCH UP MATHS YEAR 4 BOOK A © PASCAL PRESS ISBN: 9781925726145

PRACTICE

1 Fill in the missing numbers by simplifying the fractions.

Each bag has 4 of the 32 marbles, which is $\frac{1}{8}$.

a

Each bag has __ of the ___ marbles, which is —.

b

Each bag has __ of the ___ marbles, which is —.

c

Each bag has __ of the ___ marbles, which is —.

2 Simplify the fractions.

What number can BOTH the numerator and the denominator be divided by?

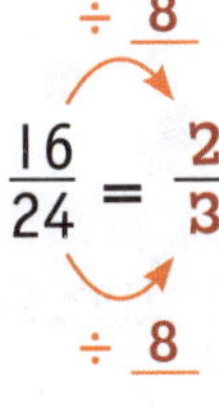

÷ 8

$\frac{16}{24} = \frac{2}{3}$

÷ 8

b

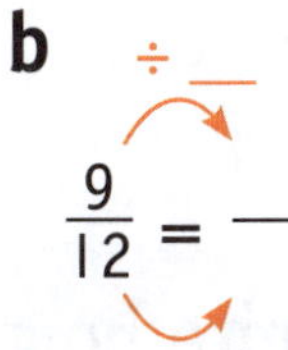

÷ __

$\frac{9}{12} =$ —

÷ __

d

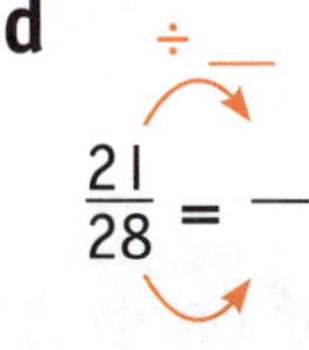

÷ __

$\frac{21}{28} =$ —

÷ __

f

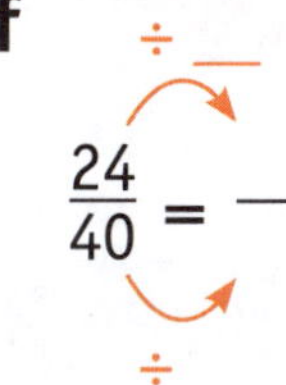

÷ __

$\frac{24}{40} =$ —

÷ __

a

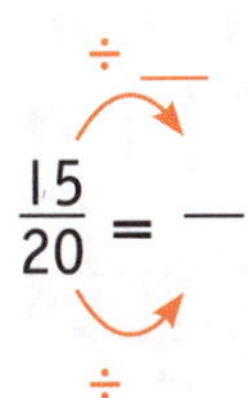

÷ __

$\frac{15}{20} =$ —

÷ __

c

÷ __

$\frac{50}{100} =$ —

÷ __

e

÷ __

$\frac{18}{30} =$ —

÷ __

g

÷ __

$\frac{40}{45} =$ —

÷ __

THIRDS AND FIFTHS OF A COLLECTION

When a group of objects is divided into three equal parts, each part is called one-third. This is written as $\frac{1}{3}$.

When a group of objects is divided into five equal parts, each part is called one-fifth. This is written as $\frac{1}{5}$.

Example 1:
These 15 marbles have been shared equally between 3 people.

Each person will get 5 marbles. One-third of 15 is 5. $\frac{1}{3}$ of 15 = 5

Example 2:
These 15 marbles have been shared equally between 5 people.

Each person will get 3 marbles. One-fifth of 15 is 3. $\frac{1}{5}$ of 15 = 3

Example 3:
These 9 marbles have been shared equally between 3 people.

Each person will get ___ marbles.

One-third of 9 is ___. $\frac{1}{3}$ of 9 = ___

Your turn

1 Circle $\frac{1}{3}$ of each group and write how many you circled.

a 1

b ___
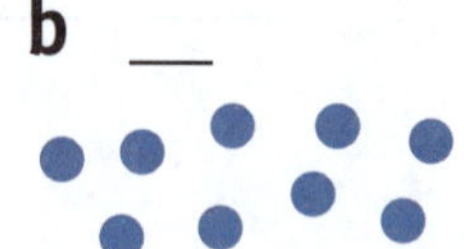

c ___
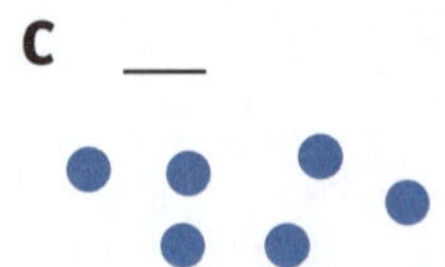

2 Circle $\frac{1}{5}$ of each group and write how many you circled.

a 2
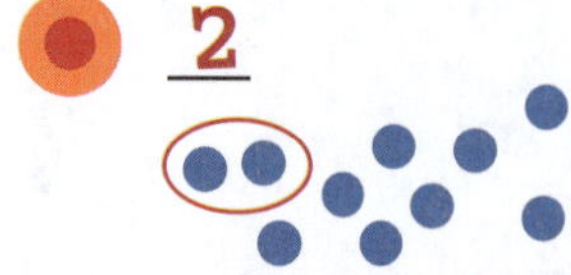

b ___

c ___
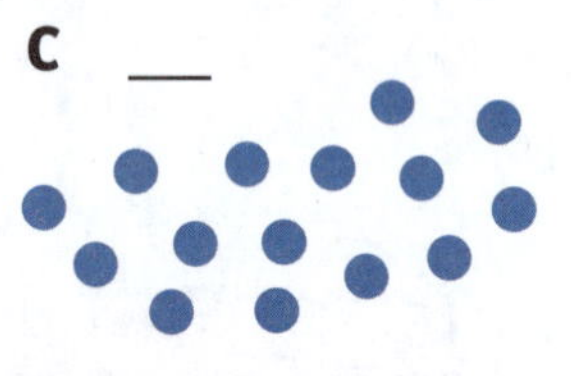

SELF CHECK Tick how you feel

Got it!	Need help...	I don't get it
 ☐	 ☐	 ☐

Check your answers
How many did you get correct? ☐

CATCH UP MATHS YEAR 4 BOOK A © PASCAL PRESS ISBN: 9781925726145

PRACTICE

1 Divide each group of objects into thirds.

 $\frac{1}{3}$ of <u>6</u> = <u>2</u> beetles

a $\frac{1}{3}$ of ___ = __ strawberries

b $\frac{1}{3}$ of ___ = __ kiwi slices

c $\frac{1}{3}$ of ___ = __ shells

2 Divide each group of objects into fifths.

$\frac{1}{5}$ of <u>10</u> = <u>2</u> orange slices

a $\frac{1}{5}$ of ___ = __ flowers

b $\frac{1}{5}$ of ___ = __ lollies

c $\frac{1}{5}$ of ___ = __ fish

3 What is one-third?

36 <u>12</u>	**b** 9 ___	**d** 21 ___	**f** 33 ___
a 3 ___	**c** 15 ___	**e** 27 ___	**g** 30 ___

4 What is one-fifth?

45 <u>9</u>	**b** 25 ___	**d** 100 ___	**f** 35 ___
a 10 ___	**c** 50 ___	**e** 30 ___	**g** 40 ___

5 Write the answer.

a One-third of 33 equals ___.

b One-third of 3 equals ___.

c One-fifth of 10 equals ___.

d One-fifth of 60 equals ___.

COMPARING FRACTIONS

Comparing fractions means deciding which fraction is bigger and which fraction is smaller.

1 whole									
$\frac{1}{2}$	$\frac{1}{2}$								
$\frac{1}{4}$	$\frac{1}{4}$	$\frac{1}{4}$	$\frac{1}{4}$						
$\frac{1}{8}$	$\frac{1}{8}$	$\frac{1}{8}$	$\frac{1}{8}$	$\frac{1}{8}$	$\frac{1}{8}$	$\frac{1}{8}$	$\frac{1}{8}$		
$\frac{1}{3}$	$\frac{1}{3}$	$\frac{1}{3}$							
$\frac{1}{5}$	$\frac{1}{5}$	$\frac{1}{5}$	$\frac{1}{5}$	$\frac{1}{5}$					
$\frac{1}{10}$	$\frac{1}{10}$	$\frac{1}{10}$	$\frac{1}{10}$	$\frac{1}{10}$	$\frac{1}{10}$	$\frac{1}{10}$	$\frac{1}{10}$	$\frac{1}{10}$	$\frac{1}{10}$

SCAN to watch video

Example 1: One-third ($\frac{1}{3}$) is larger than one-eighth ($\frac{1}{8}$).

Example 2: One-fifth ($\frac{1}{5}$) is smaller than one-half ($\frac{1}{2}$).

Example 3: Two-quarters ($\frac{2}{4}$) is equivalent to one-half ($\frac{1}{2}$).

Example 4: One-fifth ($\frac{1}{5}$) is bigger than ______________ ($\frac{}{}$).

Your turn

Colour and fill in the spaces.

● Colour one-half.

1 out of **2** = $\frac{1}{2}$

a Colour one-quarter.

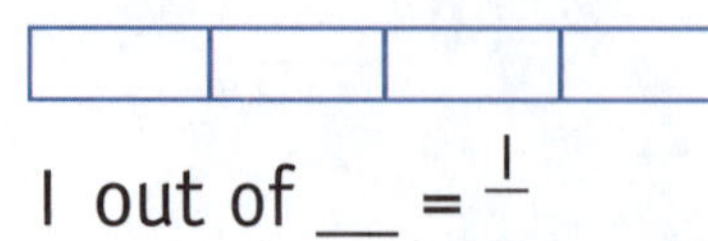

1 out of __ = $\frac{1}{}$

c Colour one-third.

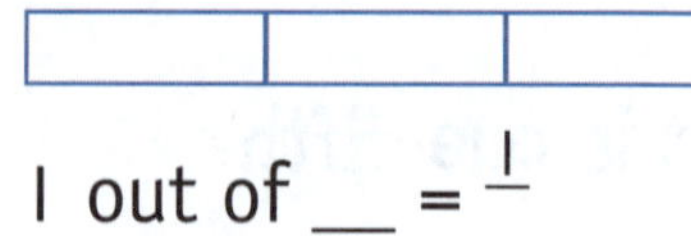

1 out of __ = $\frac{1}{}$

d Colour one-fifth.

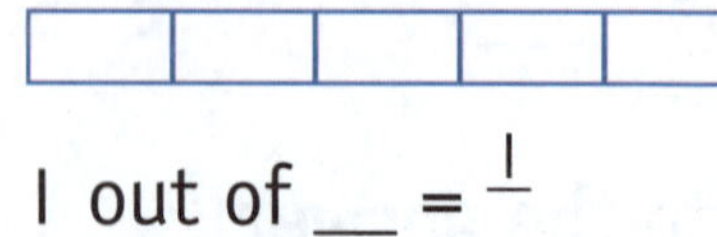

1 out of __ = $\frac{1}{}$

SELF CHECK Tick how you feel

Got it!	Need help...	I don't get it
☐	☐	☐

Check your answers
How many did you get correct?

CATCH UP MATHS YEAR 4 BOOK A © PASCAL PRESS ISBN: 9781925726145

PRACTICE

1 Number the boxes to order the fractions from largest (1) to smallest (6).

a

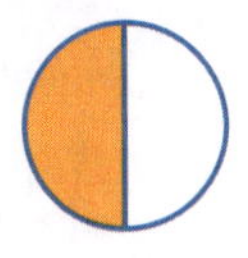 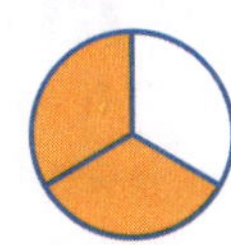   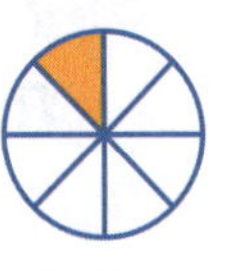 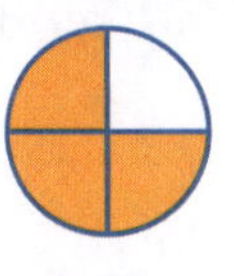

b

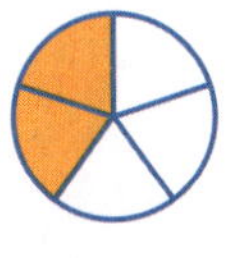 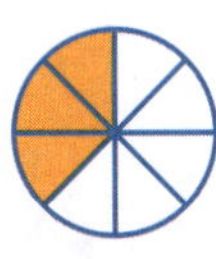    

2 Write the correct symbol: >, < or =

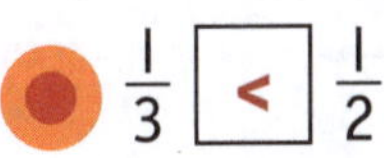

$\frac{1}{3}$ < $\frac{1}{2}$

		c	$\frac{1}{5}$ ☐ $\frac{1}{4}$	f	$\frac{3}{5}$ ☐ $\frac{2}{3}$	i	$\frac{3}{4}$ ☐ $\frac{6}{8}$
a	$\frac{1}{2}$ ☐ $\frac{4}{8}$	d	$\frac{2}{3}$ ☐ $\frac{1}{2}$	g	$\frac{4}{5}$ ☐ $\frac{3}{4}$	j	$\frac{1}{4}$ ☐ $\frac{1}{8}$
b	$\frac{1}{3}$ ☐ $\frac{1}{5}$	e	$\frac{2}{5}$ ☐ $\frac{3}{4}$	h	$\frac{2}{4}$ ☐ $\frac{1}{2}$	k	$\frac{3}{3}$ ☐ $\frac{5}{5}$

3 Draw a smaller fraction then write the fraction in the box.

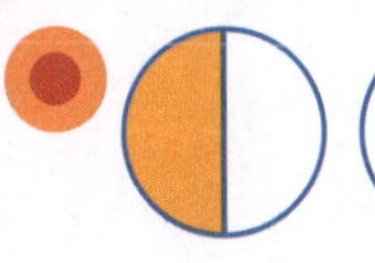

$\frac{1}{2}$

a 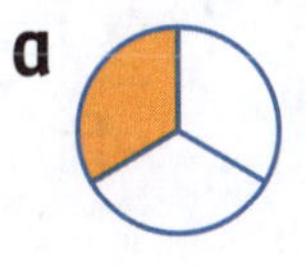   

$\frac{1}{3}$

b  

$\frac{1}{4}$

c 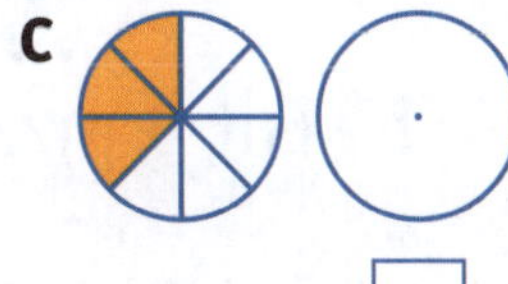

$\frac{3}{8}$

4 Draw a larger fraction then write the fraction in the box.

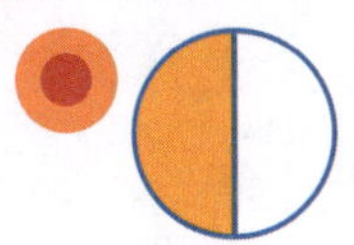
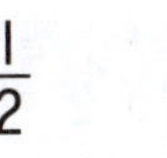

$\frac{1}{2}$ $\frac{2}{3}$

a  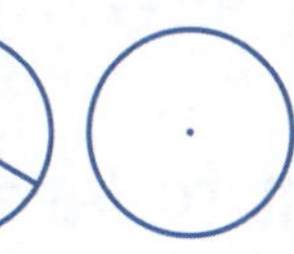

$\frac{1}{3}$

b 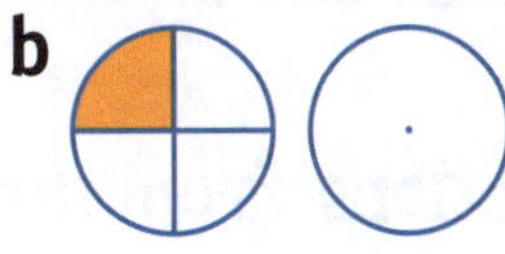

$\frac{1}{4}$

c 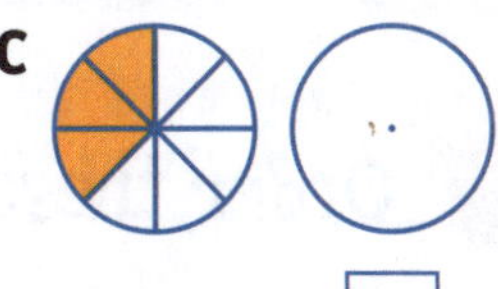

$\frac{3}{8}$

5 Draw a smaller fraction then write the fraction in the box.

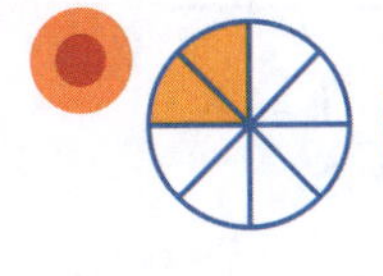

$\frac{2}{8}$ $\frac{1}{8}$

a 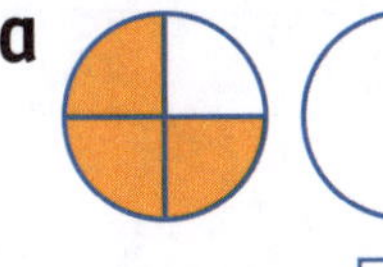 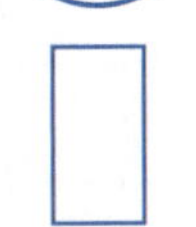

$\frac{3}{4}$

b 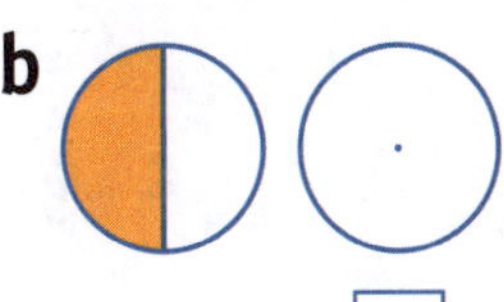 

$\frac{1}{2}$

c 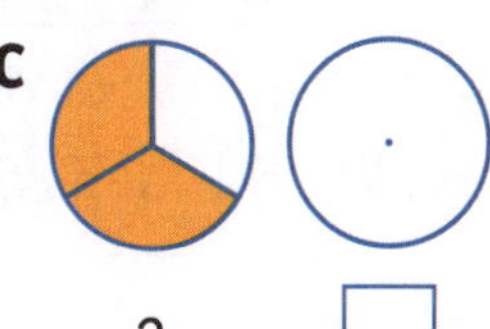

$\frac{2}{3}$

  ISBN: 9781925726145

6 Place these fractions on the number line.

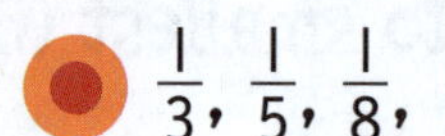

$\frac{1}{3}, \frac{1}{5}, \frac{1}{8}, 1$

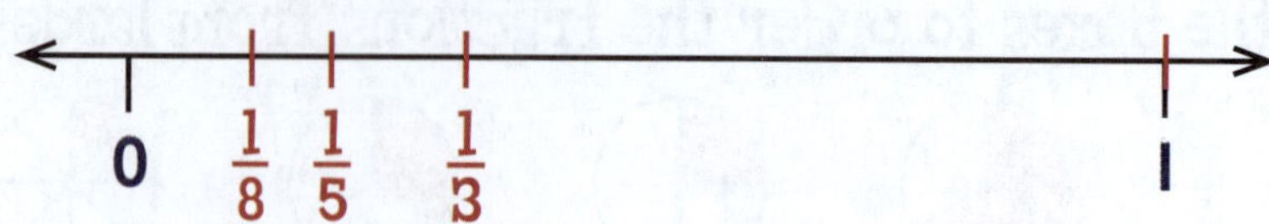

a $\frac{3}{4}, \frac{2}{3}, \frac{2}{8}, \frac{1}{2}$

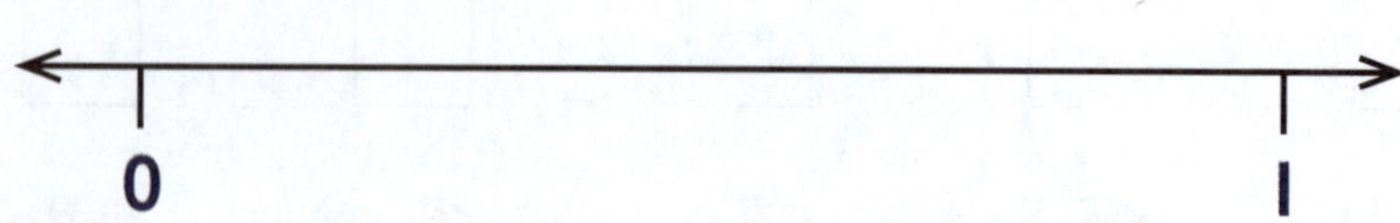

b $\frac{2}{5}, \frac{3}{4}, \frac{2}{8}, \frac{3}{5}$

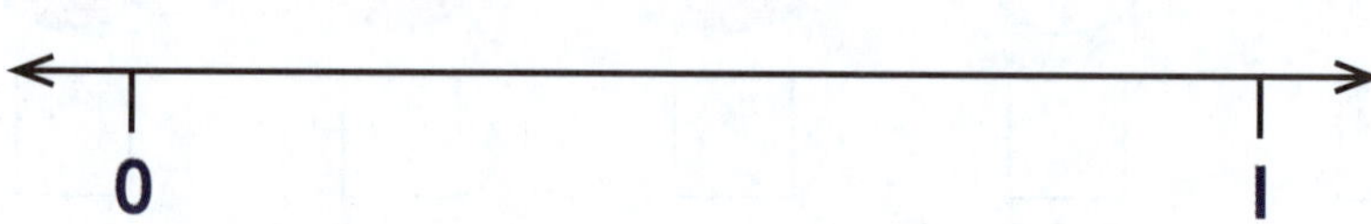

c $\frac{3}{8}, \frac{6}{8}, \frac{3}{5}, \frac{7}{8}$

7 Fill in the missing numbers.

- Eat 7 pizza slices out of 8. $\frac{7}{8}$ eaten, $\frac{1}{8}$ left

a Eat 4 pizza slices out of 8. $\frac{}{8}$ eaten, $\frac{}{8}$ left

b Eat 3 pizza slices out of 5. $\frac{}{5}$ eaten, $\frac{}{5}$ left

c Eat 2 pizza slices out of 3. $\frac{}{3}$ eaten, $\frac{}{3}$ left

d Eat 1 pizza slices out of 5. $\frac{}{5}$ eaten, $\frac{}{5}$ left

e Eat 5 pizza slices out of 5. $\frac{}{5}$ eaten, $\frac{}{5}$ left

8 Order these fractions from smallest to largest.

- $\frac{1}{4}, \frac{1}{3}, \frac{1}{8}, \frac{2}{5}$

  $\frac{1}{8}$ $\frac{1}{4}$ $\frac{1}{3}$ $\frac{2}{5}$

a $\frac{3}{5}, 1, \frac{3}{8}, \frac{2}{4}$

____

b $\frac{2}{5}, \frac{3}{4}, \frac{2}{8}, \frac{3}{5}$

____

c $\frac{5}{5}, \frac{2}{4}, \frac{6}{8}, \frac{7}{8}$

____

d $\frac{2}{3}, \frac{2}{4}, \frac{2}{8}, \frac{2}{5}$

____

e $\frac{1}{4}, \frac{3}{5}, \frac{7}{8}, \frac{1}{5}$

____

9 Circle the smallest fraction.

- ● $\frac{1}{2}$ (circled), $\frac{3}{4}$, $\frac{7}{8}$
- a $\frac{1}{4}$, $\frac{2}{3}$, $\frac{5}{8}$
- b $\frac{7}{8}$, $\frac{4}{5}$, $\frac{3}{4}$
- c $\frac{3}{4}$, $\frac{6}{8}$, $\frac{1}{4}$
- d $\frac{6}{10}$, $\frac{4}{5}$, $\frac{7}{8}$
- e $\frac{1}{2}$, $\frac{3}{4}$, $\frac{2}{5}$
- f $\frac{3}{5}$, $\frac{2}{8}$, $\frac{1}{5}$
- g $\frac{1}{4}$, $\frac{1}{3}$, $\frac{2}{5}$
- h $\frac{3}{5}$, $\frac{7}{8}$, $\frac{2}{3}$
- i $\frac{1}{4}$, $\frac{6}{8}$, $\frac{2}{3}$
- j $\frac{2}{3}$, $\frac{2}{5}$, $\frac{2}{4}$
- k $\frac{3}{4}$, $\frac{3}{5}$, $\frac{3}{8}$

10 Circle the largest fraction.

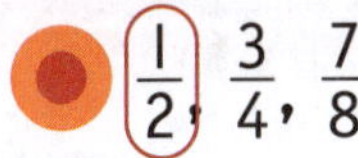

- ● $\frac{1}{2}$, $\frac{3}{4}$, $\frac{7}{8}$ (circled)
- a $\frac{1}{2}$, $\frac{3}{8}$, $\frac{2}{3}$
- b $\frac{2}{5}$, $\frac{2}{8}$, $\frac{3}{4}$
- c $\frac{2}{3}$, $\frac{3}{8}$, $\frac{3}{4}$
- d $\frac{1}{4}$, $\frac{2}{3}$, $\frac{6}{10}$
- e $\frac{3}{5}$, $\frac{7}{8}$, $\frac{2}{3}$
- f $\frac{4}{5}$, $\frac{2}{10}$, $\frac{9}{10}$
- g $\frac{3}{4}$, $\frac{2}{5}$, $\frac{3}{8}$
- h $\frac{3}{5}$, $\frac{7}{8}$, $\frac{2}{3}$
- i $\frac{1}{3}$, $\frac{2}{8}$, $\frac{1}{8}$
- j $\frac{4}{5}$, $\frac{1}{4}$, $\frac{3}{4}$
- k $\frac{3}{5}$, $\frac{7}{8}$, $\frac{2}{5}$

11 Write a smaller fraction.

- ● $\frac{3}{5}$ $\underline{\frac{1}{4}}$
- a $\frac{2}{8}$ ____
- b $\frac{1}{3}$ ____
- c $\frac{2}{3}$ ____
- d $\frac{1}{5}$ ____
- e $\frac{3}{8}$ ____
- f $\frac{4}{5}$ ____
- g $\frac{5}{8}$ ____
- h $\frac{1}{2}$ ____
- i $\frac{1}{4}$ ____
- j $\frac{3}{4}$ ____
- k $\frac{6}{8}$ ____
- l $\frac{3}{4}$ ____
- m $\frac{4}{10}$ ____
- n $\frac{2}{5}$ ____
- o $\frac{8}{10}$ ____

12 Write a larger fraction.

- ● $\frac{1}{5}$ $\underline{\frac{1}{4}}$
- a $\frac{1}{3}$ ____
- b $\frac{4}{8}$ ____
- c $\frac{3}{8}$ ____
- d $\frac{2}{3}$ ____
- e $\frac{1}{4}$ ____
- f $\frac{4}{5}$ ____
- g $\frac{3}{4}$ ____
- h $\frac{6}{8}$ ____
- i $\frac{6}{10}$ ____
- j $\frac{7}{8}$ ____
- k $\frac{2}{5}$ ____
- l $\frac{1}{10}$ ____
- m $\frac{3}{5}$ ____
- n $\frac{4}{4}$ ____
- o $\frac{1}{2}$ ____

EQUIVALENT FRACTIONS

Equivalent means equal or the same.
Equivalent fractions are fractions that are equal to or the same as each other.

$\frac{1}{4} = \frac{2}{8}$ One-quarter is equal to or the same as two-eighths.

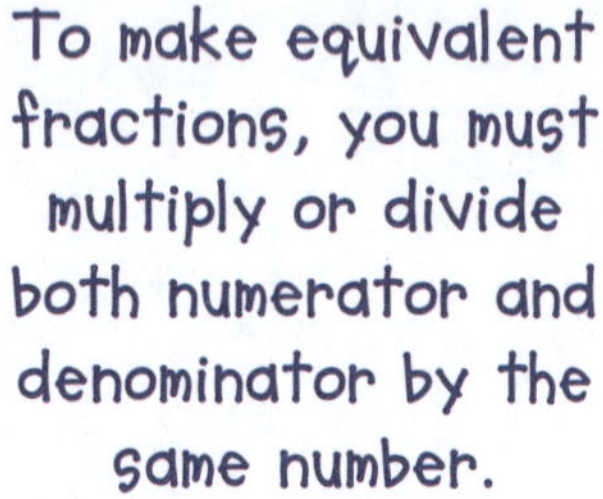

Example 1:

$\frac{1}{2} = \frac{2}{4}$ Both the 1 and the 2 in $\frac{1}{2}$ have been multiplied by 2.

Example 2:

$\frac{6}{8} = \frac{3}{4}$ Both the 6 and the 8 in $\frac{6}{8}$ have been divided by 2.

Example 3:

$\frac{10}{12} = \frac{\ }{6}$ Both the 10 and the 12 in $\frac{10}{12}$ have been divided by __.

Your turn

Colour the circle to show the fraction that is the same as the one on the left.

  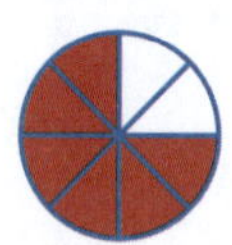

$\frac{3}{4}$ is equivalent to $\frac{6}{8}$

b  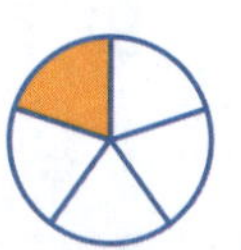 

$\frac{1}{5}$ is equivalent to $\frac{\ }{10}$

a 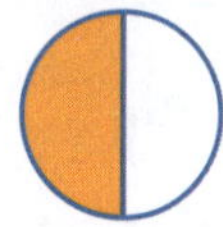 

$\frac{1}{2}$ is equivalent to $\frac{\ }{4}$

c  

$\frac{2}{8}$ is equivalent to $\frac{\ }{4}$

SELF CHECK Tick how you feel		
Got it!  ☐	Need help...  ☐	I don't get it ☐

Check your answers
How many did you get correct? ☐

CATCH UP MATHS YEAR 4 BOOK A © PASCAL PRESS ISBN: 9781925726145

PRACTICE

1 Colour these shapes to show equivalent fractions.

 $\frac{2}{5}$ is equivalent to $\frac{4}{10}$

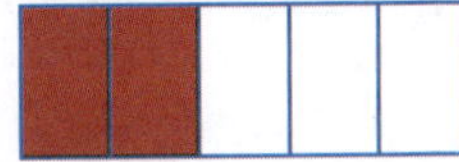 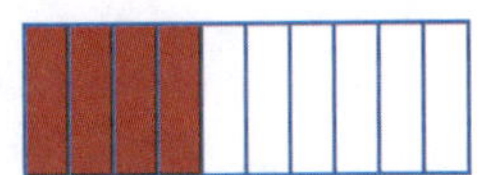

c $\frac{6}{10}$ is equivalent to $\frac{}{5}$

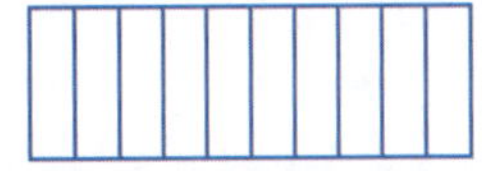 

a $\frac{3}{4}$ is equivalent to $\frac{}{8}$

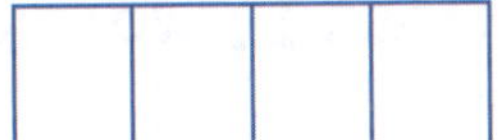 

d $\frac{4}{8}$ is equivalent to $\frac{}{4}$

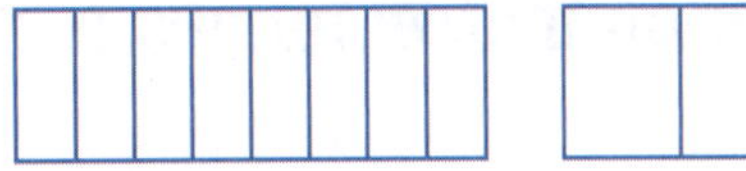

b $\frac{1}{2}$ is equivalent to $\frac{}{4}$

e $\frac{1}{4}$ is equivalent to $\frac{}{8}$

2 Create equivalent fractions.

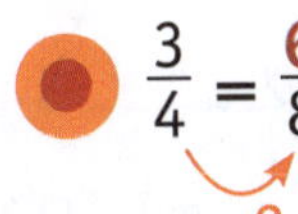 $\frac{3}{4} = \frac{6}{8}$ (× 2)

a $\frac{1}{2} = \frac{}{4}$ (× 2)

b $\frac{3}{5} = \frac{6}{10}$ (× 2)

c $\frac{2}{8} = \frac{}{4}$ (÷ 2)

d $\frac{2}{3} = \frac{}{6}$ (× 2)

e $\frac{4}{8} = \frac{}{2}$ (÷ 4)

f $\frac{2}{5} = \frac{}{10}$ (× 2)

g $\frac{2}{4} = \frac{}{8}$ (× 2)

h $\frac{1}{2} = \frac{}{8}$ (× 4)

i $\frac{1}{4} = \frac{}{8}$ (× 2)

j $\frac{1}{3} = \frac{}{6}$ (× 2)

k $\frac{1}{4} = \frac{}{20}$ (× 5)

l $\frac{10}{15} = \frac{}{3}$ (÷ 5)

m $\frac{3}{4} = \frac{}{12}$ (× 3)

n $\frac{12}{16} = \frac{}{4}$ (÷ 4)

o $\frac{3}{8} = \frac{}{24}$ (× 3)

p $\frac{4}{5} = \frac{}{30}$ (× 6)

q $\frac{20}{24} = \frac{}{6}$ (÷ 4)

r $\frac{18}{20} = \frac{}{10}$ (÷ 2)

s $\frac{12}{30} = \frac{}{10}$ (÷ 3)

FRACTIONS REVIEW

What is the numerator in these fractions?

a $\frac{1}{3}$ ___ b $\frac{2}{5}$ ___ c $\frac{3}{4}$ ___ d $\frac{2}{3}$ ___ e $\frac{1}{4}$ ___ f $\frac{3}{5}$ ___

What is the denominator in these fractions?

a $\frac{2}{3}$ ___ b $\frac{3}{8}$ ___ c $\frac{7}{8}$ ___ d $\frac{3}{4}$ ___ e $\frac{2}{5}$ ___ f $\frac{1}{2}$ ___

The denominator names a fraction. Write these fractions in words.

a $\frac{3}{5}$ ____________________

b $\frac{2}{8}$ ____________________

c $\frac{4}{5}$ ____________________

d $\frac{1}{2}$ ____________________

e $\frac{3}{4}$ ____________________

f $\frac{7}{8}$ ____________________

Cut each of these shapes in half.

a  b 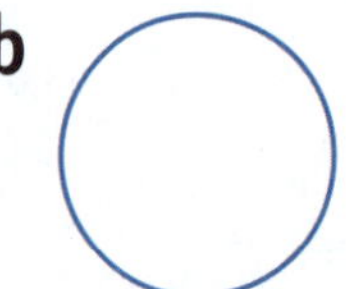 c 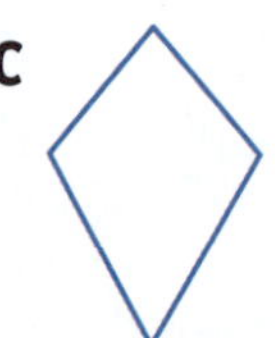 d 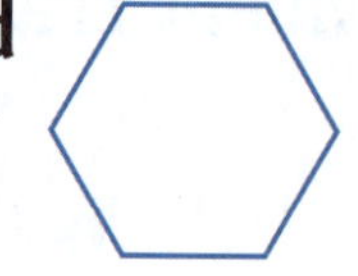 e 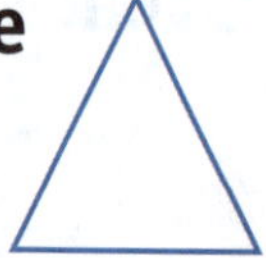

5 Cut each of these shapes into quarters.

a 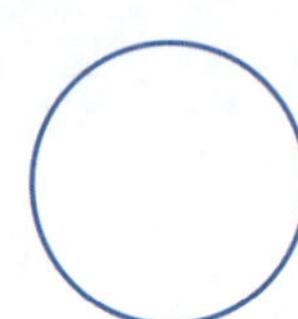 b  c  d 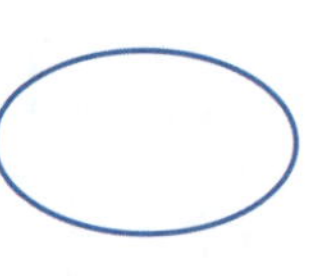 e 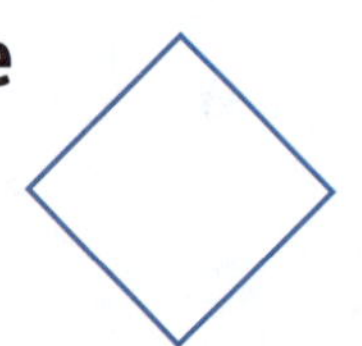

6 Circle the shapes that have been cut into eighths.

a  b  c  d  e 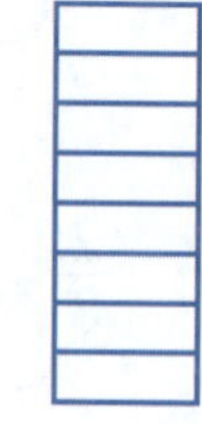

Circle with purple the shapes that have been cut into thirds and with green the shapes that have been cut into fifths.

a  b  c  d 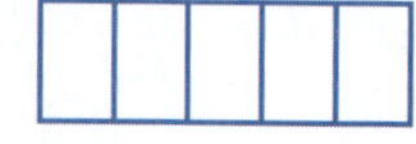

  ISBN: 9781925726145

8 Circle the shapes that have been cut into fifths.

a 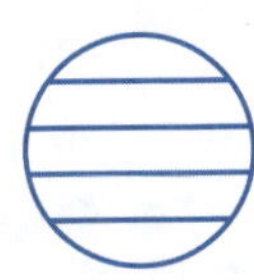

b 

c 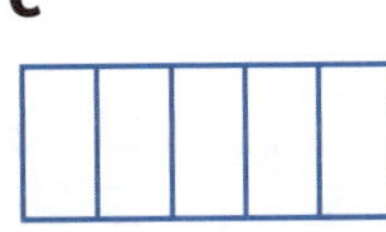

d 

e 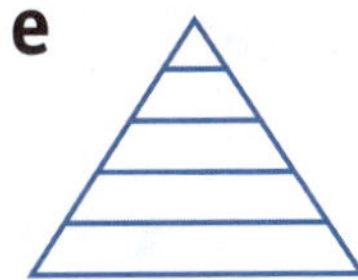

9 Circle half of each collection.

a 

Half of ___ = ___

b 

Half of ___ = ___

c 

Half of ___ = ___

d 

Half of ___ = ___

10 The groups below have been divided into halves.

a 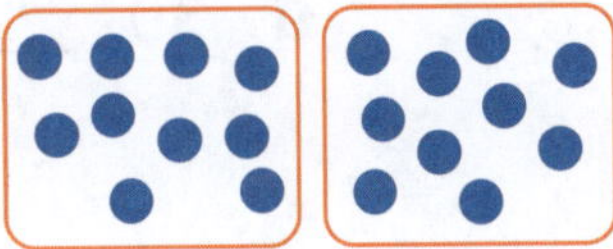

Each group has ___ of the ___ balls.

b 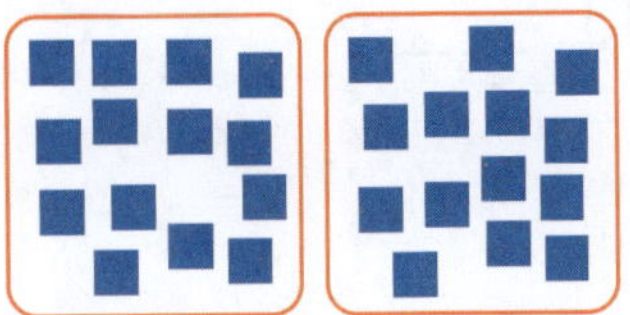

Each group has ___ of the ___ balls.

c 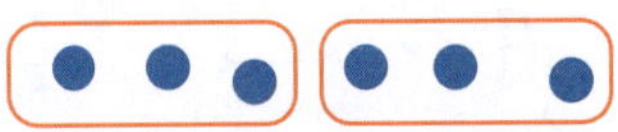

Each group has ___ of the ___ balls.

11 Circle one-quarter of each collection and then write the answers.

a 

$\frac{1}{4}$ of ___ = ___

b 

$\frac{1}{4}$ of ___ = ___

c 

$\frac{1}{4}$ of ___ = ___

REVIEW

12 What is one-eighth of each group of objects?

a ___

b ___

c ___

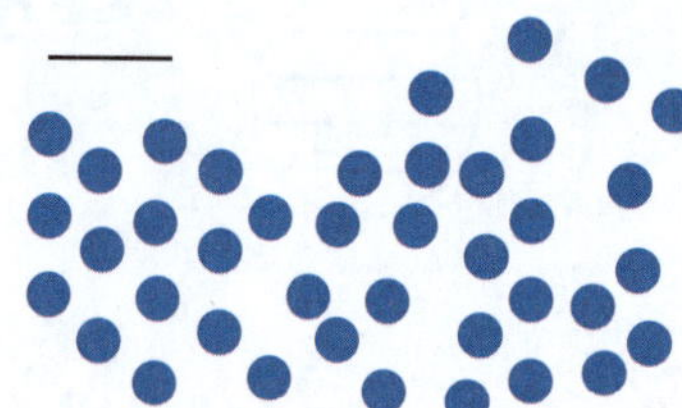

13 What is one-third of each group of objects?

a

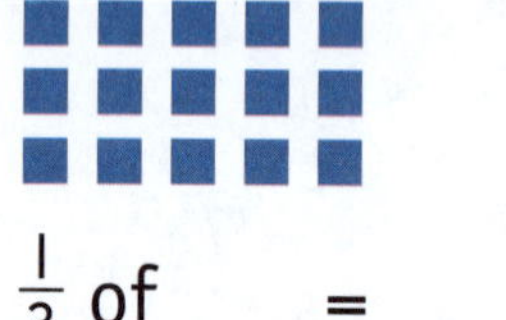

$\frac{1}{3}$ of ___ = ___

b

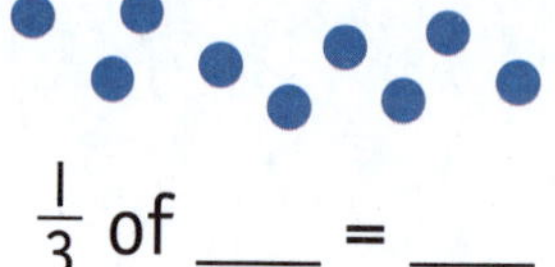

$\frac{1}{3}$ of ___ = ___

c

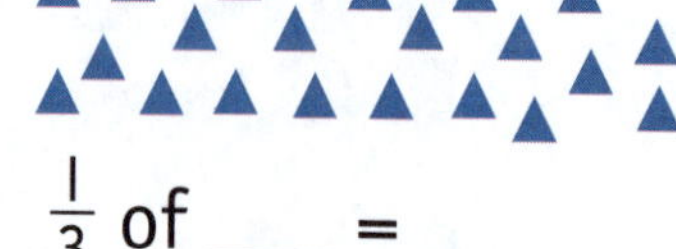

$\frac{1}{3}$ of ___ = ___

14 What is one-fifth of each group of objects?

a

$\frac{1}{5}$ of ___ = ___

b

$\frac{1}{5}$ of ___ = ___

c

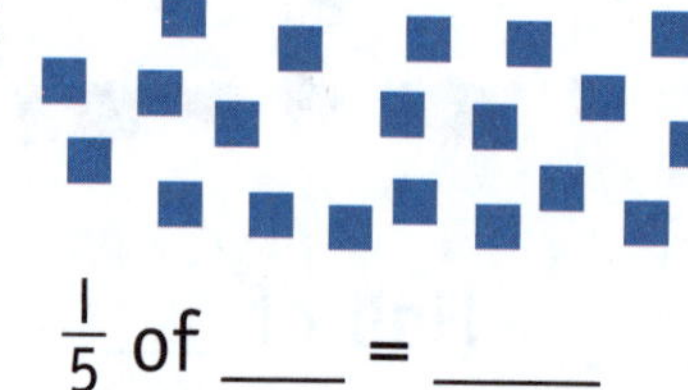

$\frac{1}{5}$ of ___ = ___

15 What is one-third?

a 3 ___ b 33 ___ c 12 ___ d 30 ___

16 What is one-fifth?

a 10 ___ b 25 ___ c 50 ___ d 35 ___

17 Write the correct symbol: >, < or =.

a $\frac{1}{2}$ ☐ $\frac{1}{3}$ c $\frac{3}{4}$ ☐ $\frac{1}{2}$ e $\frac{3}{5}$ ☐ $\frac{2}{3}$ g $\frac{4}{8}$ ☐ $\frac{1}{2}$

b $\frac{1}{3}$ ☐ $\frac{1}{8}$ d $\frac{2}{8}$ ☐ $\frac{1}{4}$ f $\frac{1}{4}$ ☐ $\frac{1}{5}$ h 1 ☐ $\frac{3}{3}$

18 Complete the following.

a Eat 6 pizza slices out of 8. $\frac{}{8}$ eaten, $\frac{}{8}$ left

b Eat 4 pizza slices out of 5. $\frac{}{5}$ eaten, $\frac{}{5}$ left

c Eat 3 pizza slices out of 3. $\frac{}{3}$ eaten, $\frac{}{3}$ left

CATCH UP MATHS YEAR 4 BOOK A © PASCAL PRESS ISBN: 9781925726145

19 Order these fractions from smallest to largest.

a $\frac{1}{8}, \frac{1}{4}, \frac{3}{5}, \frac{3}{8}, \frac{1}{2}$ ______________

c $\frac{1}{2}, \frac{1}{3}, \frac{1}{5}, \frac{1}{8}, \frac{1}{4}$ ______________

e $\frac{1}{3}, \frac{2}{5}, \frac{3}{8}, \frac{4}{4}, \frac{4}{5}$ ______________

b $\frac{3}{4}, \frac{2}{5}, \frac{1}{2}, 1, \frac{2}{8}$ ______________

d $\frac{1}{5}, \frac{3}{5}, \frac{6}{8}, \frac{2}{4}, \frac{1}{8}$ ______________

f $\frac{2}{3}, \frac{2}{5}, \frac{1}{8}, \frac{3}{10}, \frac{7}{8}$ ______________

20 Circle the smallest fraction.

a $\frac{2}{8}, \frac{1}{5}, \frac{1}{3}$
c $\frac{1}{2}, \frac{2}{3}, \frac{4}{5}$
e $\frac{1}{4}, \frac{3}{8}, \frac{2}{4}$
g $\frac{4}{5}, \frac{2}{3}, \frac{4}{8}$

b $\frac{1}{4}, \frac{3}{5}, \frac{3}{4}$
d $1, \frac{2}{5}, \frac{3}{8}$
f $\frac{1}{2}, \frac{3}{5}, \frac{7}{8}$
h $\frac{3}{4}, \frac{1}{2}, \frac{3}{8}$

21 Circle the largest fraction.

a $\frac{1}{4}, \frac{2}{3}, \frac{1}{3}$
c $\frac{3}{5}, \frac{1}{2}, \frac{3}{8}$
e $\frac{1}{2}, \frac{3}{4}, 1$
g $\frac{2}{3}, \frac{2}{4}, \frac{2}{5}$

b $\frac{3}{5}, \frac{3}{4}, \frac{7}{8}$
d $\frac{2}{5}, \frac{2}{3}, \frac{4}{8}$
f $\frac{3}{5}, \frac{2}{8}, \frac{1}{8}$
h $\frac{1}{4}, \frac{1}{3}, \frac{7}{8}$

22 Write a smaller fraction.

a $\frac{7}{8}$ ____
c $\frac{1}{4}$ ____
e $\frac{2}{3}$ ____
g $\frac{3}{4}$ ____

b $\frac{1}{3}$ ____
d $\frac{3}{5}$ ____
f $\frac{4}{8}$ ____
h $\frac{1}{2}$ ____

23 Write a larger fraction

a $\frac{6}{8}$ ____
c $\frac{3}{4}$ ____
e $\frac{1}{3}$ ____
g $\frac{1}{4}$ ____

b $\frac{2}{3}$ ____
d $\frac{2}{5}$ ____
f $\frac{5}{8}$ ____
h $\frac{4}{5}$ ____

24 Write the equivalent fraction.

a $\frac{3}{4} = \frac{}{8}$
d $\frac{2}{5} = \frac{}{10}$
g $\frac{8}{8} = \frac{}{4}$
j $\frac{1}{4} = \frac{}{8}$

b $\frac{4}{8} = \frac{}{4}$
e $\frac{1}{3} = \frac{}{6}$
h $\frac{3}{5} = \frac{}{10}$
k $\frac{4}{5} = \frac{}{10}$

c $\frac{2}{3} = \frac{}{6}$
f $\frac{1}{2} = \frac{}{8}$
i $\frac{1}{5} = \frac{}{10}$
l $\frac{1}{4} = \frac{}{8}$

DECIMALS TO HUNDREDTHS

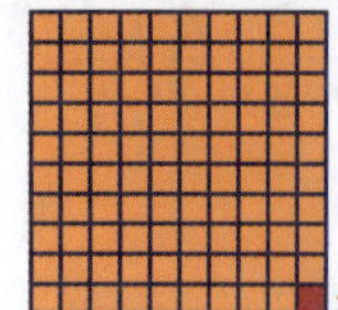

Hundredths are smaller than tenths.
This square is cut into hundredths (100 squares).

Each small square is one-hundredth of the whole.

The decimal fraction 0.01 is one-hundredth ($\frac{1}{100}$).

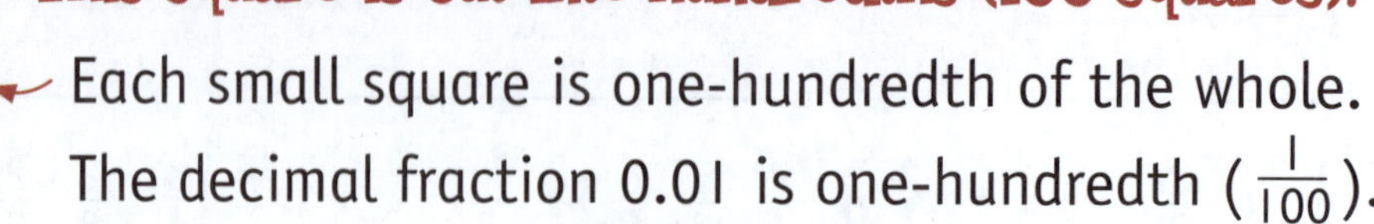

SCAN to watch video

Example 1:

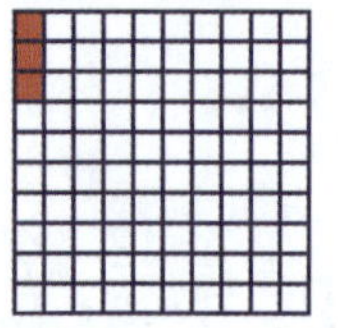

3 hundredths are coloured.

$\frac{3}{100} = 0.03$

Example 2:

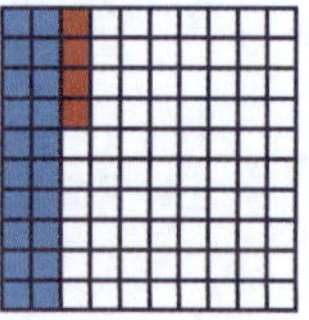

24 hundredths are coloured.

$\frac{24}{100} = 0.24$

Example 3:

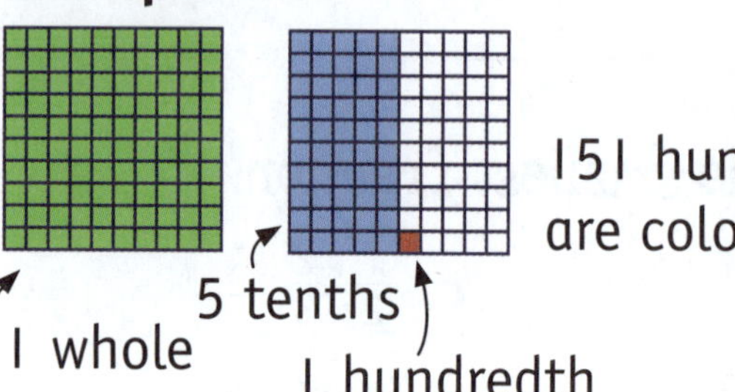

151 hundredths are coloured.

$\frac{151}{100} = 1.51$

Example 4:

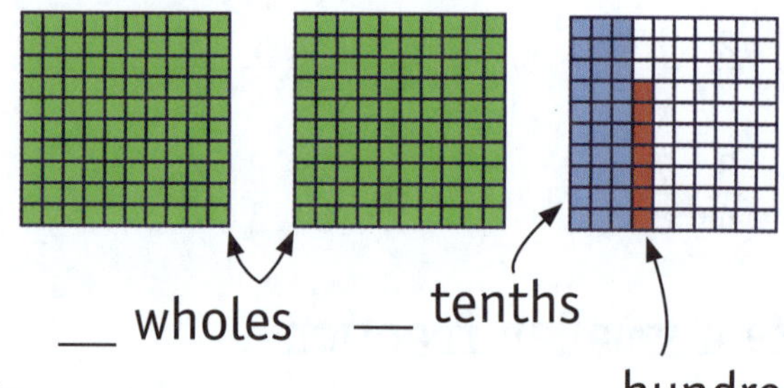

____ hundredths are coloured.

$\frac{}{100}$ = __.__

Check your answer on the video!

Your turn

What decimal fraction has been coloured?

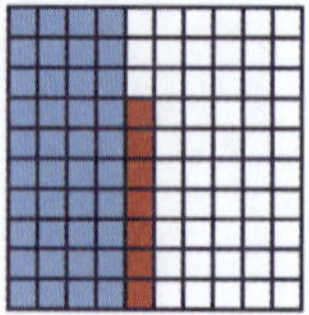

$= \frac{47}{100}$

= 0.47

a

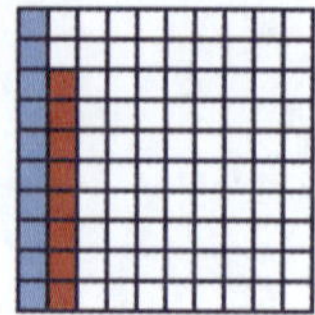

$= \frac{}{100}$

= 0.___

b

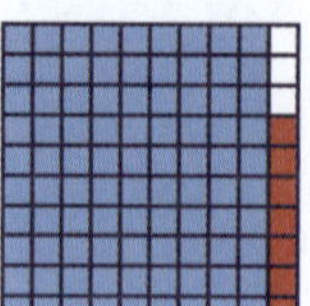

$= \frac{}{100}$

= 0.___

c

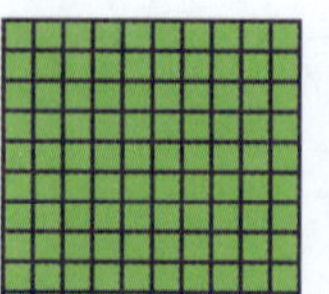

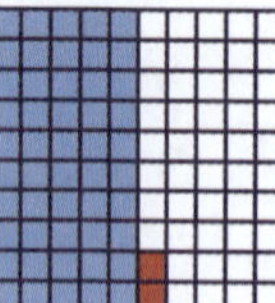

$= \frac{}{100}$

= ___.___

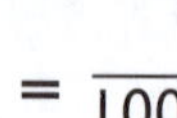

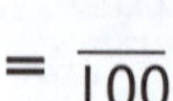

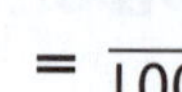

SELF CHECK Tick how you feel

Got it!	Need help...	I don't get it

Check your answers

How many did you get correct?

CATCH UP MATHS YEAR 4 BOOK A © PASCAL PRESS ISBN: 9781925726145

PRACTICE

1 Colour in the hundredths on the squares below.

● 34 hundredths

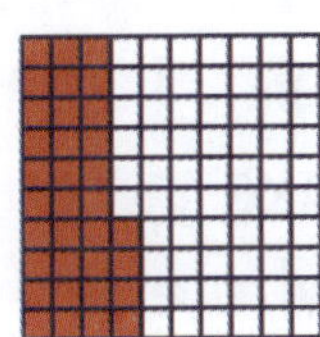

b 84 hundredths

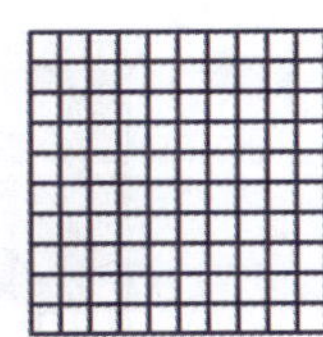

d 172 hundredths

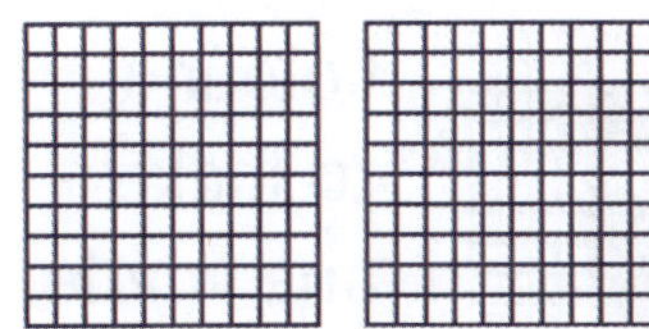

a 67 hundredths

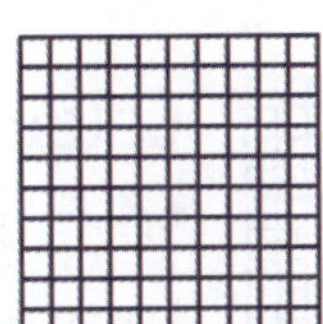

c 12 hundredths

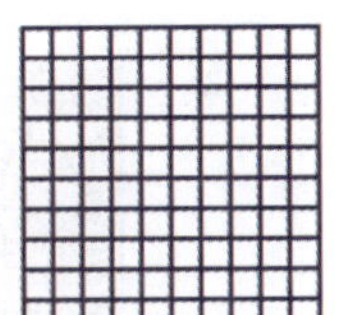

e 248 hundredths

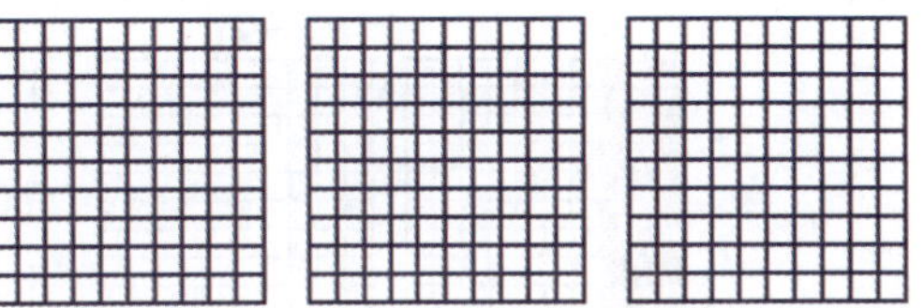

2 Write the decimal fraction for the shaded squares.

● 2.10

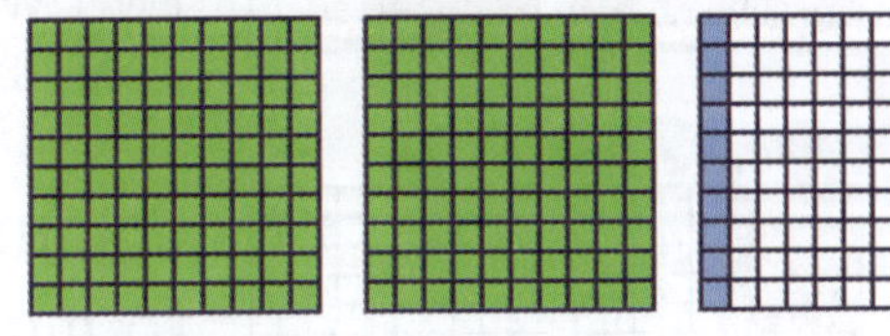

c ______

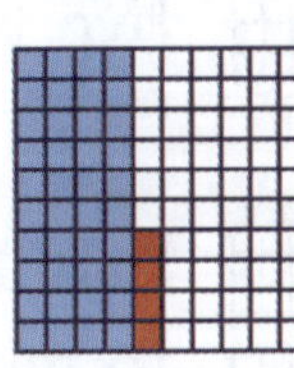

a ______

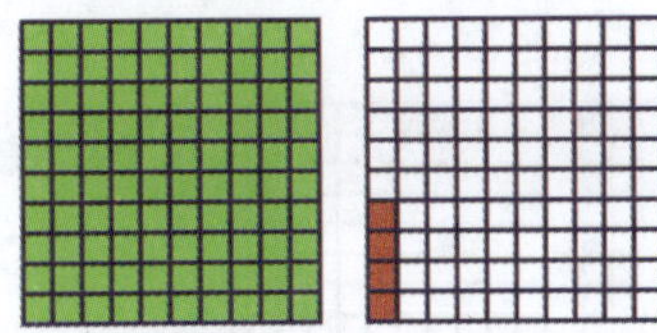

d ______

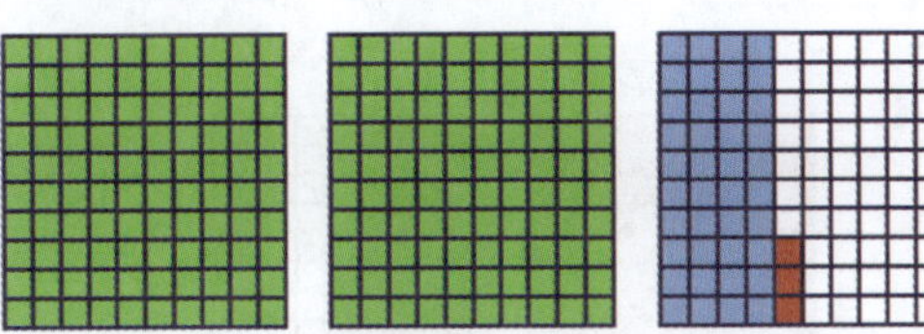

b ______

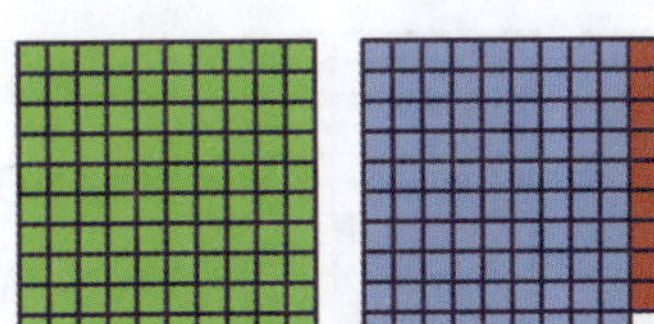

e ______

3 How many more hundredths need to be shaded to make one whole?

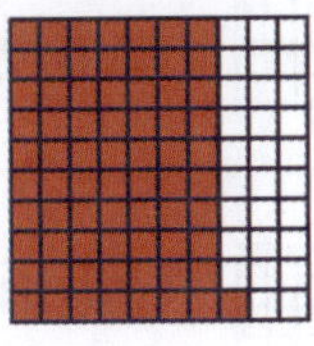

Colour 29 to make one whole.

a

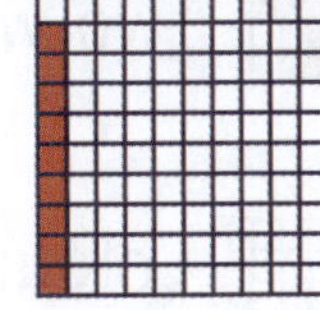

Colour ______ to make one whole.

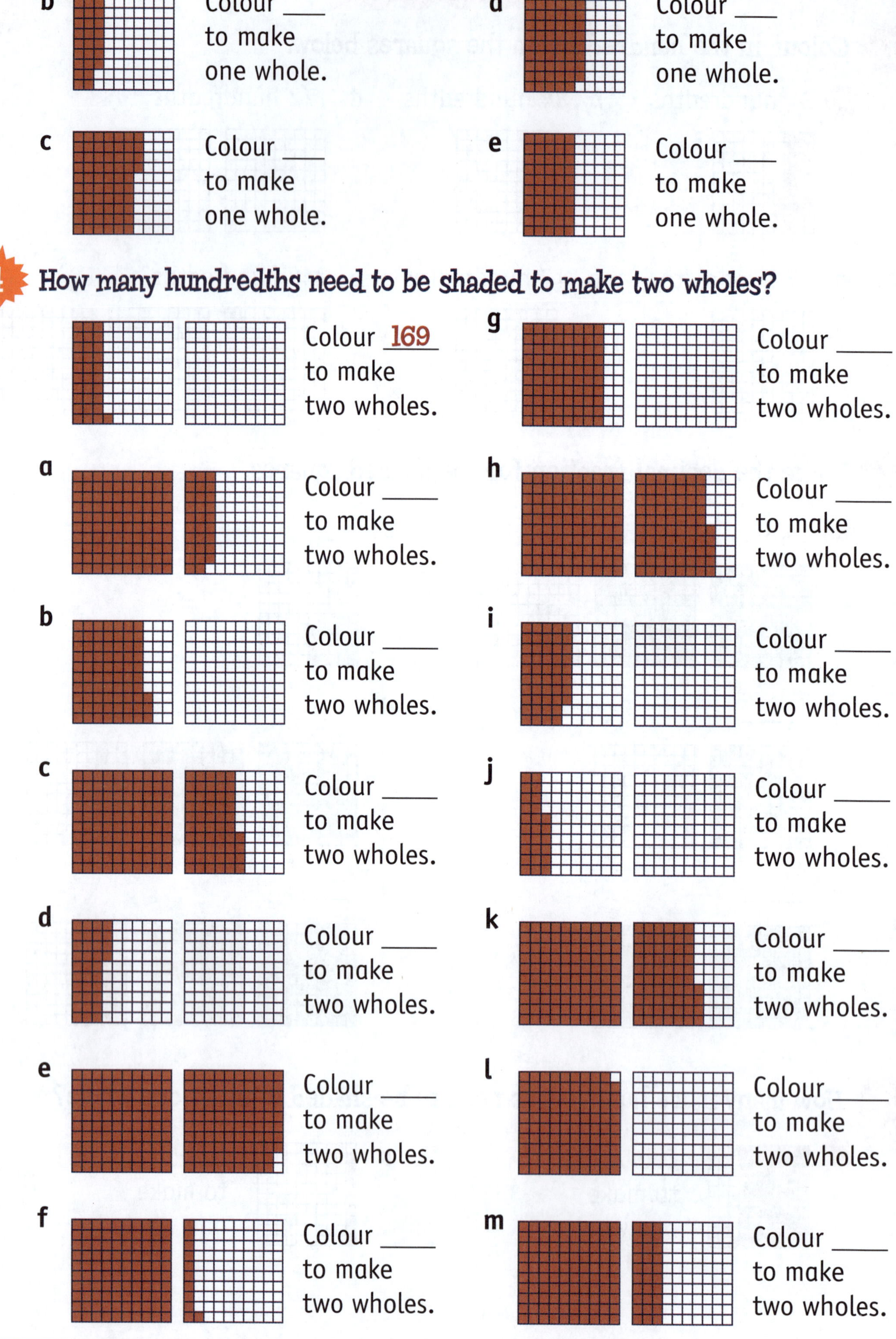

b Colour ___ to make one whole.

c Colour ___ to make one whole.

d Colour ___ to make one whole.

e Colour ___ to make one whole.

4 How many hundredths need to be shaded to make two wholes?

Colour 169 to make two wholes.

a Colour ___ to make two wholes.

b Colour ___ to make two wholes.

c Colour ___ to make two wholes.

d Colour ___ to make two wholes.

e Colour ___ to make two wholes.

f Colour ___ to make two wholes.

g Colour ___ to make two wholes.

h Colour ___ to make two wholes.

i Colour ___ to make two wholes.

j Colour ___ to make two wholes.

k Colour ___ to make two wholes.

l Colour ___ to make two wholes.

m Colour ___ to make two wholes.

CATCH UP MATHS YEAR 4 BOOK A © PASCAL PRESS ISBN: 9781925726145

RELATING TENTHS TO HUNDREDTHS

These squares are cut into tenths.

Each has 10 equal parts. Each tenth is the decimal fraction 0.10.

Example 1:

This square shows $\frac{2}{10} = 0.20$

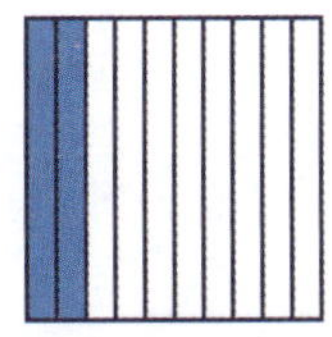

Example 2:

This square shows $\frac{7}{10} = 0.$____

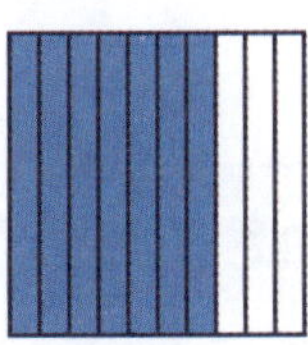

These squares are cut into hundredths.

There are 100 equal parts, each worth $\frac{1}{100} = 0.01$.

SCAN to watch video

Example 3:

This is $\frac{20}{100} = 0.20$

Check your answer on the video!

Example 4:

This is $\frac{70}{100} = 0.$____

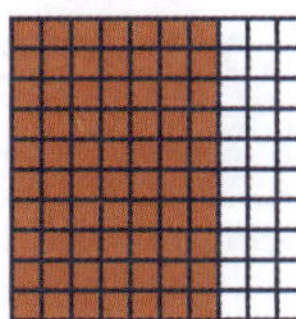

Here, one whole square and part of another square are coloured.

Example 5:

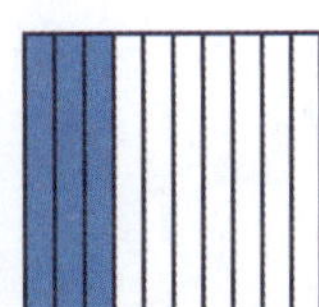

This is $\frac{13}{10} = 1.30$

Example 6:

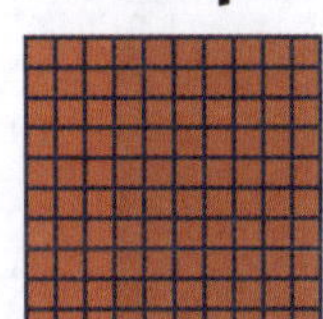

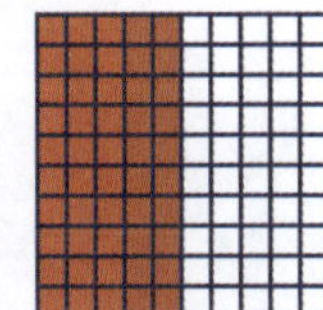

This is $\frac{150}{100} = 1.50$

Example 7:

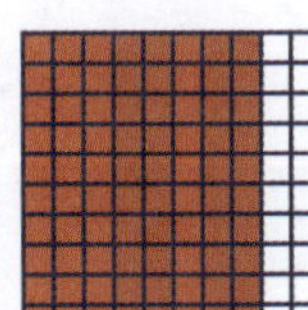

This is $\frac{}{100}$ = ______

Check your answer on the video!

Your turn

What decimal fraction is coloured?

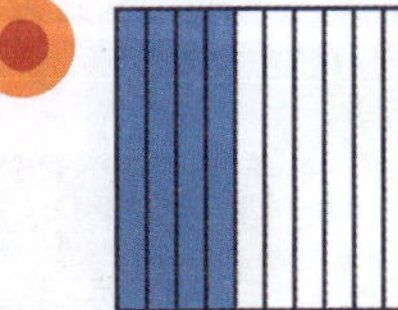

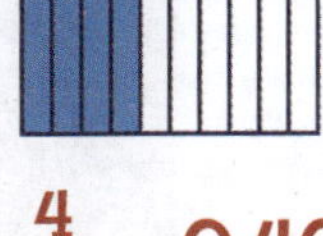

$\frac{4}{10} = 0.40$

a

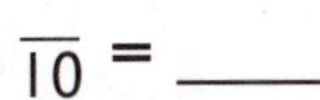

$\frac{}{10}$ = ______

b

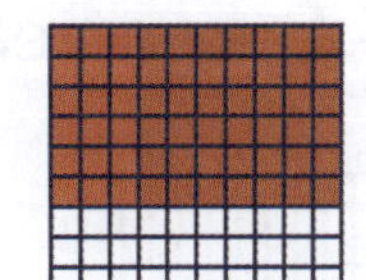

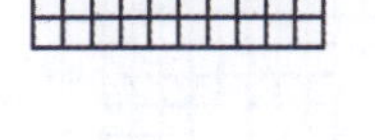

$\frac{}{100}$ = ______

c

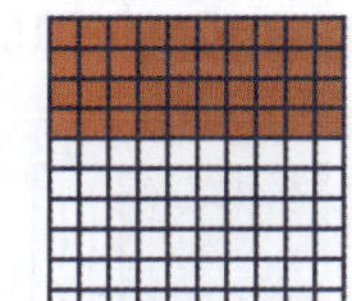

$\frac{}{100}$ = ______

SELF CHECK Tick how you feel

Got it!	Need help...	I don't get it

Check your answers

How many did you get correct?

PRACTICE

1 Shade and write the decimal fractions.

$\frac{2}{10}$ = 0.2

a 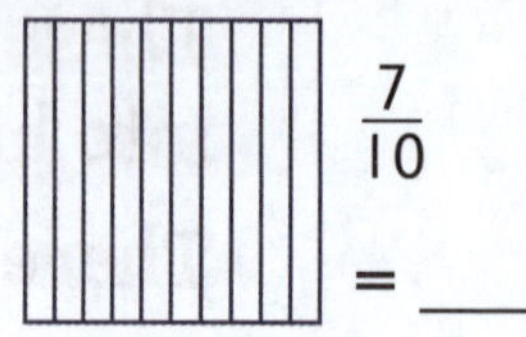

$\frac{7}{10}$ = ____

b 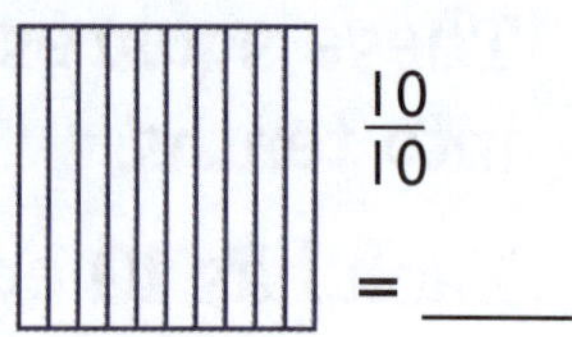

$\frac{10}{10}$ = ____

2 Shade and write the decimal fractions.

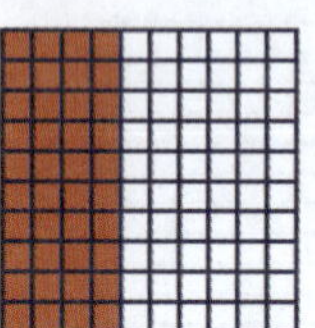

$\frac{40}{100}$ = 0.40

a 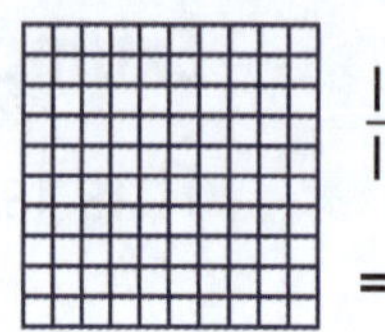

$\frac{100}{100}$ = ____

b 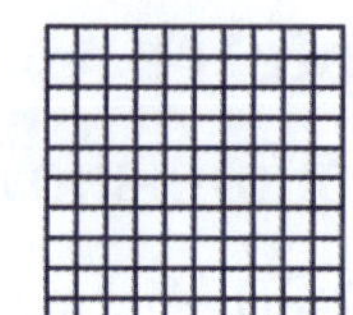

$\frac{30}{100}$ = ____

3 Complete the table.

	Words	Fraction	Decimal
●	two-tenths	$\frac{2}{10}$	0.20
a	five-tenths		
b		$\frac{6}{10}$	
c		$\frac{10}{10}$	

4 Complete the following.

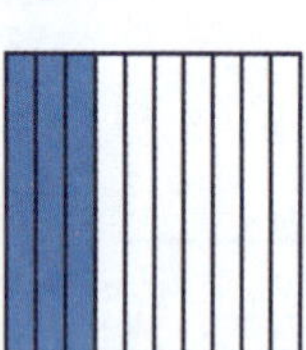 = 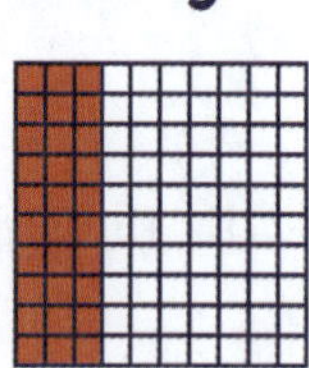

$\frac{3}{10} = \frac{30}{100}$

3 tenths = 30 hundredths

b 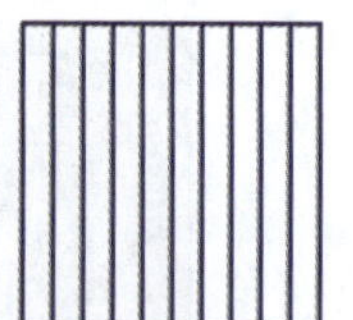 = 

$\frac{\quad}{10} = \frac{\quad}{100}$

__ tenths = ___ hundredths

a 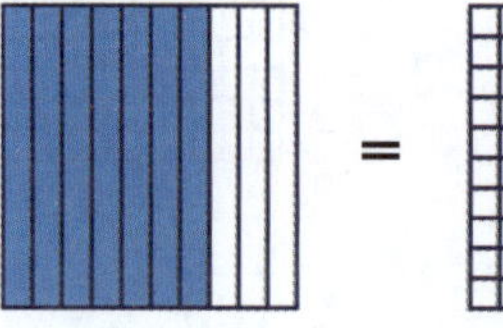

$\frac{\quad}{10} = \frac{\quad}{100}$

__ tenths = ___ hundredths

c 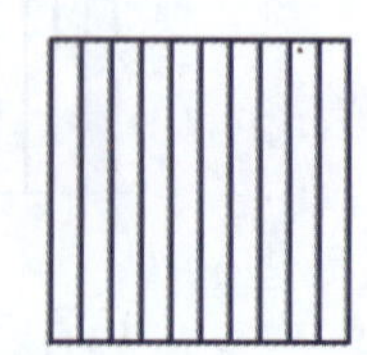 = 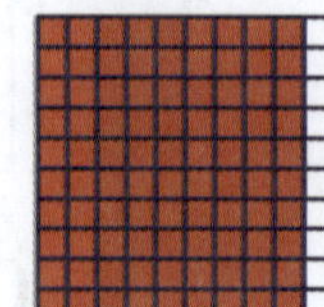

$\frac{\quad}{10} = \frac{\quad}{100}$

__ tenths = ___ hundredths

CATCH UP MATHS YEAR 4 BOOK A © PASCAL PRESS ISBN: 9781925726145

WRITING DECIMALS

Fractions can be written as decimals.

The 0 in the ones position means there are no whole numbers.

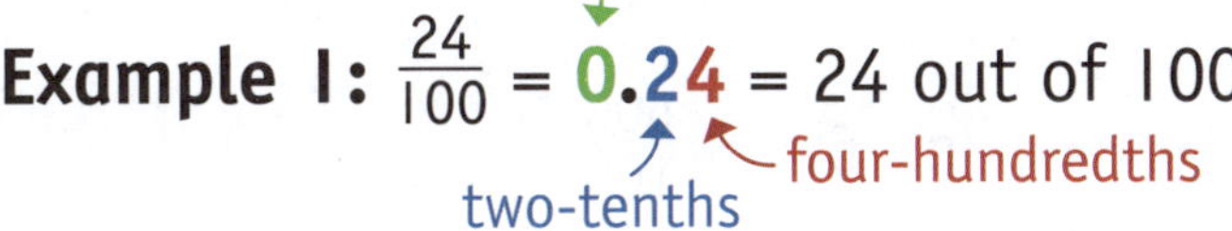

Example 1: $\frac{24}{100}$ = 0.24 = 24 out of 100

(2 = two-tenths; 4 = four-hundredths)

no whole numbers

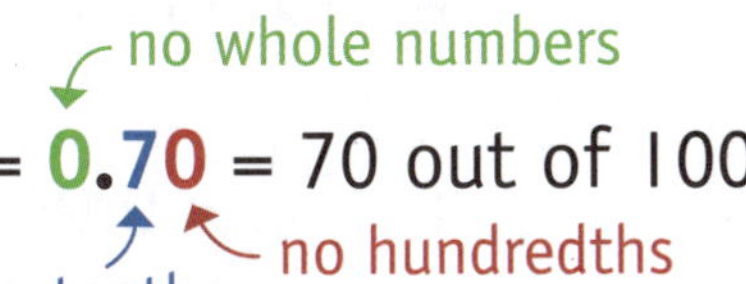

Example 2: $\frac{70}{100}$ = 0.70 = 70 out of 100

(7 = seven-tenths; 0 = no hundredths)

Remember to put the decimal point between the ones place and the tenths place.

The 1 in the ones position means there is 1 whole number.

Example 3: $\frac{137}{100}$ = 1.37 = 137 out of 100

(3 = three-tenths; 7 = seven-hundredths)

The ___ in the ones position means there are 2 whole numbers.

Example 4: $\frac{249}{100}$ = ___.___ ___ = ____ out of 100

(___ tenths; ___ hundredths)

Complete the table.

	Fraction	Decimal fraction	Out of 100
●	$\frac{16}{100}$	0.16	16 out of 100
a	$\frac{28}{100}$		___ out of ___
b			10 out of 100
c		0.93	___ out of ___
d			312 out of 100

SELF CHECK Tick how you feel

Got it!	Need help...	I don't get it
☐	☐	☐

Check your answers

How many did you get correct? ☐

  ISBN: 9781925726145

PRACTICE

1 Answer True or False.

- 19 hundredths > 0.24 **False**

a $\frac{72}{100} < \frac{198}{100}$ ________

b 0.34 = 34 hundredths ________

c $\frac{127}{100}$ = 172 hundredths ________

d $\frac{23}{100}$ = 23 hundredths ________

e 0.06 = 6 hundredths ________

f 7 hundredths > $\frac{70}{100}$ ________

g 6.6 > $\frac{600}{100}$ ________

h 12.20 = $\frac{1220}{100}$ ________

i $\frac{40}{100} < 0.04$ ________

2 Complete the table.

	Diagram	Fraction	Decimal
•		$\frac{69}{100}$	0.69
a		$\frac{281}{100}$	
b		$\frac{}{100}$	0.03
c		$\frac{}{100}$	

3 Write the decimal fraction.

- $\frac{136}{100}$ = 1.36

a $\frac{73}{100}$ = ______

b $\frac{14}{100}$ = ______

c $\frac{277}{100}$ = ______

d $\frac{403}{100}$ = ______

e $\frac{710}{100}$ = ______

f $\frac{8}{100}$ = ______

g $\frac{10}{100}$ = ______

h $\frac{983}{100}$ = ______

CATCH UP MATHS YEAR 4 BOOK A © PASCAL PRESS ISBN: 9781925726145

DECIMAL FRACTIONS IN WORDS

A decimal fraction is a fraction with a denominator (bottom number) that is 10, or a power of 10 like 100, 1000 and 10 000.

SCAN to watch video

Examples

Fraction	Fraction in words	Decimal fraction	Decimal in words
$\frac{34}{100}$	thirty-four hundredths	0.34	zero point three four
$\frac{86}{100}$	eighty-six hundredths	0.86	zero point eight six
$\frac{349}{100}$	three hundred and forty-nine hundredths	3.49	three point four nine
$\frac{111}{100}$	______________ ________ hundredths	__ . __ __	one point one ______
___	two hundred and sixty-nine hundredths	__ . __ __	

Check your answer on the video!

Your turn

Complete the table.

	Fraction	Fraction in words	Decimal fraction	Decimal in words
●	$\frac{73}{100}$	seventy-three hundredths	0.73	zero point seven three
a			0.58	
b	$\frac{197}{100}$			one point nine seven

SELF CHECK Tick how you feel		
Got it! ☐	Need help... ☐	I don't get it ☐

Check your answers
How many did you get correct? ☐

PRACTICE

1 Write these decimals in words.

- 0.38 zero point three eight

a 0.95 ______

b 0.20 ______

c 0.71 ______

d 4.33 ______

e 6.12 ______

2 Write these fractions as a number out of 100.

- sixteen hundredths = 16 out of 100

a two hundredths = ____ out of 100

b forty-eight hundredths = ____ out of 100

c ninety-three hundredths = ____ out of 100

d seventy hundredths = ____ out of 100

e one hundred and two hundredths = ____ out of 100

3 Write these fractions as decimal fractions.

- twelve hundredths = 0.12

a eighty-two hundredths = ______

b seven hundredths = ______

c thirty hundredths = ______

d one hundred and seventy-nine hundredths = ______

e two hundred and sixty-four hundredths = ______

CATCH UP MATHS YEAR 4 BOOK A © PASCAL PRESS ISBN: 9781925726145

PLACE VALUE

Place value is the value of each digit in a number.
It means how much the digit is worth.

3 tens, 2 ones, 4 tenths, 7 hundredths

32.47

Tens	Ones	Point	Tenths	Hundredths
3	2	.	4	7

Examples

Decimal	Hundreds	Tens	Ones	Point	Tenths	Hundredths
13.59		1	3	.	5	9
48.68		4		.	6	8
171.26	1		1	.	2	
257.99						

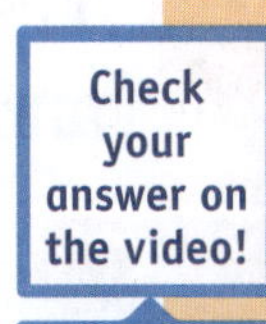

Your turn

1 Use blue to circle the tenths.

● 36.34 **a** 243.62 **b** 120.32 **c** 40.84

2 Use red to circle the hundredths.

● 34.33 **a** 341.26 **b** 203.21 **c** 83.04

3 Use green to circle the ones.

● 62.34 **a** 14.36 **b** 324.23 **c** 436.24

4 Use yellow to circle the tens.

● 29.23 **a** 70.34 **b** 123.58 **c** 542.63

SELF CHECK Tick how you feel

Got it!	Need help...	I don't get it
 ☐	 ☐	 ☐

Check your answers
How many did you get correct? ☐

  ISBN: 9781925726145

PRACTICE

1 Complete the table.

	Decimal	Tens	Ones	Point	Tenths	Hundredths
●	17.42	1	7	.	4	2
a	63.81					
b	57.90					
c	40.75					
d	38.69					
e	94.21					
f	83.06					

2 Trace the tens with yellow, the ones with green, the tenths with blue and the hundredths with red.

● 36.24

a 13.09

b 47.36

c 72.09

d 83.95

e 57.60

f 92.72

g 64.58

3 Place these decimals in ascending order.

● 2.73, 2.82, 2.14, 2.08, 2.37 2.08, 2.14, 2.37, 2.73, 2.82

a 5.05, 5.50, 5.15, 5.03, 5.58 ____________

b 8.73, 8.30, 8.70, 8.37, 7.38 ____________

c 9.16, 6.19, 9.61, 6.91, 19.6 ____________

d 6.21, 12.62, 2.61, 1.26, 6.12 ____________

e 7.57, 7.75, 5.75, 7.55, 5.57 ____________

f 10.35, 10.53, 13.10, 13.35, 5.03 ____________

g 1.23, 4.56, 7.89, 0.34, 1.07 ____________

h 14.36, 14.96, 14.69, 16.34, 13.46 ____________

CATCH UP MATHS YEAR 4 BOOK A © PASCAL PRESS ISBN: 9781925726145

PARTITIONING DECIMALS

A decimal is made up of parts. These parts can be whole numbers and fractions of whole numbers.

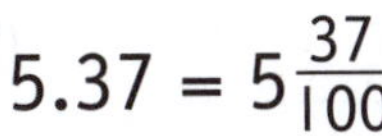

$$5.37 = 5\frac{37}{100}$$

This mixed number is made up of 5 wholes, $\frac{3}{10}$ and $\frac{7}{100}$.

It could also be written as $5 + \frac{37}{100}$.

Example 1:

$6.23 = 6\frac{23}{100} = 6\frac{2}{10} + \frac{3}{100} = 6 + \frac{23}{100}$

Example 2:

$2.75 = 2\frac{}{100} = 2\frac{}{10} + \frac{}{100} = 2 + \frac{}{100}$

Example 3:

$3.96 = 3\frac{}{100} = 3\frac{}{10} + \frac{}{100} = 3 + \frac{}{100}$

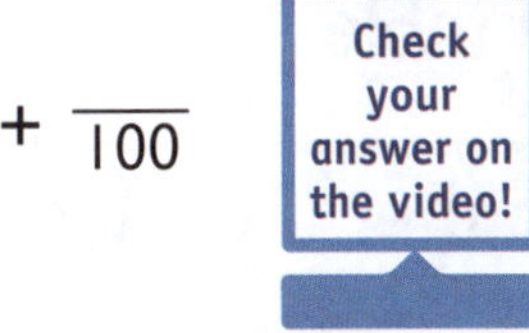

Complete the table.

	Mixed number	Decimal	Wholes	Tenths	Hundredths	Wholes	Hundredths
●	$3\frac{42}{100}$	3.42	3	$\frac{4}{10}$	$\frac{2}{100}$	3	$\frac{42}{100}$
a	$7\frac{26}{100}$	7.26					
b	$8\frac{14}{100}$	8.14					
c	$6\frac{25}{100}$	6.25					
d	$9\frac{70}{100}$	9.70					
e	$2\frac{16}{100}$	2.16					

SELF CHECK Tick how you feel

Got it! ☐ Need help... ☐ I don't get it ☐

Check your answers

How many did you get correct? ☐

  ISBN: 9781925726145

PRACTICE

1 Complete the table.

	Mixed Number	Wholes	Tenths	Hundredths
●	$1\frac{5}{100}$	1	$\frac{0}{10}$	$\frac{5}{100}$
a		3	$\frac{7}{10}$	$\frac{6}{100}$
b	$4\frac{15}{100}$			
c		7	$\frac{4}{10}$	$\frac{8}{100}$
d	$6\frac{38}{100}$			
e		5	$\frac{9}{10}$	$\frac{9}{100}$
f	$9\frac{23}{100}$			

2 Complete the table.

	Mixed number	Wholes	Hundredths
●	$3\frac{73}{100}$	3	$\frac{73}{100}$
a		2	$\frac{4}{100}$
b	$5\frac{12}{100}$		
c		6	$\frac{82}{100}$
d	$2\frac{4}{100}$		
e		8	$\frac{16}{100}$

3 Look at the number expander, then expand the decimal fraction.

● | 7 | 3 . 4 | tenths | 6 | hundredths |

= 73 + 4 tenths + 6 hundredths = $73 + \frac{4}{10} + \frac{6}{100}$

CATCH UP MATHS YEAR 4 BOOK A © PASCAL PRESS ISBN: 9781925726145

a

5	4 . 9	8	hundredths

= ________________________________ = ______________

b

8	2 . 5	3	hundredths

= ________________________________ = ______________

c

6	1 . 7	tenths	5	hundredths

= ________________________________ = ______________

d

2	8	4 . 1	tenths	8	hundredths

= ________________________________ = ______________

Australian money uses decimals. Whole dollars are written to the left of the decimal point and cents (parts of a dollar) are written to the right. For example, \$4.56 has 4 whole dollars and $\frac{56}{100}$ of a dollar (56 cents).

4 Fill in the missing numbers.

- ● \$10.45 = \$10 + $\frac{45}{100}$
- **a** \$8.57 = \$8 + $\frac{}{100}$
- **b** \$4.36 = \$4 + $\frac{}{100}$
- **c** \$3.98 = \$3 + $\frac{}{100}$
- **d** \$12.64 = \$12 + $\frac{}{100}$
- **e** \$6.22 = \$6 + $\frac{}{100}$
- **f** \$7.51 = \$7 + $\frac{}{100}$
- **g** \$12.35 = \$12 + $\frac{}{100}$

5 Write these fractions as decimals.

- ● \$17 + $\frac{53}{100}$ = \$17.53
- **a** \$39 + $\frac{42}{100}$ = ________
- **b** \$54 + $\frac{68}{100}$ = ________
- **c** \$1 + $\frac{36}{100}$ = ________
- **d** \$2 + $\frac{49}{100}$ = ________
- **e** \$5 + $\frac{78}{100}$ = ________
- **f** \$2 + $\frac{93}{100}$ = ________
- **g** \$16 + $\frac{65}{100}$ = ________

DECIMALS REVIEW

Complete the fractions to match the shading.

a

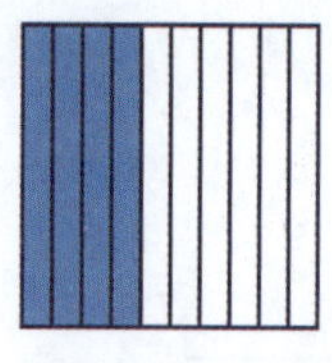

$\frac{\quad}{10}$ = 0.___

c

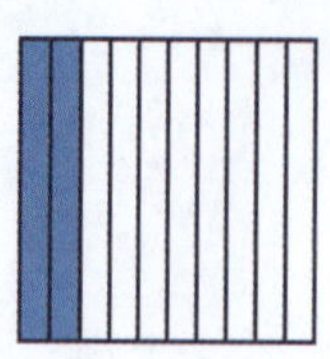

$\frac{\quad}{\quad}$ = ___.___

e

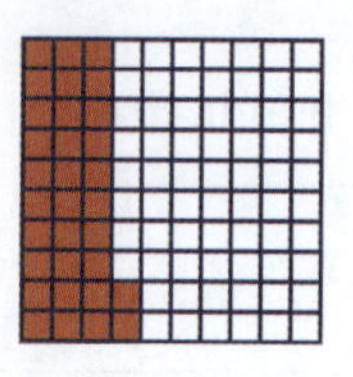

$\frac{\quad}{\quad}$ = ___.___

g

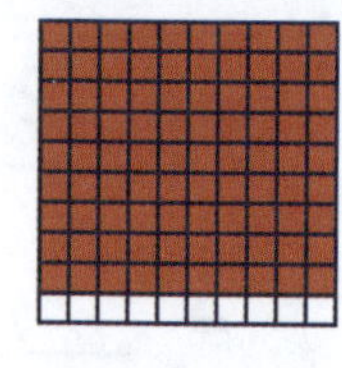

$\frac{\quad}{\quad}$ = ___.___

b

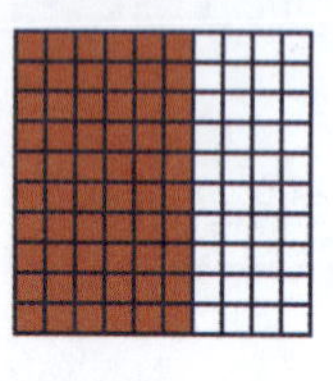

$\frac{\quad}{100}$ = 0.___

d

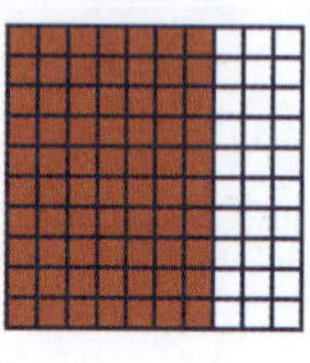

$\frac{\quad}{\quad}$ = ___.___

f

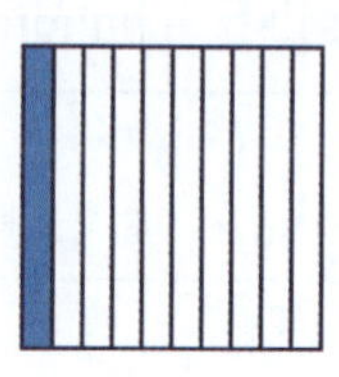

$\frac{\quad}{\quad}$ = ___.___

h

$\frac{\quad}{\quad}$ = ___.___

Complete the fractions to match the shading.

a

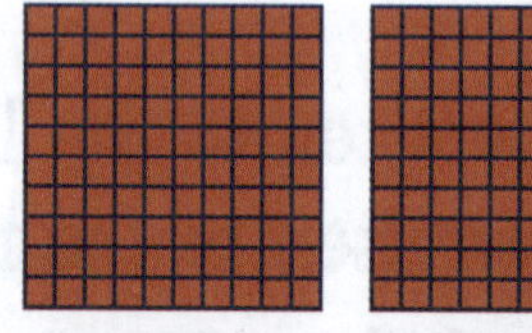

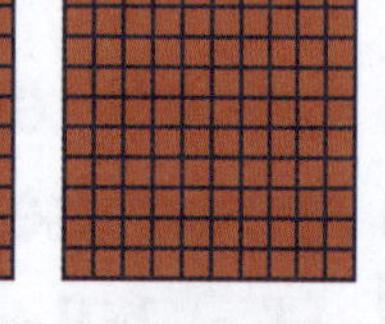

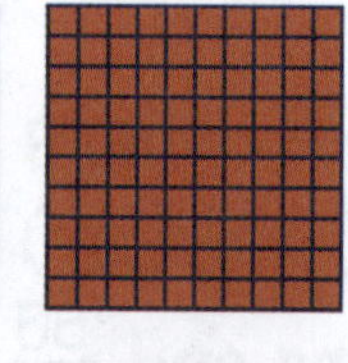

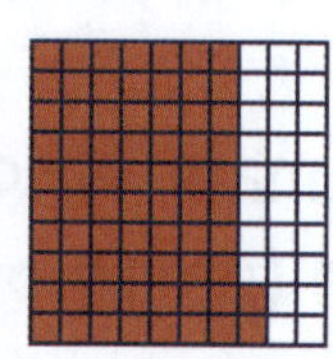

___ $\frac{\quad}{\quad}$ = ___.___

b

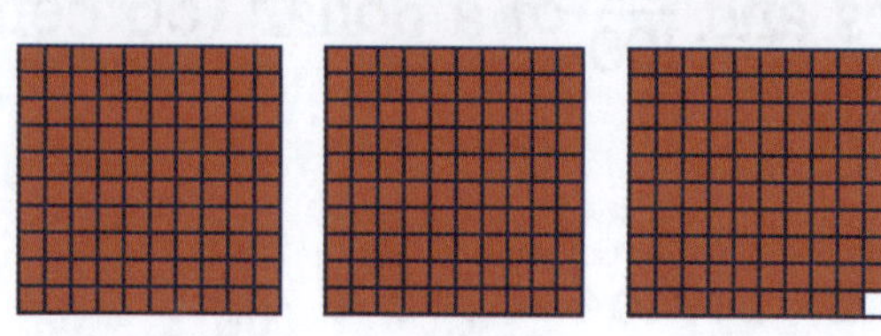

___ $\frac{\quad}{\quad}$ = ___.___

Complete the table.

	Words	Fraction	Decimal
a	three-tenths		0.___
b		$\frac{8}{10}$	
c	seven-tenths		
d		$\frac{10}{10}$	
e			0.10
f		$\frac{5}{10}$	

CATCH UP MATHS YEAR 4 BOOK A © PASCAL PRESS ISBN: 9781925726145

g	six-tenths		
h	nine-tenths		
i	sixty-seven hundredths		
j		$\frac{22}{100}$	
k			0.89
l	one hundred and twenty-one hundredths		
m		$\frac{361}{100}$	
n			5.37
o		$\frac{809}{100}$	
p	seven hundred and sixteen hundredths		

4 How many more hundredths need to be shaded to make one whole?

a ____

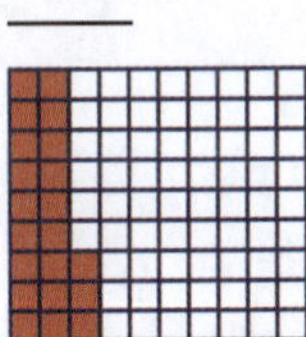

b ____

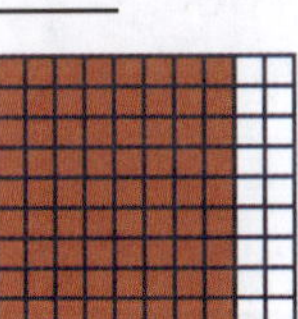

c ____

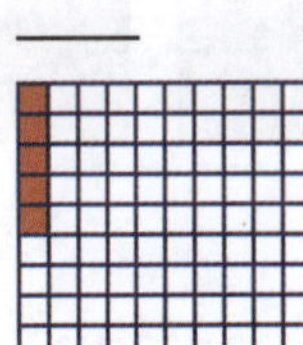

d ____

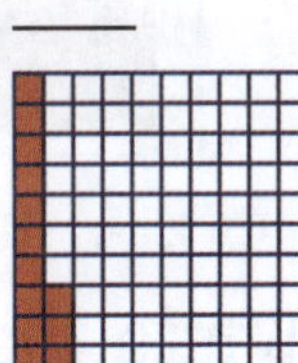

5 How many more hundredths need to be shaded to make two wholes?

a ____

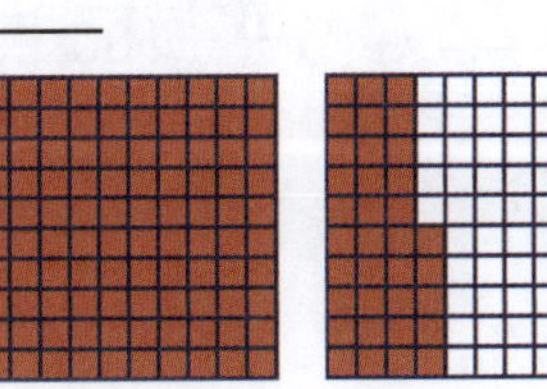

c ____

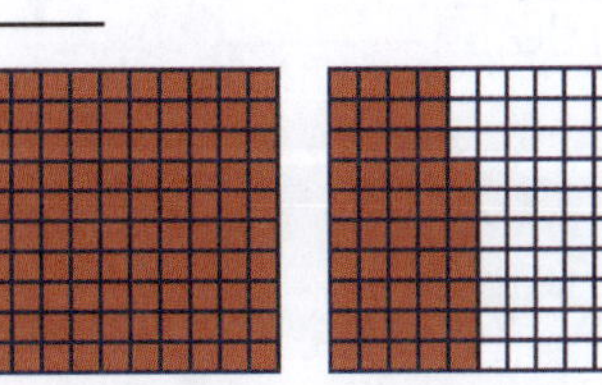

e ____

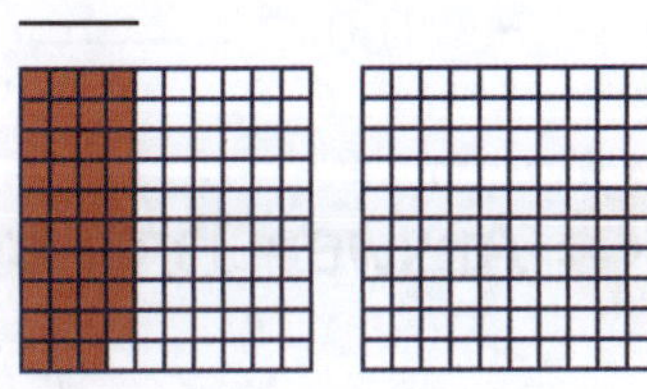

b ____

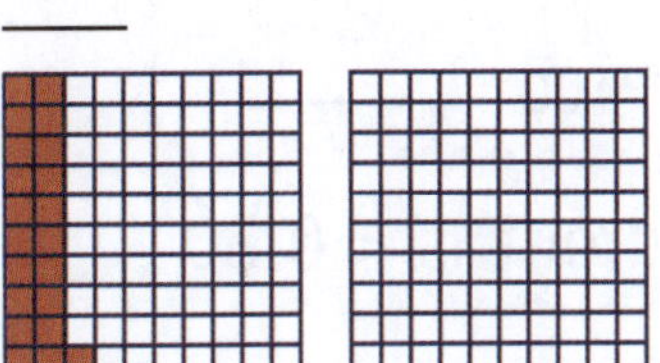

d ____

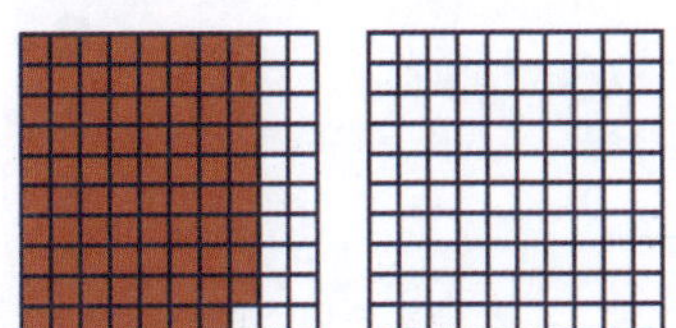

f ____

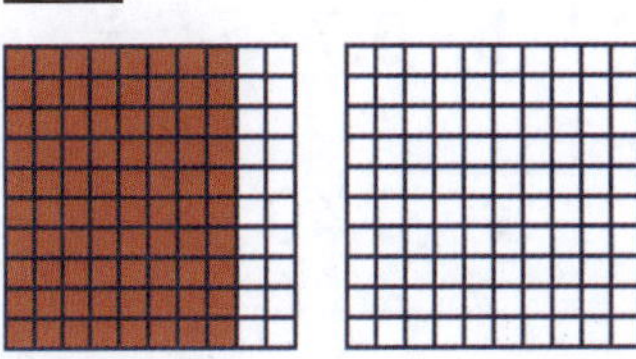

REVIEW

6 Complete the table.

	Coloured Squares	Fraction	Decimal
a	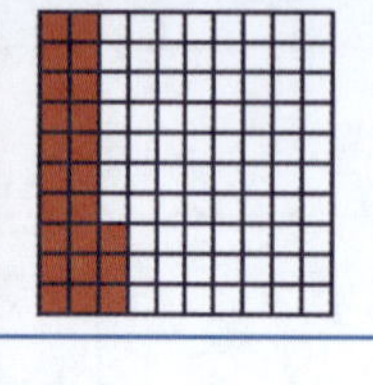 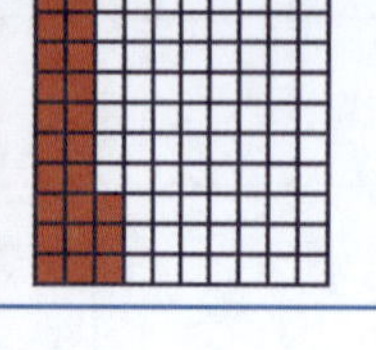 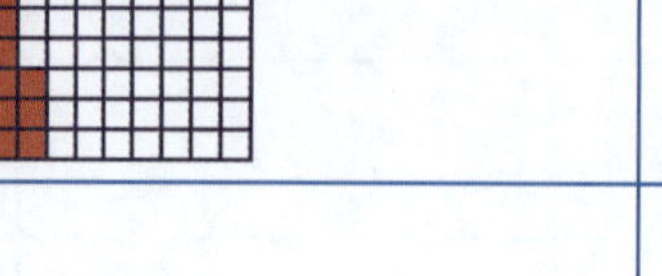	$\frac{\quad}{100}$	___.___
b	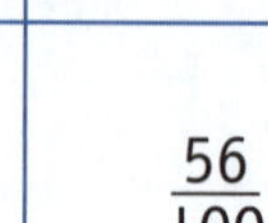	$\frac{56}{100}$	___.___
c	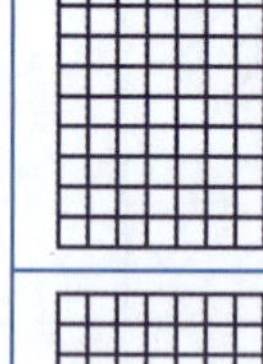	$\frac{\quad}{100}$	2.29
d		$\frac{143}{100}$	___.___
e	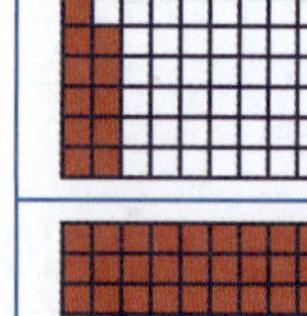	$\frac{\quad}{100}$	___.___
f	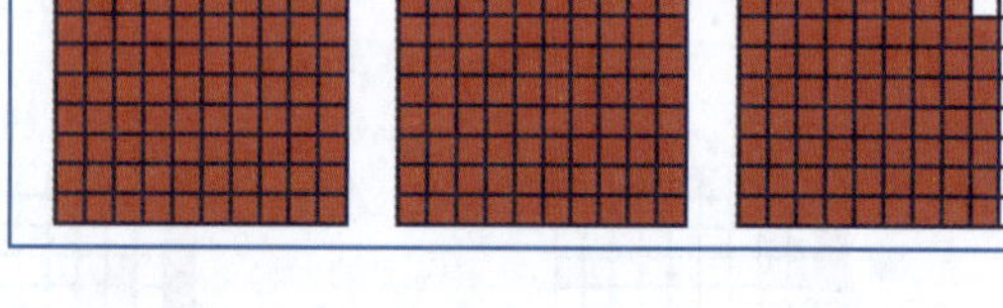	$\frac{\quad}{100}$	___.___

7 Write the decimal fraction.

a $\frac{127}{100}$ = ________ c $\frac{356}{100}$ = ________ e $\frac{495}{100}$ = ________ g $\frac{230}{100}$ = ________

b $\frac{73}{100}$ = ________ d $\frac{109}{100}$ = ________ f $\frac{9}{100}$ = ________ h $\frac{17}{100}$ = ________

8 Answer True or False.

a 17 hundredths > $\frac{71}{100}$ ________________

b $0.72 > \frac{72}{100}$ ________________

c $0.40 = \frac{40}{100}$ ________________

d $8.2 < \frac{82}{100}$ ________________

e $0.81 = \frac{18}{100}$ ________________

f 8 hundredths < 0.80 ________________

CATCH UP MATHS YEAR 4 BOOK A © PASCAL PRESS ISBN: 9781925726145

9 Complete the table.

	Fraction	Fraction in words	Decimal fraction	Decimal in words
a		ninety-one hundredths		
b			0.37	
c	$\frac{77}{100}$			
d				one point eight three
e		three hundred and thirty-nine hundredths		
f			4.78	
g				six point four six
h		eight hundred and sixty-two hundredths		

10 Write these decimals in words.

a 0.42 ______________________

b 1.40 ______________________

c 1.63 ______________________

d 1.06 ______________________

e 4.82 ______________________

f 3.95 ______________________

g 9.99 ______________________

REVIEW

Write these fractions as a number out of 100.

a sixty-one hundredths = ___ out of 100

b ninety-two hundredths = ___ out of 100

c seventy-eight hundredths = ___ out of 100

d one hundred and two hundredths = _____ out of 100

e three hundred and eighty-one hundredths = _____ out of 100

f two-hundred and three hundredths = _____ out of 100

12 Write as decimal fractions.

a seventy-two hundredths = ______

b one hundred and fifty-three hundredths = ______

c ninety-four hundredths = ______

d three hundred and ninety-two hundredths = ______

e two hundred and seventy seven hundredths = ______

f one hundred and forty-eight hundredths = ______

13 Circle the tenths in the following.

a 3.82 **b** 0.46 **c** 38.95 **d** 132.86 **e** 493.89

14 Circle the hundredths in the following.

a 27.41 **b** 137.25 **c** 0.93 **d** 1.92 **e** 437.20

Circle the ones in the following.

a 92.37 **b** 0.95 **c** 24.29 **d** 147.38 **e** 87.40

Circle the tens in the following.

a 10.36 **b** 92.37 **c** 143.73 **d** 107.24 **e** 71.31

17 **Place these decimals in ascending order.**

a 3.47, 7.43, 4.37, 7.34, 3.74 ______________________

b 1.30, 1.03, 1.24, 1.68, 1.06 ______________________

c 8.26, 6.28, 2.86, 8.62, 6.82 ______________________

18 **Complete the table.**

	Mixed number	Wholes	Tenths	Hundredths	Mixed number	Wholes	Hundredths
a	$8\frac{16}{100}$				$8\frac{16}{100}$		
b		2	$\frac{6}{10}$	$\frac{3}{100}$			
c						1	$\frac{33}{100}$
d	$7\frac{46}{100}$				$7\frac{46}{100}$		
e		5	$\frac{4}{10}$	$\frac{7}{100}$			
f						9	$\frac{78}{100}$
g	$3\frac{56}{100}$				$3\frac{56}{100}$		
h		8	$\frac{8}{10}$	$\frac{6}{100}$			
i						4	$\frac{24}{100}$
j	$5\frac{6}{100}$				$5\frac{6}{100}$		
k		10	$\frac{7}{10}$	$\frac{1}{100}$			
l	$12\frac{36}{100}$				$12\frac{36}{100}$		

19 **How many wholes are in these mixed numbers?**

a $5\frac{25}{100}$ ___ b $7\frac{37}{100}$ ___ c $1\frac{24}{100}$ ___

REVIEW

20 How many tenths are in these mixed numbers?

a $6\frac{26}{100}$ ______ b $4\frac{54}{100}$ ______ c $7\frac{22}{100}$ ______

21 How many hundredths are in these mixed numbers?

a $7\frac{34}{100}$ ______ b $5\frac{41}{100}$ ______ c $3\frac{27}{100}$ ______

22 Write the decimal fraction on the number expander in expanded form.

a

4	9 . 3	tenths	8	hundredths

= ____ + __ tenths + __ hundredths = ____ + $\frac{\quad}{10}$ + $\frac{\quad}{100}$

b

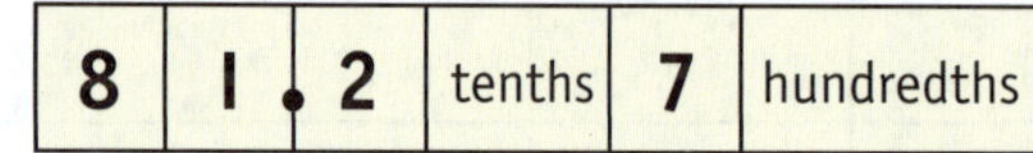

5	6 . 5	9	hundredths

= ____ + ____ hundredths = ____ + $\frac{\quad}{100}$

c

8	1 . 2	tenths	7	hundredths

= ____ + __ tenths __ hundredths = ____ + $\frac{\quad}{10}$ + $\frac{\quad}{100}$

23 Write these amounts of money as fractions.

a \$3.52 = \$___ + $\frac{\quad}{100}$

b \$1.57 = \$___ + $\frac{\quad}{100}$

c \$14.50 = \$___ + $\frac{\quad}{100}$

d \$82.45 = \$___ + $\frac{\quad}{100}$

e \$20.80 = \$___ + $\frac{\quad}{100}$

f \$143.26 = \$____ + $\frac{\quad}{100}$

24 Write these fractions as decimals.

a \$10 + $\frac{75}{100}$ = \$______

b \$2 + $\frac{53}{100}$ = \$______

c \$1 + $\frac{40}{100}$ = \$______

d \$28 + $\frac{54}{100}$ = \$______

e \$6 + $\frac{7}{100}$ = \$______

f \$102 + $\frac{5}{100}$ = \$______

CATCH UP MATHS YEAR 4 BOOK A © PASCAL PRESS ISBN: 9781925726145

PATTERNS

A pattern is when a sequence of objects, letters or numbers is following a rule.

Object patterns

Example 1:

is repeated

Example 2:

is repeated

Example 3:

is repeated

Example 4:

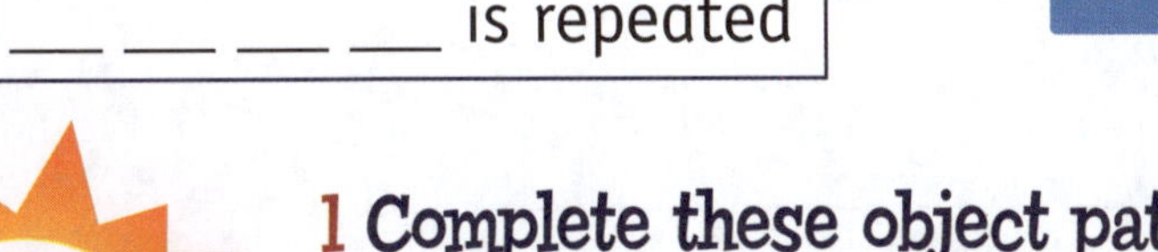
___ ___ ___ ___ is repeated

Number patterns

Example 5:
2, 4, 6, 8, 10, 12

Rule: + 2

Example 6:
5, 10, 15, 20, 25, 30

Rule: + 5

Example 7:
63, 54, 45, 36, 27, 18

Rule: – 9

Check your answer on the video!

Example 8:
27, 24, 21, 18, 15, 12

Rule: ___

Your turn

1 Complete these object patterns.

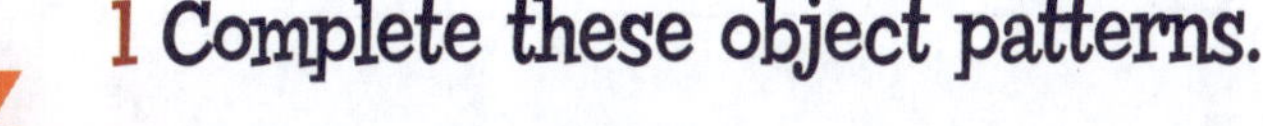

a  ___ ___ ___ ___ ___

b  ___ ___ ___ ___ ___

2 Complete these number patterns and write the rule.

● 7, 14, 21, 28, 35, 42, 49, 56, 63 Rule: + 7

a 4, 8, 12, 16, ___, ___, ___, ___, ___ Rule: ___

b 110, 100, 90, 80, ___, ___, ___, ___, ___ Rule: ___

SELF CHECK Tick how you feel

Got it!	Need help...	I don't get it

Check your answers
How many did you get correct?

  ISBN: 9781925726145

PRACTICE

1 Complete the patterns and write the rule.

- 21, 19, 17, 15, 13, 11 Rule: – 2

a 30, 25, 20, ___, ___, ___ Rule: ___

b 48, 44, 40, ___, ___, ___ Rule: ___

c 81, 72, 63, ___, ___, ___ Rule: ___

d 2, 4, 6, ___, ___, ___ Rule: ___

e 40, 50, 60, ___, ___, ___ Rule: ___

f 7, 17, 27, ___, ___, ___ Rule: ___

g 91, 81, 71, ___, ___, ___ Rule: ___

2 Continue these shape patterns.

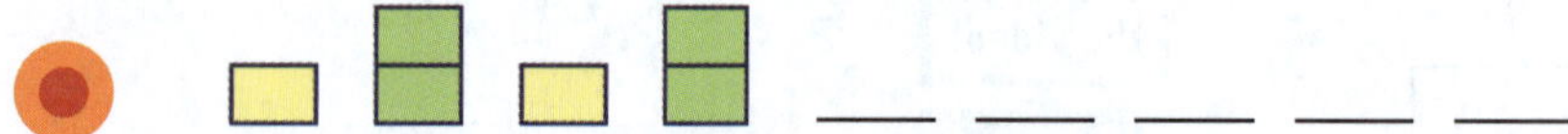

- ___ ___ ___ ___ ___

a ___ ___ ___ ___ ___

b ___ ___ ___ ___ ___

c P A D P A ___ ___ ___ ___ ___

d ___ ___ ___ ___ ___

e ___ ___ ___ ___ ___

f ___ ___ ___ ___ ___

g ___ ___ ___ ___ ___

CATCH UP MATHS YEAR 4 BOOK A © PASCAL PRESS ISBN: 9781925726145

3 Write the missing numbers in the patterns below.

- 26, 28, 30, 32, 34, 36
- a 12, 22, 32, ___, ___, ___
- b 90, ___, ___, 60, 50, ___
- c 12, 15, ___, ___, ___, 27
- d 14, ___, 24, ___, 34, ___
- e ___, 90, 85, ___, 75, ___
- f ___, 75, ___, 71, 69, ___
- g 56, 63, ___, ___, 84, ___
- h 36, 30, ___, ___, 12, ___
- i 9, ___, 29, ___, 49, ___
- j ___, 16, 24, ___, ___, 48
- k 125, ___, 115, ___, 105, ___
- l 7, 12, ___, ___, 27, 32, ___
- m 4, 9, 14, ___, ___, ___

4 Draw what is missing to complete the repeating shapes.

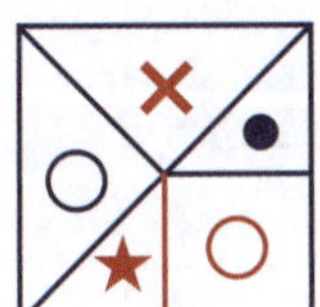

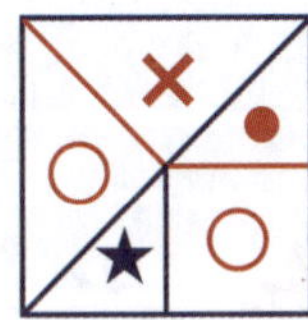

a

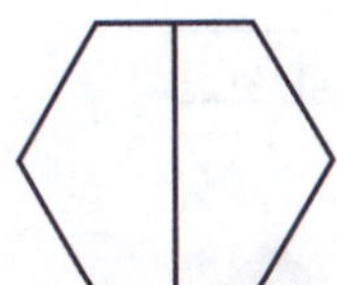
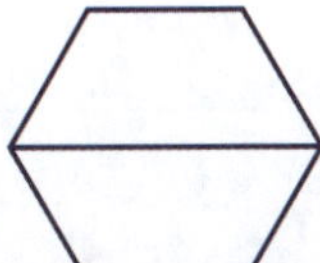
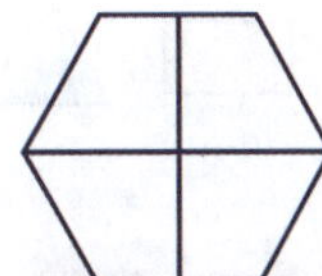
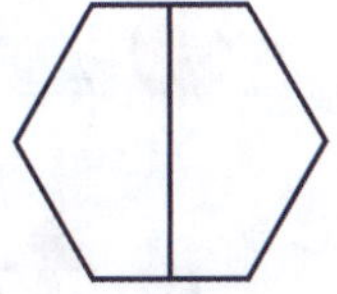

b
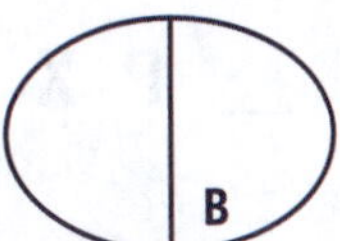

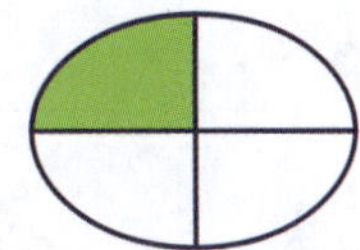

c
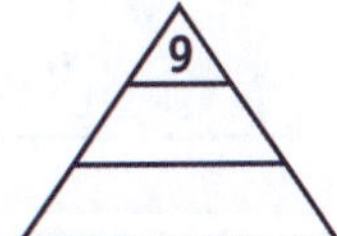

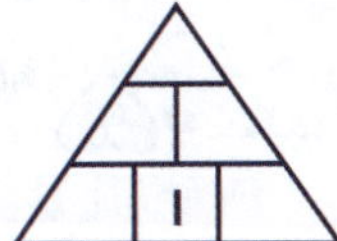

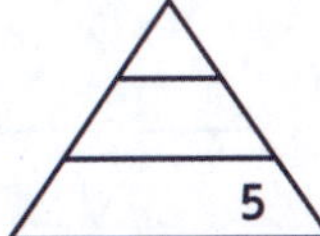

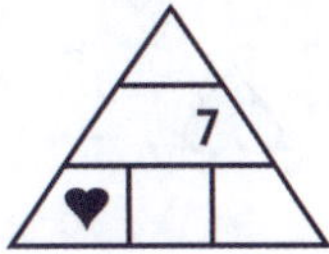

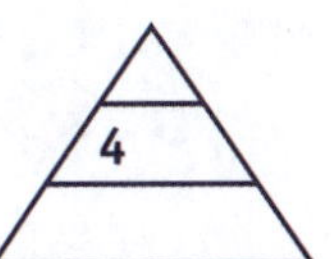

d
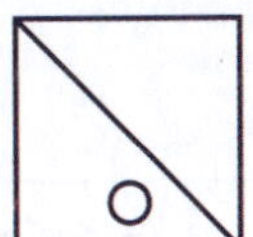
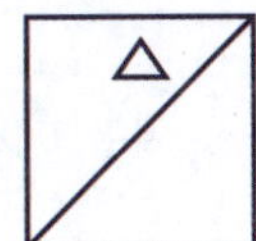
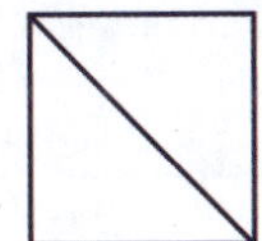
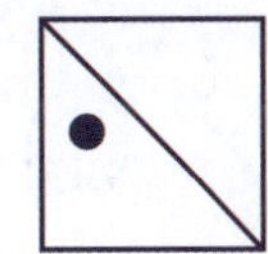

e

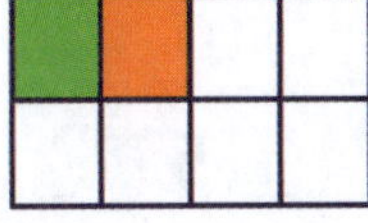
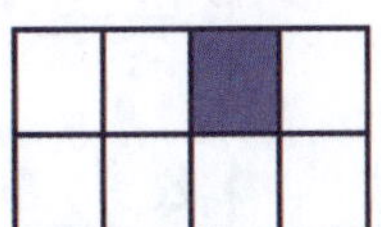
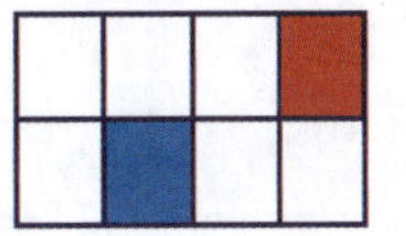
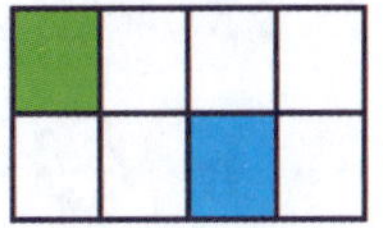

f 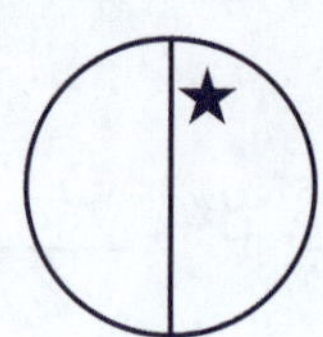 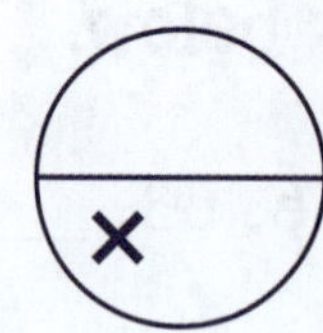 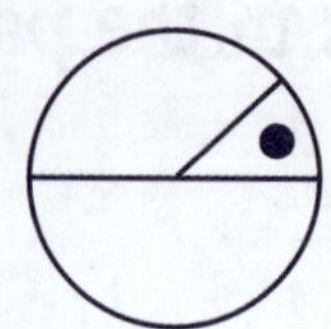 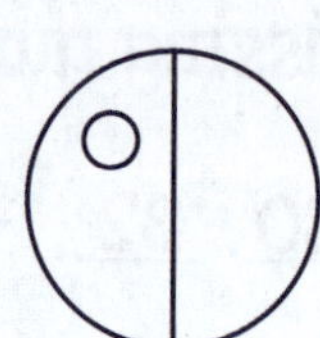 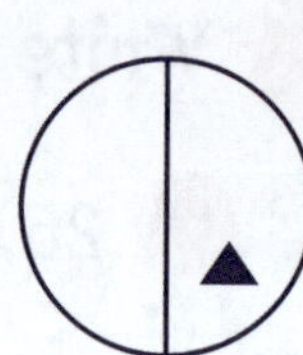

5 Write the rule for each pattern.

● 160, 80, 40, 20 Divide by 2

a 46, 43, 40, 37 ________

b 18, 24, 30, 36 ________

c 93, 91, 89, 87 ________

d 4, 7, 10, 13, 16 ________

e 29, 25, 21, 17, 13 ________

f 163, 158, 153, 148, 143 ________

6 Complete these object patterns.

● ▲ ■ ▬ ▲ __ __ __ ■ ▬ __ __ ▬

a ↑ ↓ ← → __ __ ← __ ↑ __ ← → __ __

b X O Y P __ __ __ __ X __ Y __ __ O __ P X

c ▲ ● ◗ __ ● __ ▲ __ ◗ __ __ __ ▲ ● ◗ __

d ♥ × ○ ♠ __ × ○ __ __ × __ ♠ ♥ × __ __ ♥

e ⊞ ◫ ⊟ ⊠ __ __ __ ⊠ ⊞ __ __ __ ⊞ ◫ __

f ⦶ ⊖ ⊗ ⊕ ⦶ __ __ __ __ ⊖ ⊗ __ __ __ __

g B A ■ ● B __ __ ● B __ ■ ● __ A __ __ __ A ■

h 4 5 7 9 __ __ 7 9 __ 5 __ 9 4 __ __ __ __ 5 7 __ __ __ 7 9

CATCH UP MATHS YEAR 4 BOOK A © PASCAL PRESS ISBN: 9781925726145

PATTERN TABLES

You can use tables to show patterns.
The rows of numbers are related by a rule.

SCAN to watch video

Example 1:

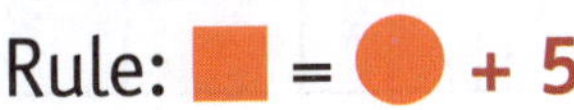

Rule: ■ = ● + 5

●	1	2	3	4	5
■	6	7	8	9	10

■ = ● + 5
■ = 1 + 5
■ = 6

Example 2:

Rule: ▲ = ★ × 2

★	2	4	6	8	10
▲	4	8	12	16	20

▲ = ★ × 2
▲ = 2 × 2
▲ = 4

Example 3:

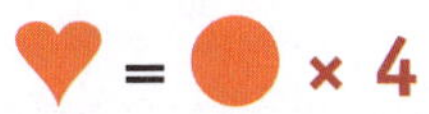

♥ = ● × 4

●	1	3	5	7	9
♥	4		20		

Example 4:

★ = ▲ − 3

▲	5	10	15	20	25
★		7			22

Check your answer on the video!

Your turn

Complete the pattern tables.

● ▲ = ■ ÷ 3

■	27	24	21	18
▲	9	8	7	6

a ★ = ● × 3

●	3	5	7	9
★				

b ★ = ■ + 6

■	4	5	6	7
★				

c ● = 15 − ▲

▲	2	4	6	8
●				

SELF CHECK Tick how you feel

Got it!	Need help...	I don't get it
☐	 ☐	 ☐

Check your answers
How many did you get correct? ☐

  ISBN: 9781925726145

PRACTICE

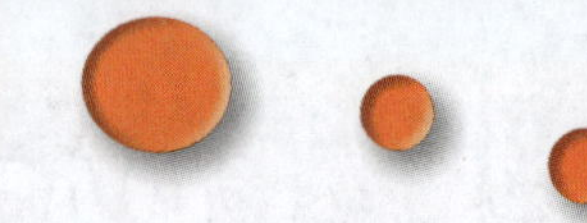

1 Complete the pattern tables.

● = ■ ÷ 2

■	24	22	20	18	16
●	12	11	10	9	8

a ■ = 5 + ▲

▲	19	18	17	16	15
■					

b ★ = ▲ × 8

▲	6	7	8	9	10
★					

c ⬬ = ● – 8

●	32	30	28	26	24
⬬					

2 Complete the tables.

a

▲	3	4	5	6	7
× 2	6	8	10	12	14
× 4					
× 6					
× 7					
× 8					

b

●	1	2	3	4	5
× 3					
× 5					
× 9					
× 10					
× 12					

3 Now solve these two-step patterns.

Rule: multiply by 3, plus 1. Pattern: 2, 7, 22, 67, 202, 607

a Rule: subtract 3, multiply by 2. Pattern: 12, 18, 30, ____, ____, ____

b Rule: add 3, multiply by 2. Pattern: 8, 22, 50, ____, ____, ____

4 Complete the table.

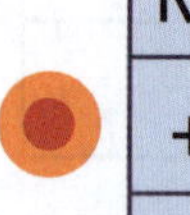

	Rule	2	4	6	12	15	10	11
	+ 5	7	9	11	17	20	15	16
a	× 9							
b	+ 6							

CATCH UP MATHS YEAR 4 BOOK A © PASCAL PRESS ISBN: 9781925726145

EQUIVALENT NUMBER SENTENCES

Number sentences that are equal to each other are called equivalent number sentences. One side of the number sentence equals the other.

Example 1:

□ + 55 = 83

□ = 28

Example 2:

this side = this side

□ − 15 = 19

□ = 34

Example 3:

24 + 32 = □

□ = ___

Equivalent means 'equal in value'.

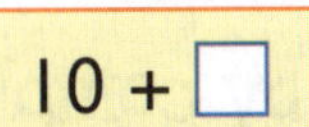

Example 4:

10 + 2	30 − □

□ = 18

Example 5:

10 + □	8 + 9

□ = 7

Example 6:

18 + 6	3 × □

□ = ___

Your turn

Complete these equivalent number sentences.

● □ × 4 = 6 × 2

□ = 3

a □ − 12 = 3 × 6

□ = ___

b

30 ÷ 2	□ × 5

□ = ___

c

3 × 12	4 × □

□ = ___

SELF CHECK Tick how you feel

Got it!	Need help...	I don't get it
□	□	□

Check your answers

How many did you get correct? □

PRACTICE

1 Find the missing number.

- 23 + 6 = **35** – 6
- **a** 13 + ☐ = 10 + 5
- **b** 100 – 25 = 17 + ☐
- **c** 42 ÷ 7 = 3 × ☐
- **d** ☐ + 10 = 5 × 11
- **e** 3 × ☐ = 42 – 6
- **f** 27 ÷ 3 = 20 – ☐
- **g** 110 ÷ 10 = ☐ + 6
- **h** 50 ÷ 5 = 2 × ☐
- **i** 48 ÷ ☐ = 36 ÷ 3
- **j** 25 × 2 = 100 ÷ ☐
- **k** 64 ÷ 8 = 24 ÷ ☐

2 Balance the scales by completing the equivalent number sentences.

- 88 ÷ 11 | 4 × ☐ — ☐ = ____
- **a** 3 × 9 | 53 – ☐ — ☐ = ____
- **b** 4 × 12 | 60 – ☐ — ☐ = ____
- **c** 10 + 22 | 4 × ☐ — ☐ = ____
- **d** 108 ÷ 12 | 20 – ☐ — ☐ = ____
- **e** 60 – ☐ | 3 × 10 — ☐ = ____
- **f** 100 – 50 | 25 × ☐ — ☐ = ____
- **g** 3 × ☐ | 70 – 40 — ☐ = ____
- **h** 5 × 7 | 50 – ☐ — ☐ = ____
- **i** 4 × 9 | 36 ÷ ☐ — ☐ = ____

CATCH UP MATHS YEAR 4 BOOK A © PASCAL PRESS ISBN: 9781925726145

TERMS IN PATTERNS

A term is one of the parts or members of a pattern.

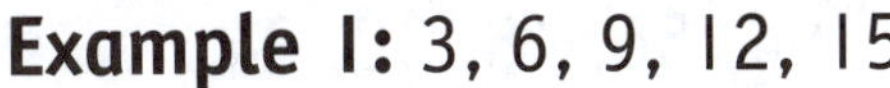

Example 1: 3, 6, 9, 12, 15

Each number is a term.
6 is the second term in this pattern.

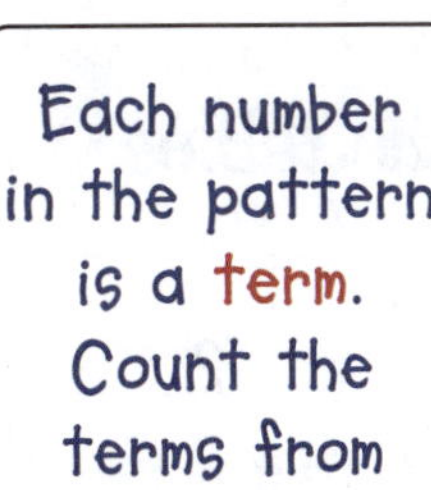

Example 2: 5, 10, 15, 20

In this number pattern,
the first term is 5
and the fourth term is 20.

Example 3: 16, 20, 24, 28, 32, 36

The first term is ___.

The fifth term is ___.

Example 4: 81, 72, 63, 54, 45, 36

The third term is ___.

The sixth term is ___.

Your turn

What is the third term in each number pattern?

- ● 2, 4, 6, 8, 10 [6]
- **a** 3, 6, 9, 12, 15 [ ]
- **b** 5, 10, 15, 20, 25 [ ]
- **c** 4, 8, 12, 16, 20 [ ]
- **d** 1, 3, 5, 7, 9 [ ]
- **e** 1, 11, 21, 31, 41 [ ]
- **f** 81, 72, 63, 54, 45 [ ]
- **g** 100, 90, 80, 70, 60 [ ]

SELF CHECK Tick how you feel

Got it!	Need help...	I don't get it
☐	☐	☐

Check your answers
How many did you get correct? ☐

PRACTICE

1 Write the fourth term of each number pattern.

● 1, 3, 5, 7, 9 7

a 3, 6, 9, 12, 15 ___

b 7, 14, 21, 28, 35 ___

c 64, 56, 48, 32, 24 ___

2 Complete the tables.

●

Order of term	1	2	3	4	5	6
Term	1	4	9	16	25	36

a

Order of term	1	2	3	4	5	6
Term	5	10	15			

b

Order of term	1	2	3	4	5	6
Term	3	5	7			

c

Order of term	1	2	3	4	5	10
Term	4	8	12			

d

Order of term	1	2	3	4	5	6
Term	48	40	32			

e

Order of term	1	2	3	4	5	10
Term	2	4	6			

3 Look at the dot pattern, then complete the table.

1 3 6 10 15

a

Order of term	1	2	3	4	5	6	7	8	9	10
Term										

b How many dots would be in the 12th term? ____

4 Look at the dot pattern, then complete the table.

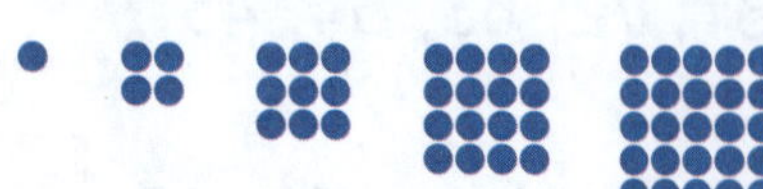

a

Order of term	1	2	3	4	5	6	7	8	9	10
Term	1	4	9	16	25					

b How many dots would be in the 12th term? ____

CATCH UP MATHS YEAR 4 BOOK A © PASCAL PRESS ISBN: 9781925726145

PATTERNS AND ALGEBRA REVIEW

Complete the patterns.

a B, A, ☆, B, A, ___, ___, ___, ___

b ♡, ○, □, ♡, ○, ___, ___, ___, ___

c ○, ∘, △, ○, ∘, ___, ___, ___, ___

d ○, ⬯, △, ○, ⬯, ___, ___, ___, ___

e 8, 16, 24, 32, ___, ___, ___, ___

f 100, 90, 80, 70, ___, ___, ___, ___

g 42, 36, 30, 24, ___, ___, ___, ___

h 27, 36, 45, 54, ___, ___, ___, ___

2 Write the missing numbers in the patterns below.

a 5, 10, ___, ___, ___, 30, 35

b 8, 18, ___, ___, ___, 58, 68

c 6, 11, 16, ___, ___, 31, ___

d 7, 16, ___, ___, 43, ___, 61

e ___, 100, ___, 80, 70, ___, 50

f 83, 79, ___, 71, 67, ___, ___

g 130, ___, 122, ___, 114, ___

h 96, ___, ___, 66, 56, ___, 36

Draw what is missing to complete the repeating shapes.

a

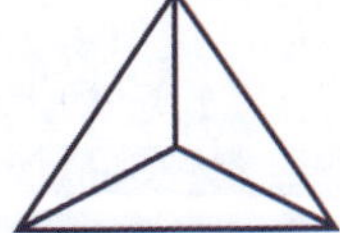  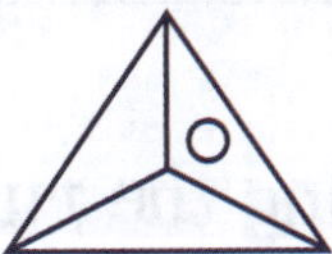 

b

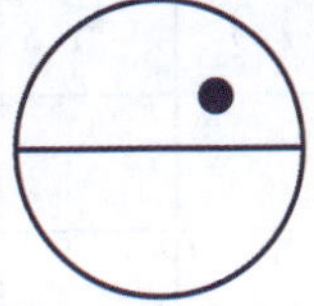 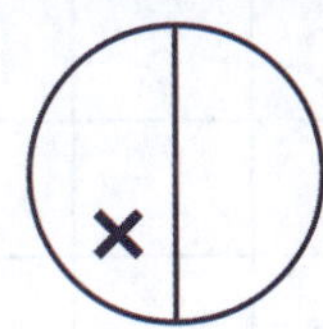 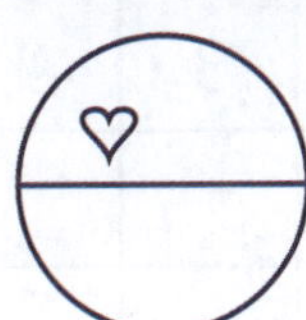 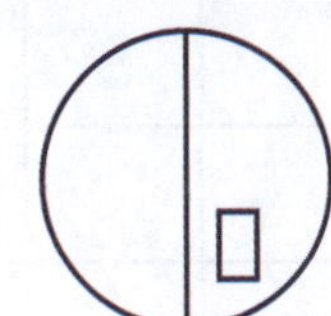

c

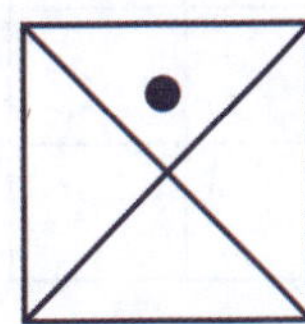 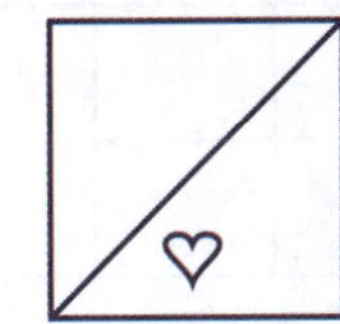 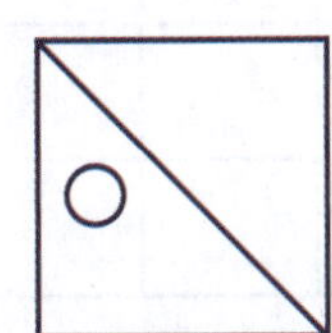 

REVIEW

d 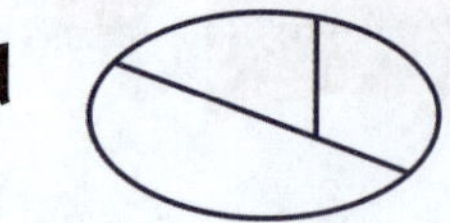  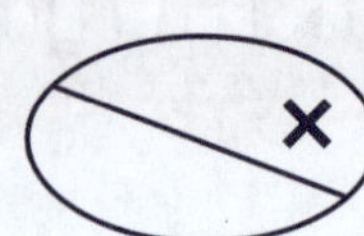 

e   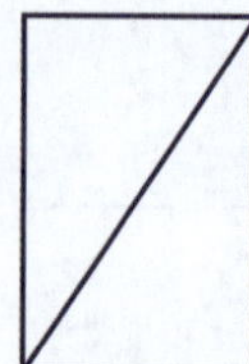 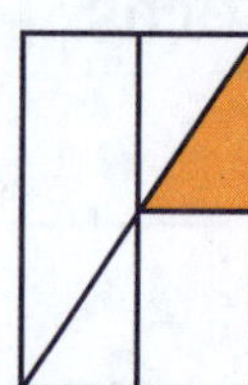

f 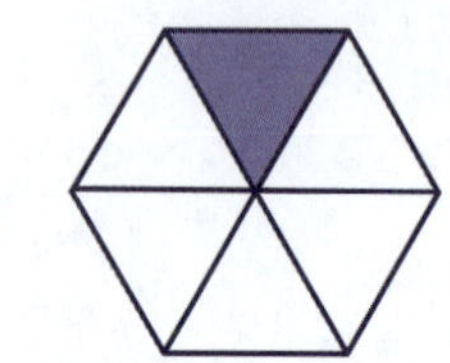 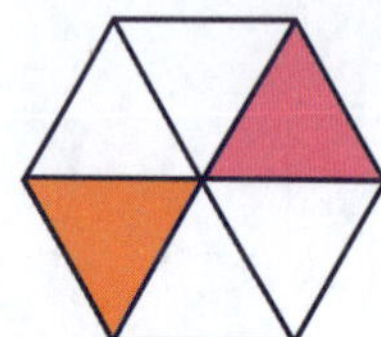 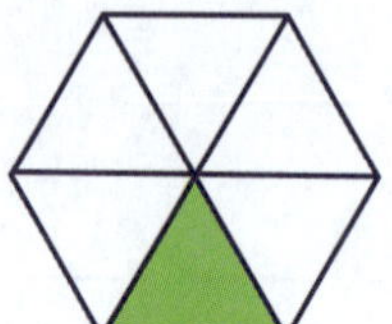 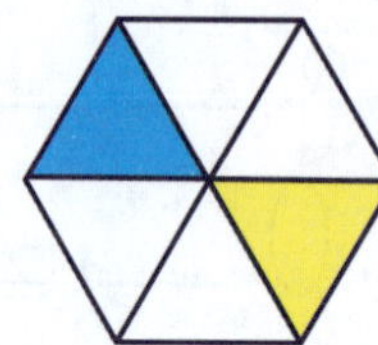

4 Write the rule for each pattern in words and numbers.

a 5, 8, 11, 14, 17 ______________

b 91, 87, 83, 79, 75 ______________

5 Complete the pattern tables.

a ★ = ▲ × 4

▲	2	3	4	5	6	7	8
★							

b ■ = ● − 5

●	15	14	13	12	11	10	9
■							

6 Complete the table by following the rules.

Rule	1	4	8	7	3	9	10	6	11	12	5
+ 6											
× 5											
× 2, + 3											
× 3, + 1											

CATCH UP MATHS YEAR 4 BOOK A © PASCAL PRESS ISBN: 9781925726145

7 Solve these two-step patterns.

a Rule: multiply 2, add 3. Pattern: 5, 13, ______, ______, ______, ______

b Rule: multiply 5, minus 2. Pattern: 3, 13, ______, ______, ______, ______

c Rule: minus 2, times 3. Pattern: 6, 12, ______, ______, ______, ______

8 Complete the equivalent number sentences.

a 23 + 5 = 4 × ☐

b 5 × ☐ = 40 − 10

c 4 × 12 = 36 + ☐

d 60 ÷ 10 = 20 − ☐

e 4 × ☐ = 100 ÷ 5

f ☐ + 16 = 6 × 5

9 Balance the scales.

a 8 × 3 | 36 − ☐

☐ = ___

b 4 × ☐ | 50 − 18

☐ = ___

c ☐ + 20 | 5 × 5

☐ = ___

d 6 × 2 | 24 − ☐

☐ = ___

10 What is the second term in the number patterns?

a 4, 8, 12, 16 ___

b 5, 10, 15, 20 ___

c 3, 7, 11, 15 ___

d 9, 19, 29, 39 ___

11 Answer the questions.

Order of term	1	2	3	4	5	6	7	8	9
Term	4	8	12	16	20	24	28	32	36

What would be the:

a 10th term? ____

b 12th term? ____

c 20th term? ____

CHANCE

The chance of something is how possible it is that it will or will not happen.

If the chance is that an event **might happen** it means that the event could happen but it is not definite that it will. It is only possible that it might happen.

For example:

- If a die is rolled it lands on a six.
- If a ball is chosen out of a bag of red and green balls, it is red.

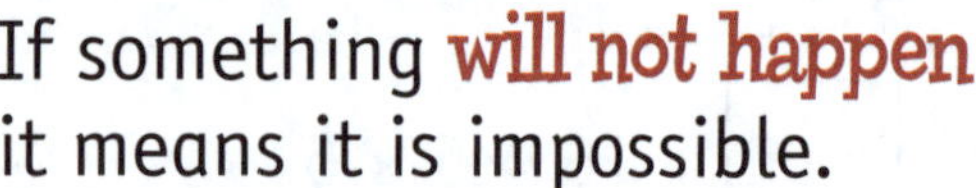

If something **will not happen** it means it is impossible.

For example:

- It will not rain jellybeans.
- Your school will not go on an excursion to Saturn.

If something **will happen** it means it will definitely happen.

For example:

- The sun will rise in the morning.
- Out of a jar of red jellybeans, a red jellybean is chosen.

Example 1:
What is another event that will not happen?

____

Example 2:
What is another event that will happen?

____

Circle with red if the event will happen, with blue if the event won't happen and with green if it might happen.

a If today is Wednesday, tomorrow is Tuesday.

b A dog learns a new trick.

c If today is Monday tomorrow is Tuesday.

Check your answers
How many did you get correct?

CATCH UP MATHS YEAR 4 BOOK A © PASCAL PRESS ISBN: 9781925726145

PRACTICE

1 Colour in the chance words that match the event.

You will read at school today.

will happen
won't happen
might happen

c You will play with your friend after school.

will happen
won't happen
might happen

a You will grow wings.

will happen
won't happen
might happen

b Your teacher will turn into a kangaroo.

will happen
won't happen
might happen

d If today is Friday, tomorrow will be Saturday.

will happen
won't happen
might happen

2 Draw in the boxes below and write on the lines about what you draw.

Something that might happen	Something that will not happen	Something that will happen
______________	______________	______________
______________	______________	______________
______________	______________	______________

CERTAIN AND UNCERTAIN EVENTS

A certain event means that the event will happen. There can be no other result.

Here are some certain events.

Example 1:

March is the month after February.

Example 2:

There are four seasons in the year.

Example 3:

Tuesday is the day before

__________________.

Check your answer on the video!

An uncertain event means there could be another possible result.

SCA to wat vide

Here are some uncertain events.

Example 4:

It will be a sunny day today.

Example 5:

I will eat pasta for dinner.

Example 6:

I will ______________________

______________________ after school.

Your turn

Explain why.

a **March is the month after February.**
Why is this a certain event?

____

b **I will eat pasta for dinner.**
Why is this an uncertain event?

____

SELF CHECK Tick how you feel		
Got it! ☐	Need help... ☐	I don't get it ☐

Check your answers

How many did you get correct? ☐

CATCH UP MATHS YEAR 4 BOOK A © PASCAL PRESS ISBN: 9781925726145

PRACTICE

1 Circle the certain events with green and uncertain events with blue.

		c	I will go to school today.	f	I was born on 2 May. My birthday is 2 May.
a	We will get takeaway for dinner.	d	My parents will cook dinner.	g	New Year's Day is 1 January.
b	Christmas Eve is 24 December.	e	Australia Day is 26 January.	h	My grandma will call me at 1 pm.

2 Draw a picture of something certain and something uncertain.

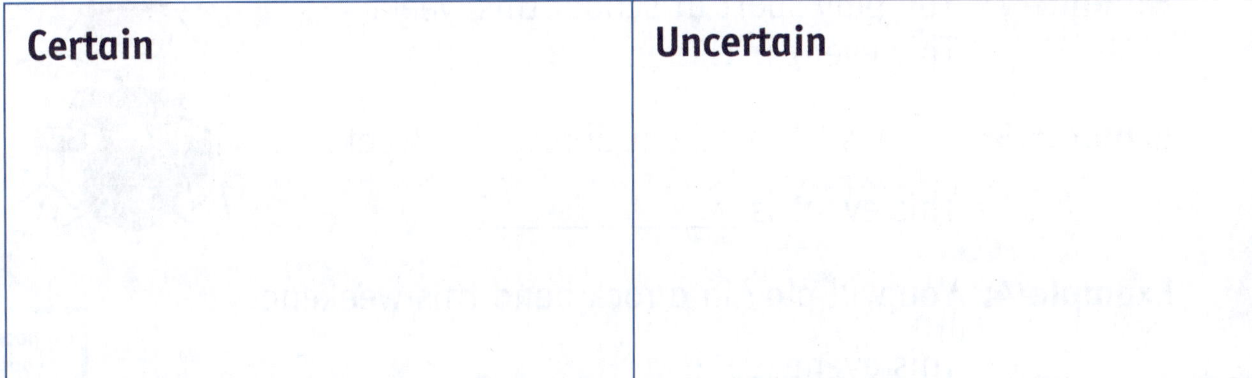

3 Write certain or uncertain next to the statements.

a From a deck of cards, I will choose a card with numbers or pictures.

____

b If I throw a die, it will land on ⚁. ________________

c If I throw a die, I will throw a number between 1 and 6.

____

d I will choose a red counter out of a jar of coloured counters.

____

LIKELY AND UNLIKELY EVENTS

SCAN to watch video

A likely event is something that would usually happen.

For example:

- You play sport on the weekend.
- You go to school.
- You eat lunch.

Likely

An unlikely event is something that would not usually happen.

For example:

- You live in New Zealand for a year.
- You swim when it is snowing.
- You sing in front of the whole school.

Unlikely

Example 1: You will climb Mount Everest tomorrow.
This event is **unlikely**.

Example 2: You play sport at school this week.
This event is **likely**.

Example 3: You visit the school library this week.

This event is ____________.

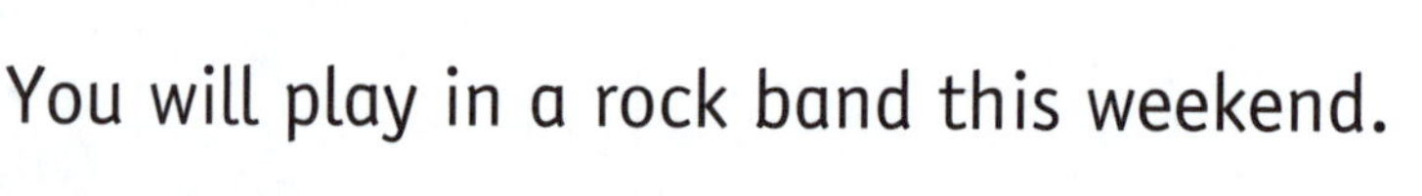

Example 4: You will play in a rock band this weekend.

This event is ____________.

Your turn

Join the events with the correct label

- You go on a school excursion. — Likely

a You eat dinner.

b You have a party at school.

c You swim at the beach in summer.

Likely
Unlikely

SELF CHECK	Tick how you feel	
Got it! ☐	Need help... ☐	I don't get it ☐

Check your answers

How many did you get correct? ☐

CATCH UP MATHS YEAR 4 BOOK A © PASCAL PRESS ISBN: 9781925726145

PRACTICE

1 Join each event to the correct label.

It will rain on a cloudy day.

a I will eat waffles for breakfast every day.

b There is a giraffe outside my bedroom window.

Likely

Unlikely

g I will get a cat for my birthday.

c I will go to school on Saturday.

f I will wear a jumper in winter.

e In summer the weather is hot.

d The sun will be shining during the day.

2 Think of three likely and three unlikely events and write them below.

Likely events

Unlikely events

  ISBN: 9781925726145

PROBABILITY

Sometimes, if one event happens, another event cannot happen. Some events cannot happen at the same time.

Example 1:

If a baby boy is born,
the baby cannot be a girl.

Example 2:

If a coin is tossed and it lands on heads,
it cannot land on tails at the same time.

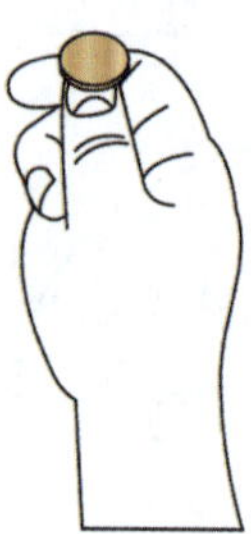

Example 3:

If I roll a die and it shows six, I can't roll a __
on the same die at the same time.

Example 4:

If I win a race, I cannot __________ that race.

Example 5:

If I go up the stairs, I cannot go __________
the stairs at the same time.

Can the two events happen at the same time?
Write yes or no.

- The sink will overflow. There is no plug in the sink. No

a I draw a red flower. I only have a blue pencil. ______

b I choose a red card.
I have a full deck of black and red cards. ______

SELF CHECK Tick how you feel		
Got it! ☐	Need help... ☐	I don't get it ☐

Check your answers
How many did you get correct? ☐

CATCH UP MATHS YEAR 4 BOOK A © PASCAL PRESS ISBN: 9781925726145

PRACTICE

Match events in Box A with the event in Box B that cannot happen at the same time.

Box A	Box B
• If it is raining	• I draw a green car.
a If a girl is born	• The coin lands on heads.
b If I have a yellow crayon to draw a yellow car	• The shoe shop only sells high heels.
c I buy sports shoes.	• We didn't have our spelling test this week.
d I get all my spelling words right in my spelling test.	• It is a boy.
e I buy a black motorbike.	• I look down at the ground.
f I turn left.	• There are only red motorbikes for sale.
g I look up at the sky	• It cannot be dry.
h If I toss a coin and it lands on tails	• I turn right.
i At sunrise	• The sun sets.

RELATED EVENTS

Some events are related.
If one occurs, it is more likely another event will occur.

If you try ice-skating for the first time, there [is] [~~is not~~] a chance you will fall over.

If you ride a skateboard for the first time, you [might] [~~will not~~] fall off.

SCAN to watch video

Sometimes the chance of something happening IS NOT affected by another event.
The events are not related.

If Pete cannot swim, it [~~does~~] [does not] affect the chance of him learning to ride a bicycle.

If your favourite food is pizza, it [~~does~~] [does not] affect the chance of you going swimming.

Example 1: If I roll a die, it [does] [does not] affect the result when I flip a coin.

Example 2: If I go out in the rain, there [is] [is not] a chance I will get wet.

Tick the event that cannot affect the chance of the first event happening.

- Ben got in trouble at school yesterday.
  - ☑ Ben enjoy`s swimming.
  - ☐ Ben didn't attend school yesterday.

a I will go and watch a movie with my friend Annie.

- ☐ The movie cinema is closed today.
- ☐ I like eating popcorn.

SELF CHECK Tick how you feel		
Got it! ☐	Need help... ☐	I don't get it ☐

Check your answers
How many did you get correct? ☐

CATCH UP MATHS YEAR 4 BOOK A © PASCAL PRESS ISBN: 9781925726145

PRACTICE

1 **Tick the event that cannot affect the chance of the first event happening.**

- Bailey's bike chain fell off today.
  - ☑ Bailey learnt to ride his bike in the holidays.
  - ☐ Bailey's mum taught him how to fix bike chains.

a Roisin will go skiing tomorrow.
- ☐ The forecast tomorrow is heavy snowfall.
- ☐ Roisin's sister Ann won a medal in the Olympics.

b Tonight I will eat dinner at 7 pm.
- ☐ My train was late and I got home at 7:30 pm.
- ☐ My favourite dinner is curry.

c Antonia baked cookies for her brother Christian.
- ☐ Antonia loves baking cookies.
- ☐ Christian loves mountain biking.

d Elise will go to the beach for a swim tomorrow.
- ☐ Elise's mum is a pro-surfer.
- ☐ The forecast is for a thunderstorm tomorrow.

e Danny decided to cook a barbecue for dinner.
- ☐ Danny decided he would cook kebabs on the barbecue.
- ☐ Danny likes painting.

f Chris loves to dance.
- ☐ Chris decorates cakes.
- ☐ Chris enjoys hip-hop dancing.

OUTCOMES

An outcome is any possible result that can happen in a chance experiment.

Example 1:

What are all the possible outcomes when three jellybeans are chosen from a jar that has the same number of red and blue jellybeans?

Possible outcomes:

- All three jellybeans are red.
- All three jellybeans are blue.
- One jellybean is red and two jellybeans are blue.
- Two jellybeans are red and one jellybean is blue

An example of an impossible outcome is choosing a green jellybean because there are no green jellybeans in the jar.

Example 2:

A box has 3 red balls, 1 green ball and 2 blue balls.

If I take out one ball, what are all the possible outcomes?

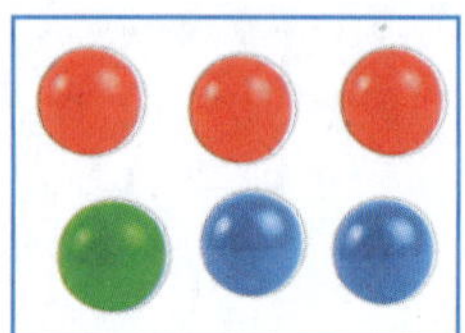

R	R	R	G		

There are ___ possible outcomes.

List the possible outcomes when three jellybeans are chosen from a jar that has the same number of green (G) and orange (O) jellybeans.

G G G ________

________ ________

____

SELF CHECK Tick how you feel

Got it!	Need help...	I don't get it
☐	☐	☐

Check your answers

How many did you get correct? ☐

CATCH UP MATHS YEAR 4 BOOK A © PASCAL PRESS ISBN: 9781925726145

PRACTICE

BETTY'S BIG SCOOP ICE CREAM

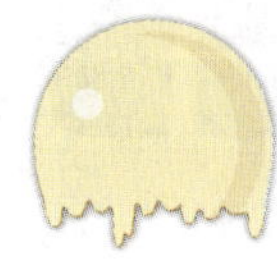

Chocolate Vanilla Strawberry Mint Bubblegum

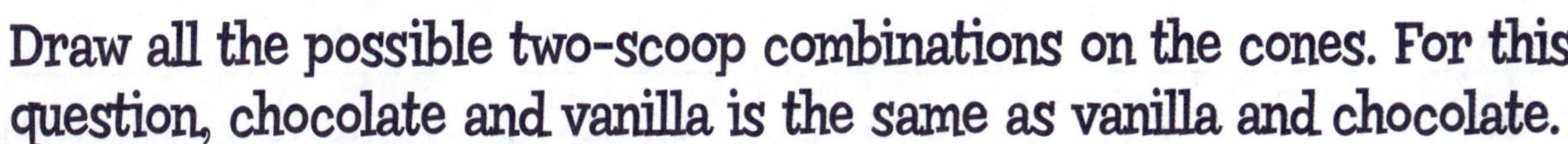

1 Draw all the possible two-scoop combinations on the cones. For this question, chocolate and vanilla is the same as vanilla and chocolate.

chocolate
vanilla

List all the possible outcomes for the combination of a shirt and a pair of shorts.

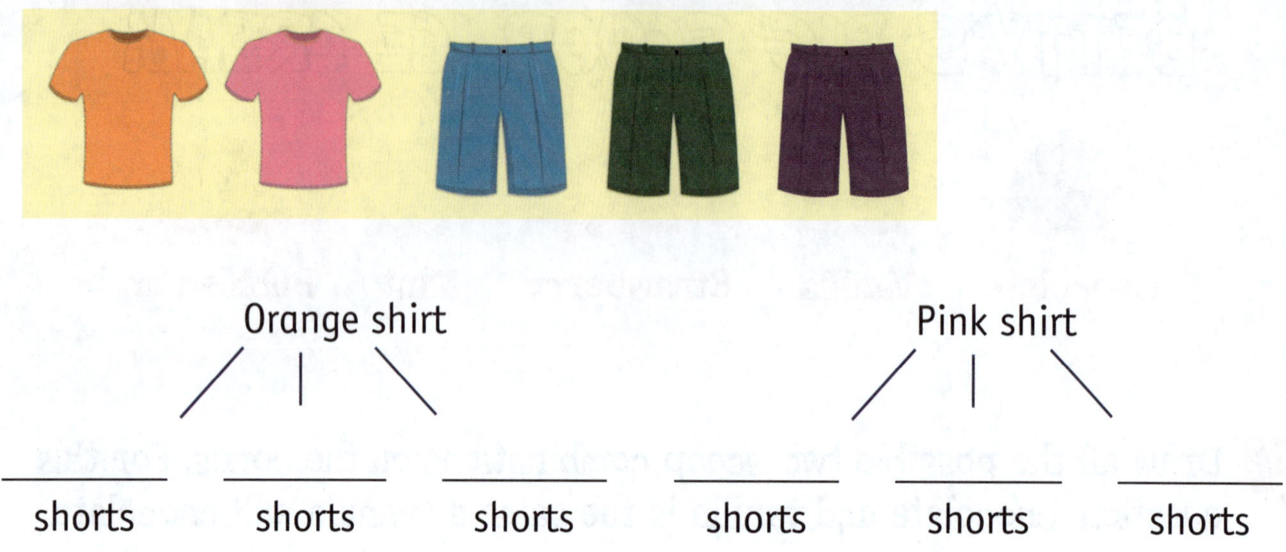

There are __ possible outcomes for this experiment.

Allan made up lolly bags using one of each of these types of lollies.

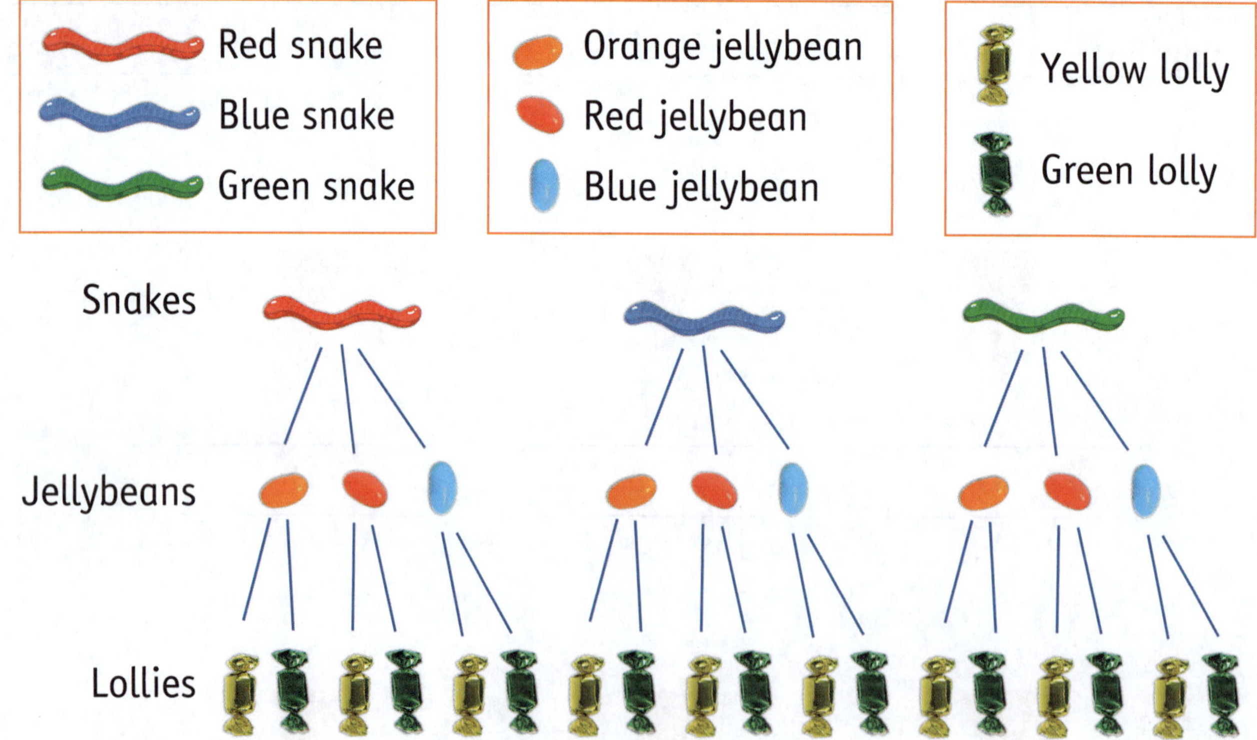

How many possible outcomes are there? ___

CATCH UP MATHS YEAR 4 BOOK A © PASCAL PRESS ISBN: 9781925726145

When a coin is tossed, there are two outcomes: a head or a tail.

The coin has a one in two chance of landing on a head.

The coin has a one in two chance of landing on a tail.

What are the three outcomes if two coins are tossed? For this question, order is not important.

____

____

____

5 Len's basketball club has a new uniform.

Sweatbands are in three different colours: blue, white and yellow.
T-shirts are in two different colours: white and blue.
Shorts are in two different colours: yellow and blue.

What are the possible uniform combinations Len can make?

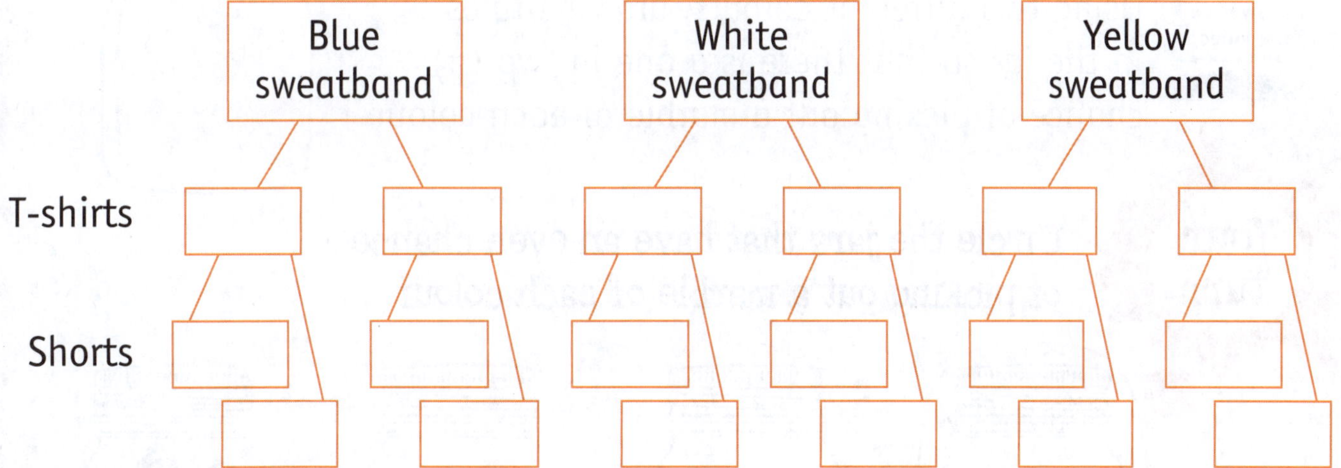

EVEN CHANCE

When a coin is tossed, it lands on heads (H) or tails (T). The chance of either outcome is even. There are ONLY two possible outcomes.

SCAN to watch video

There is a one in two ($\frac{1}{2}$) chance of getting heads.

There is a one in two ($\frac{1}{2}$) chance of getting tails.

Example 1:

If six green marbles and six blue marbles are in a jar, there is a one in two ($\frac{1}{2}$) chance that when one marble is chosen, it will be green or blue.

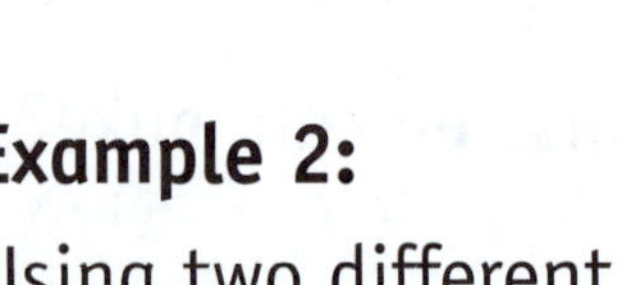

There is an even chance of taking a green or blue marble out of the jar because there are six marbles of each colour.

Check your answer on the video!

Example 2:

Using two different colours, draw marbles in the jar so that there is a one in two ($\frac{1}{2}$) chance of picking out a marble of each colour.

Your turn

Circle the jars that have an even chance of picking out a marble of each colour.

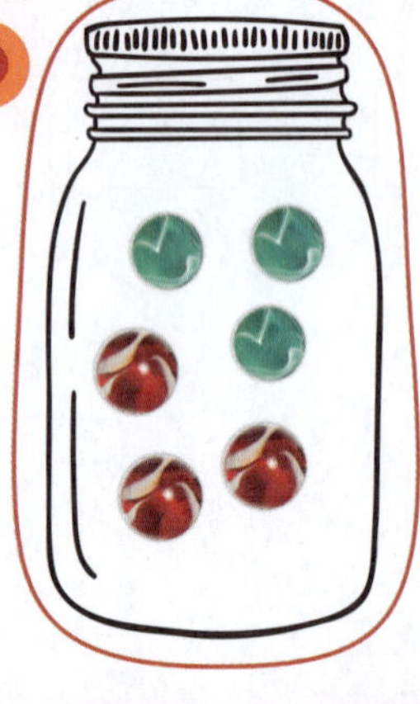

a

b

c

SELF CHECK Tick how you feel		
Got it! ☐	Need help... ☐	I don't get it ☐

Check your answers
How many did you get correct? ☐

CATCH UP MATHS YEAR 4 BOOK A © PASCAL PRESS ISBN: 9781925726145

PRACTICE

1 Draw extra marbles in each jar so that there is a one in two ($\frac{1}{2}$) chance outcome of the two different colours.

b
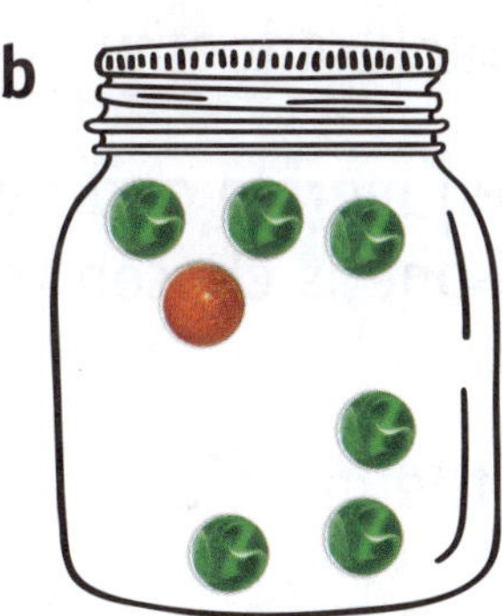

d

f

a

c

e

g

2 Which spinners have a one in two ($\frac{1}{2}$) chance outcome? Circle them.

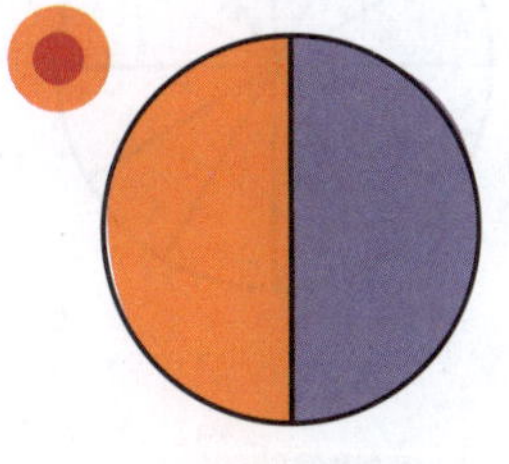

b

d
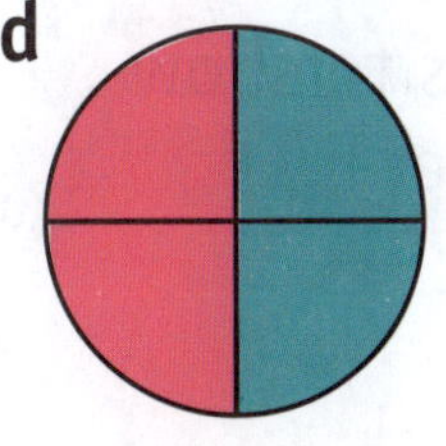

f
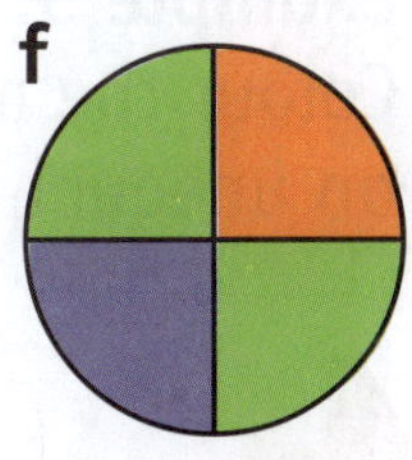

a
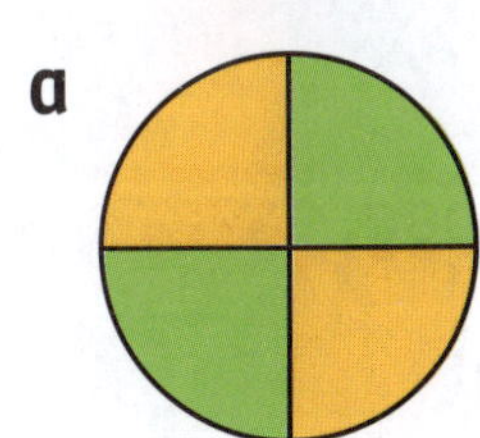

c
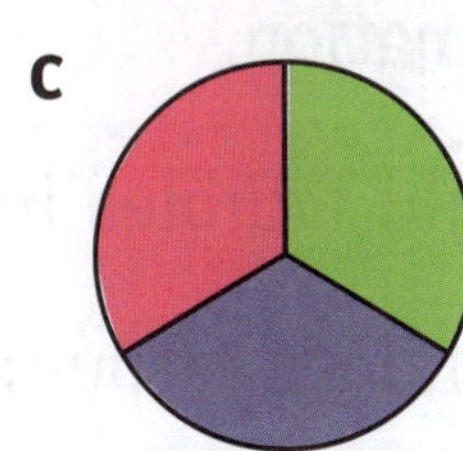

e
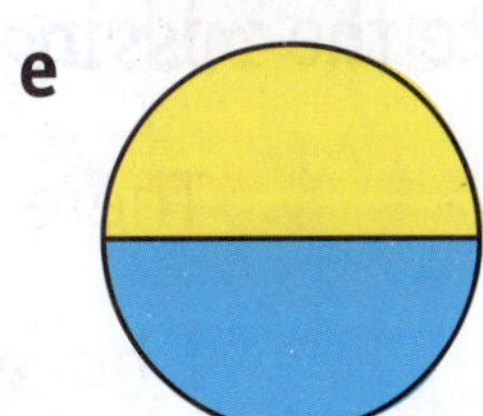

g
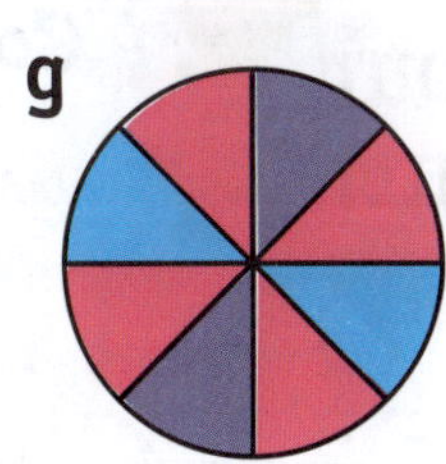

3 Fill in the gaps.

a If I put five black and five blue cards in a bag,
I will have a __ in __ ($\frac{1}{\ }$) chance of picking out a black card.

b If I place one red and one green peg in a bag,
I will have a __ in __ ($\frac{1}{\ }$) chance of picking out a green peg.

UNEQUAL CHANCE

When there is more of one thing than of another thing, there is an unequal chance of each outcome occurring.

Example 1: If a bag holds one green marble and five red marbles, what colour would the marble be if one is chosen from the bag?

Possible outcomes:

- There is a one in six ($\frac{1}{6}$) chance the marble will be green.
- There is a five in six ($\frac{5}{6}$) chance the marble will be red.

It is an unequal chance because there are more red than green marbles. It is NOT fair.

Example 2:
Colour the spinner to show an unequal chance.

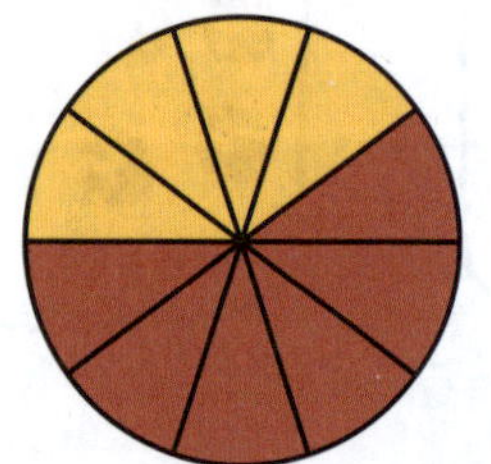

Example 4:
Colour the spinner to show an unequal chance.

Example 3:
Colour the marbles to show an unequal chance.

Check your answer on the video!

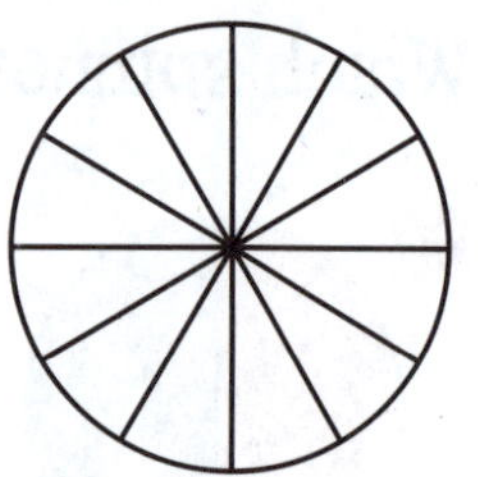

Your turn

Complete the missing information.

There are eight marbles in this bag.

I have a $\frac{\ }{8}$ chance of pulling out a purple marble and I have a $\frac{\ }{8}$ chance of pulling out a blue marble.

I have an equal/unequal chance of pulling out a blue marble.

Check your answers
How many did you get correct?

CATCH UP MATHS YEAR 4 BOOK A © PASCAL PRESS ISBN: 9781925726145

PRACTICE

1 There are 12 jellybeans in this jar. Answer the questions below.

● How many jellybeans are red?

4 out of 12

a ___ out of 12 jellybeans are blue.

b The ___________ and ___________ jellybeans have a 3 in 12 chance of being picked.

c The least likely colour of jellybean to be picked is ___________.

2 Use this jar of jellybeans to answer the questions.

a There are ___ jellybeans.

b There is a ___ in ___ chance ($\frac{}{8}$) the jellybean will be green.

c There is a ___ in ___ chance ($\frac{}{8}$) the jellybean will be red.

d It is an equal/unequal chance.

3 Use this spinner to answer the questions.

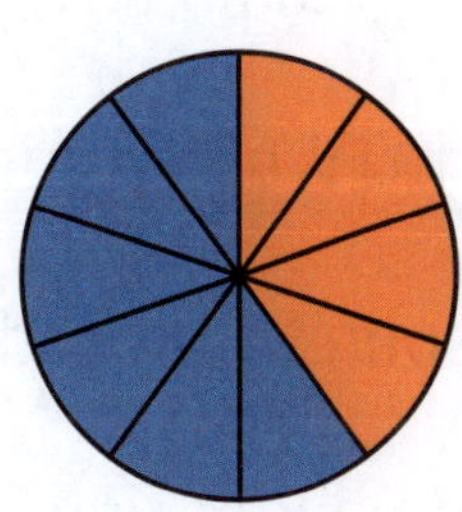

a There are ___ sections on the spinner.

b There is a ___ in ___ chance ($\frac{}{10}$) the spinner will land on blue.

c There is a ___ in ___ chance ($\frac{}{10}$) the spinner will land on orange.

d It is an equal/unequal chance.

CHANCE REVIEW

1 Join the events with the correct label.

a It will rain on a cloudy day.

b My mum will turn into a horse.

c The day after Thursday is Friday.

will happen
won't happen
might happen

d My dad will make a cake for dessert.

e You will go swimming in summer.

f My pet chicken will learn to sing.

2 Colour in certain or uncertain for each event.

a	There are seven days in one week.	Certain	Uncertain
b	I will have training tonight.	Certain	Uncertain
c	I will win my soccer game on the weekend.	Certain	Uncertain
d	There are 12 months in one year.	Certain	Uncertain

3 Colour the box green if the event is likely and red if the event is unlikely.

a ☐ You eat pizza on Friday night.

b ☐ You dance with your teacher at assembly.

c ☐ You don't go to school.

d ☐ You eat breakfast.

e ☐ You ride a horse to school.

f ☐ You fly an aeroplane after school.

g ☐ You go to sleep at night.

h ☐ You turn blue if you eat too many blueberries.

i ☐ You have a shower.

j ☐ You swim in summer.

k ☐ You watch TV after midnight.

l ☐ You eat something today.

m ☐ There is a cow in your classroom.

n ☐ You use an umbrella when it rains.

  ISBN: 9781925726145

4 Match the events that cannot happen at the same time.

a	A puppy is born.	• I draw a yellow flower.
b	It is raining.	• It is a cat.
c	I have only a red pencil.	• It is red or amber.
d	A light switch is switched on.	• It is dry.
e	The traffic light is green.	• The light switch is off.

5 Write whether the statements are true or false.

a A woman has a baby girl, so the next baby she has must be a boy.

____

b A coin is tossed and lands on tails, so the next time the coin is tossed it will be heads.

____

c If it is sunny, you may get sunburnt.

____

d If one potato is rotten in a sack of potatoes, they must all be rotten.

____

e If it is raining, you can use an umbrella to help you stay dry.

____

6 List all the combinations of T-shirts and shorts.

REVIEW

7 List all possible outcomes if I take two gummy bears out of the jar at the same time.

__________ and __________

__________ and __________

__________ and __________ __________ and __________

__________ and __________ __________ and __________

__________ and __________ __________ and __________

__________ and __________ __________ and __________

8 Jan's netball club has a new uniform

The hats are three different colours: white (WH), red (RH), blue (BH).

The skirts are three different colours: white (WS), red (RS), blue (BS).

The tops are two different colours: red (RT), blue (BT).

Write all the possible uniform combinations Jan can make.

9 Colour the marbles to show the chance.

a Even chance

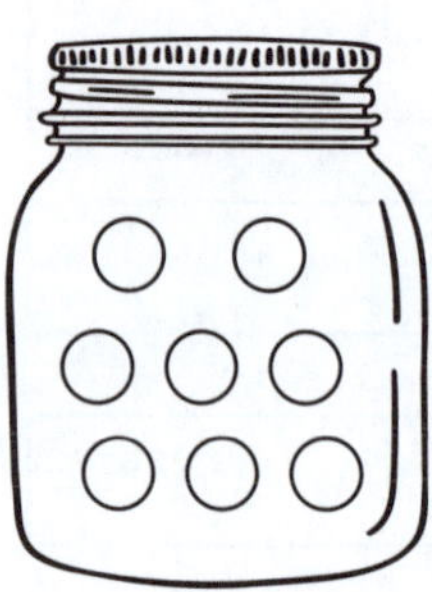

b Unequal chance

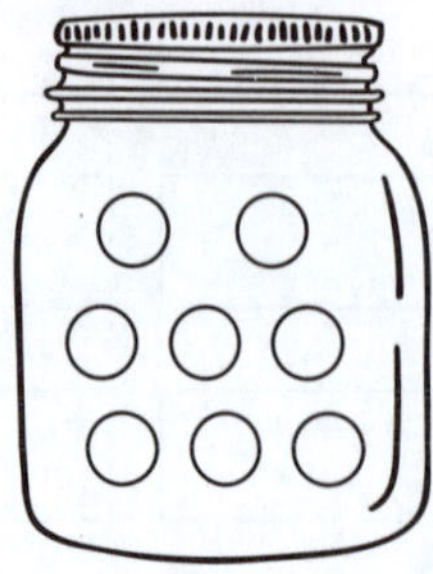

c One in two chance

d Not fair chance

CATCH UP MATHS YEAR 4 BOOK A © PASCAL PRESS ISBN: 9781925726145

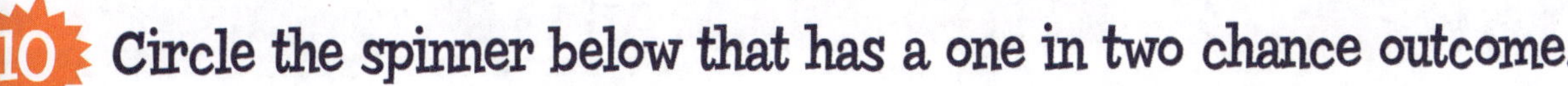

10 Circle the spinner below that has a one in two chance outcome.

a
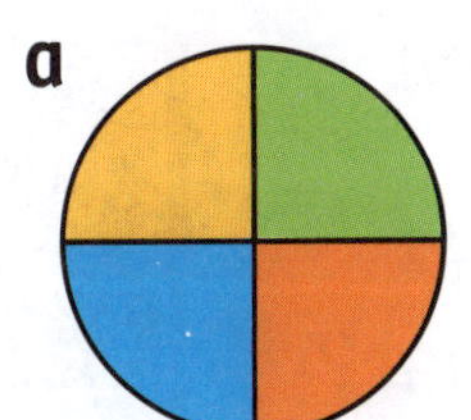
b
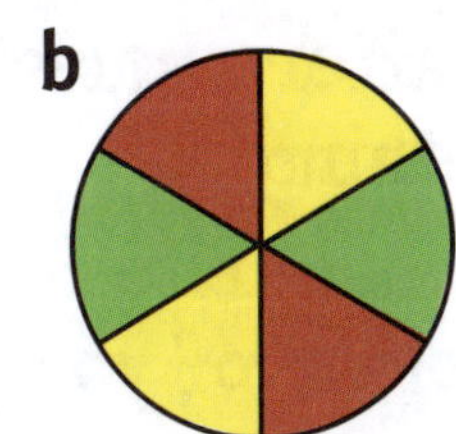
c
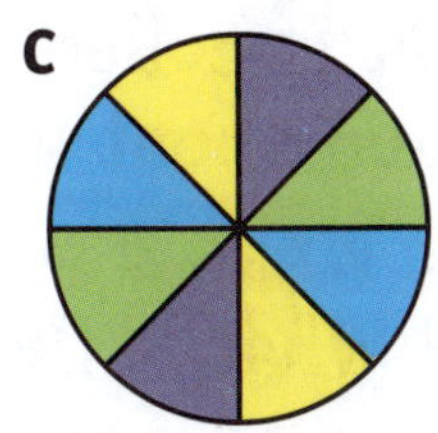
d
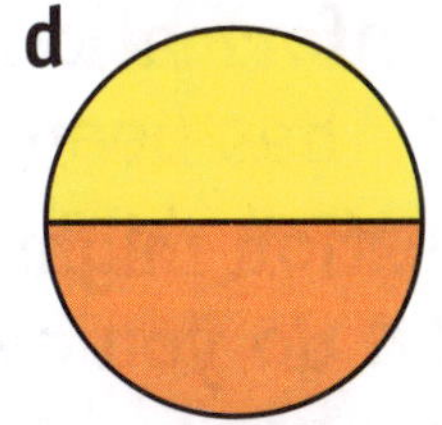
e
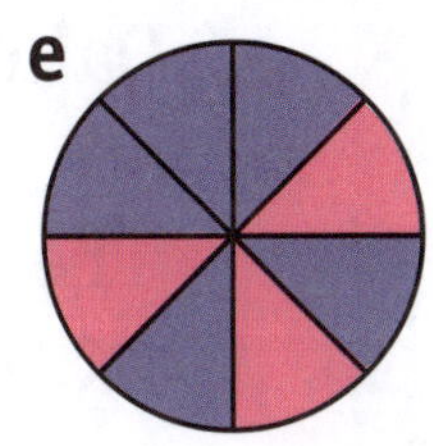

11 Draw balls in each jar to have a one in two ($\frac{1}{2}$) chance

a
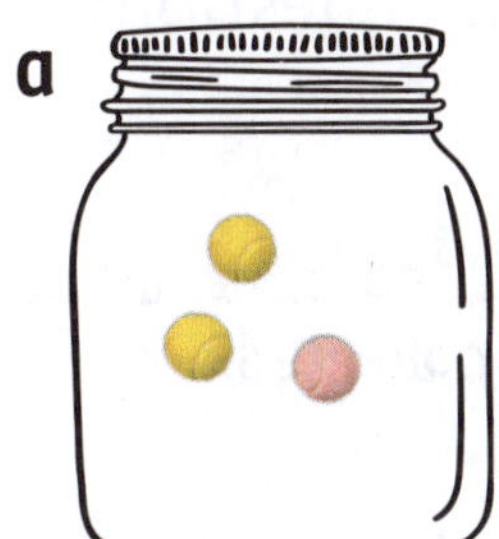
b
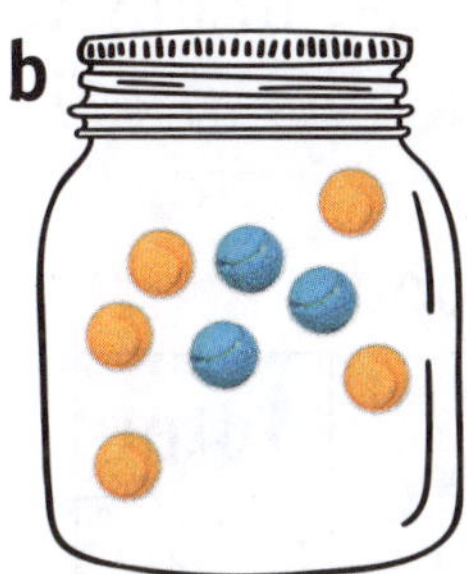
c

d

12 Circle the bags of marbles that show an unequal chance.

a
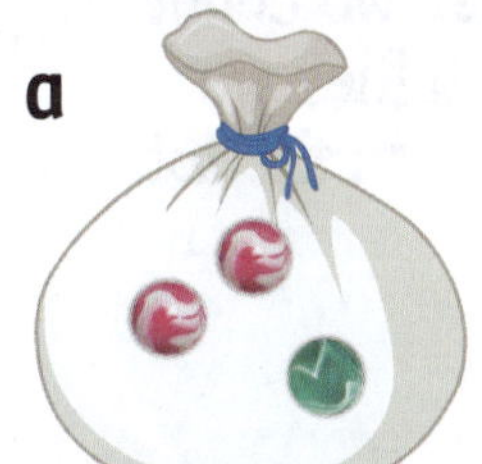
b
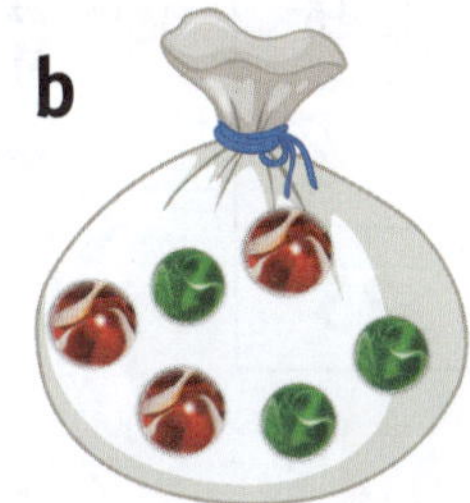
c
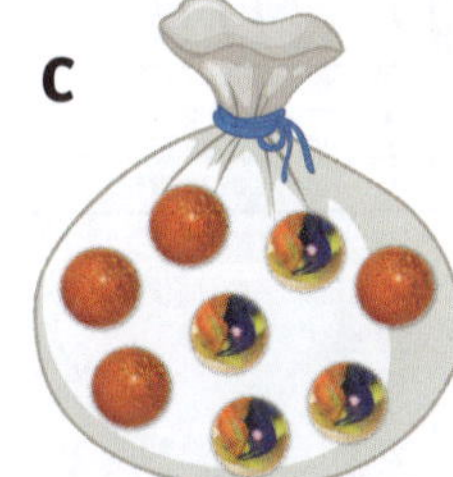
d
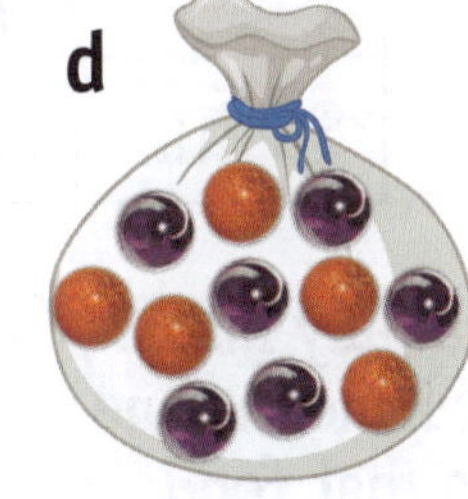
e
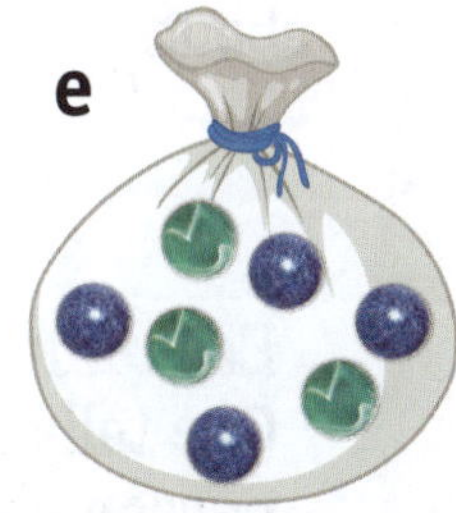

13 Complete the missing information.

a
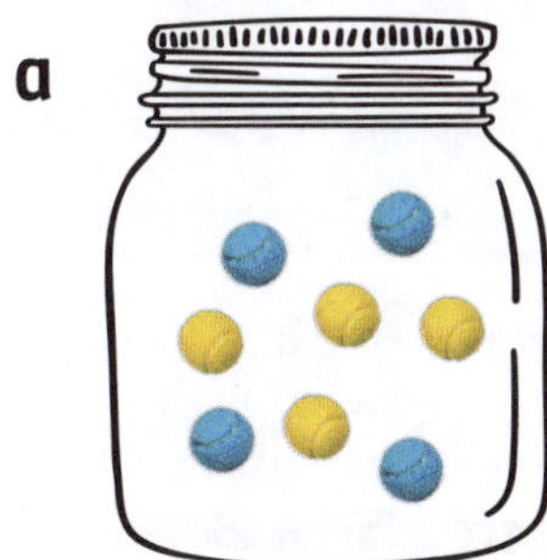

I have a $\frac{}{8}$ chance of taking out a yellow ball and I have a $\frac{}{8}$ chance of taking out a blue ball.

I have an equal/unequal chance of pulling out a yellow ball.

b

I have a $\frac{}{12}$ chance of taking out a green ball and I have a $\frac{}{12}$ chance of taking out an orange ball.

I have an equal/unequal chance of pulling out a green ball.

DATA AND TABLES

Data is information that is collected after a question has been asked about something.

The question might be: What is your favourite food? What pet(s) do you own? What is the colour of your car?

A table is often used to display the data.

Example 1:

The table below shows the data collected in answer to the question: *What are the favourite foods of the students in 4T?*

4T's Favourite Foods ← This is the information being collected.

Food	Tally	Total
Pizza	||||	4
Pasta	~~||||~~ |||	8
Tacos	~~||||~~ ||||	9
Sushi	||||	4
		25

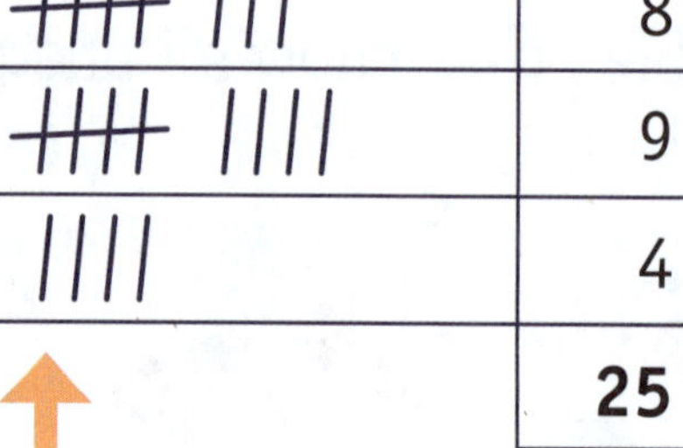

These are the foods data is being collected about. → (Food column)

After we count the tallies we write in the total. → (Total column)

Each line is a tally mark. One tally mark means one person chose that food.

The total number of students asked what their favourite food is (all the totals added together).

Lots of information can be obtained from this table.

- The most popular food is tacos.
- Twenty-five students were surveyed (asked what their favourite food is).

Use the table to answer the questions.

We count the tallies or look at the total in the table.

a How many students chose pasta? 8

b How many students chose pizza? 4

c Which two foods were chosen by an equal number of students?

pizza and sushi

CATCH UP MATHS YEAR 4 BOOK A © PASCAL PRESS ISBN: 9781925726145

Example 2:

Count the tally marks and record the totals.

Type of Boats Seen on Sunday

Type of Boat	Tally	Total
Sailboat	𝍸 𝍸 ||	
Speedboat	𝍸 𝍸 𝍸 ||||	
House boat	𝍸 ||	
Pontoon boat	𝍸 𝍸 𝍸 𝍸 |	

Your turn

Use the table to complete the following sentences.

Drinks Sold at Caringbar Public Canteen

Day	Tally	Total
Monday	𝍸 𝍸 ||||	14
Tuesday	𝍸 𝍸 𝍸 ||||	19
Wednesday	𝍸 𝍸 𝍸 𝍸	20
Thursday	𝍸 𝍸 𝍸 |	16
Friday	𝍸 𝍸 𝍸 𝍸 𝍸 |	26
		95

a The most drinks were sold on ______________.

b The fewest drinks were sold on ______________.

c The difference in the numbers of drinks sold on Thursday and Friday is ____.

d The total number of drinks sold is ____.

SELF CHECK Tick how you feel		
Got it! ☐	Need help... ☐	I don't get it ☐

Check your answers

How many did you get correct? ☐

PRACTICE

1 **Fill in the missing information in the tables below.**

a **Lolly Colour in Packet**

Colour	Tally	Total
Red	𝍷𝍷	
Blue		4
Yellow	𝍸 𝍷	
Green		5

b **4S's Favourite Colour**

Colour	Tally	Total
Black		10
Pink		5
Orange	𝍷𝍷𝍷	
Blue		8
Yellow	𝍷	

Michael asked 20 people what their favourite sport was and he organised the data he collected into this table.

Favourite Sport

Colour	Tally	Total
Football	𝍸 𝍷𝍷	7
Soccer	𝍸 𝍷	6
Cricket	𝍷𝍷𝍷	3
Tennis	𝍷𝍷𝍷𝍷	4
		20

2 **Answer these questions using the data in Michael's table.**

- What was the most popular sport? football

a What was the least popular sport? ______________

b How many people chose tennis? ___

c How many people chose soccer? ___

3 **Work out the difference. The difference is the amount that one number is bigger than another. It is found by subtracting.**

- What is the difference between the number of people who chose football and tennis? $7 - 4 = 3$

a What is the difference between the number of people who chose soccer and cricket? __________

b What is the difference between the number of people who chose tennis and cricket? __________

CATCH UP MATHS YEAR 4 BOOK A © PASCAL PRESS ISBN: 9781925726145

PICTURE GRAPHS

A picture graph uses pictures to represent data.
A key is used to explain what the pictures mean.

Example 1:

Use the graph to answer the following questions.

Number of books read in Term 3

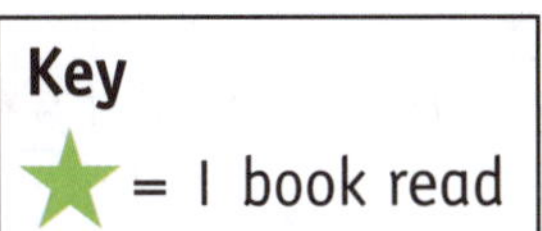

a How many books altogether were read? 20

b Who read the most books?

Rashid

c Who read the least number of books?

Robert

d What was the difference between the number of books read by Rashid and Robert? 5

e Which two people read the same number of books?

Sofia and Julie

f How many books did Karen read? 3

  ISBN: 9781925726145

Example 2: Use the graph to fill in the table, and then answer the following questions.

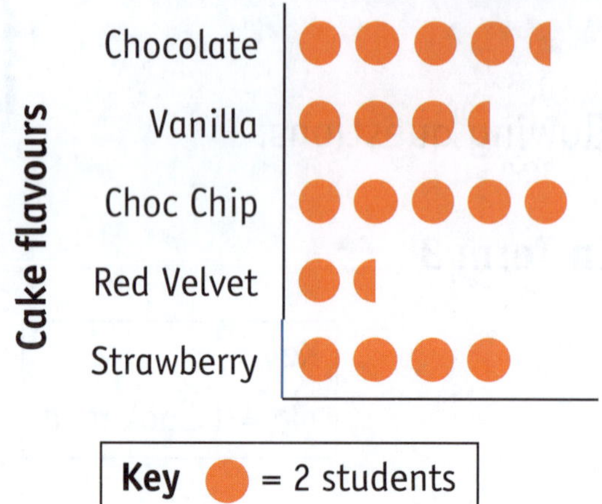

4D's Favourite Cake Flavour

Cake Flavour	Total
Chocolate	
Vanilla	
Choc Chip	
Red Velvet	
Strawberry	
Total:	

a What was the least popular cake flavour? ___

b How many students chose Red Velvet? ___

c ___ students chose Vanilla.

d What is the difference between the number of votes for Chocolate and for Strawberry? ___

e How many students were surveyed? ___

Your turn

Use the graph to fill in the table.

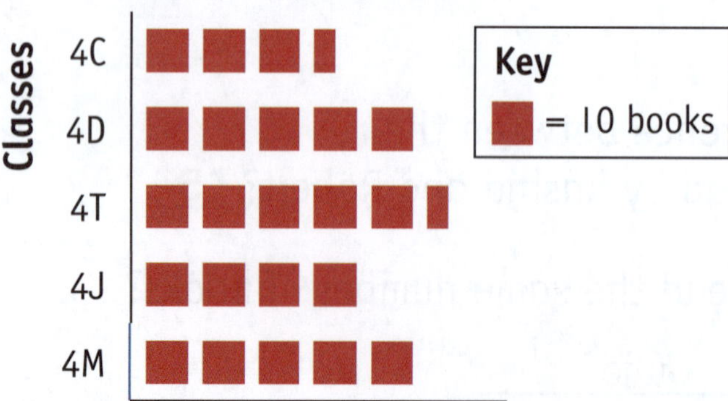

Number of Books Read in Book Week

Class	Total
4C	
4D	
4T	
4J	
4M	
Total:	

SELF CHECK Tick how you feel

Got it!	Need help...	I don't get it
☐	☐	☐

Check your answers
How many did you get correct? ☐

CATCH UP MATHS YEAR 4 BOOK A © PASCAL PRESS ISBN: 9781925726145

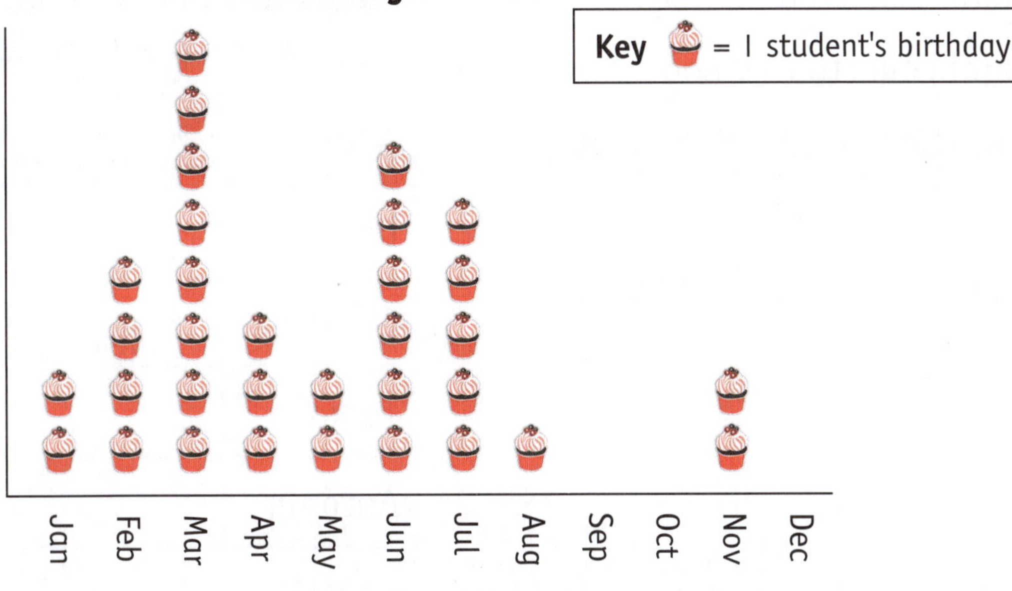

1 Answer the questions from the data in the picture graph.

a How many students are in 4C? ___

b Which months have the same number of students' birthdays?

____

c How many students have a birthday in July?___

d Which month has the most birthdays? _______

e Which months have no birthdays? _______

____

f How many students have a birthday in February? ___

g What is the difference between the number of birthdays in March and May? ___

h What is the difference between the number of birthdays in June and September? ___

Using the symbol = 5 bags of rubbish collected, complete the picture graph to show the following information.

- Allan collected 10 bags.

a Dani collected 15 bags.

b Christian collected 20 bags.

c Tarka collected 15 bags.

d Antonia collected 10 bags.

Bags of Rubbish Collected on Clean-up Australia Day

Allan	
Dani	
Christian	
Tarka	
Antonia	

3 Use the picture graph to answer the questions.

- How many houses were built in 2017? 20

a How many houses were built in 2020? ____

b Which year had the least number of houses built?

____

c What is the difference between the number of houses built in 2021 and 2019?

____

Number of Houses Built by I-Build Houses

2017	
2018	
2019	
2020	
2021	

Key

= 10 houses built

= 5 houses built

d Which year had 40 houses built? ________

e How many houses were built altogether in 2020 and 2021? ________

f Which year had double the number of houses built as in 2017? ________

CATCH UP MATHS YEAR 4 BOOK A © PASCAL PRESS ISBN: 9781925726145

The children in 4K made a graph of the pets they own. The girls' data is recorded in orange and the boys' in green. Use the graph to answer the following questions.

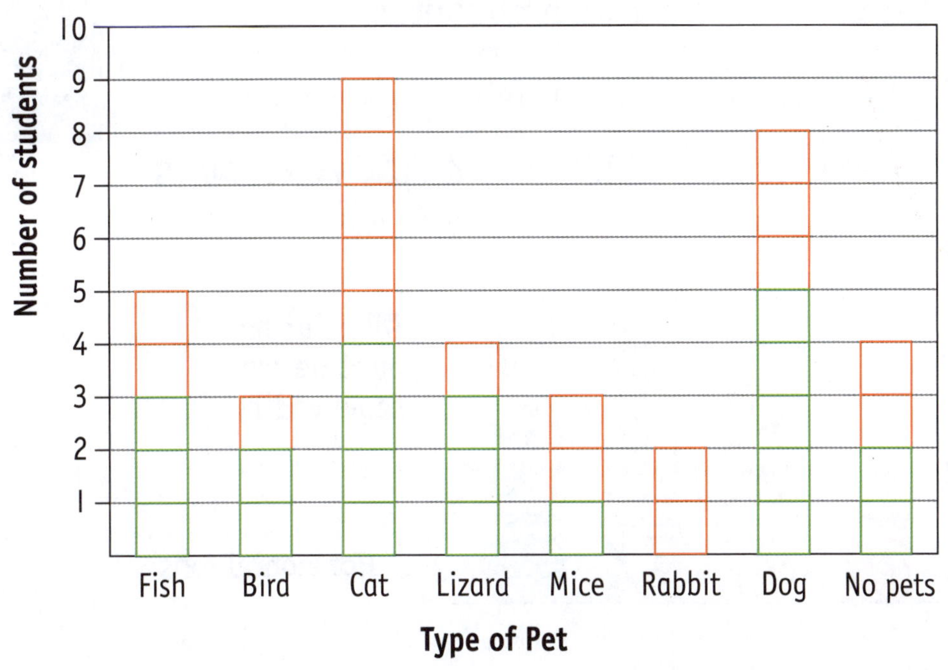

- How many girls had birds? 1

a How many students had fish as pets? ____

b How many boys owned lizards? ____

c What was the most popular pet? ________________

d How many students did not own a pet? ____

e Which two pets were equally as popular?

________________ and ________________

f What is the difference between the numbers of the most popular and least popular pets? ____

g How many boys are in the class? ____

h How many students owned a lizard or a rabbit? ____

i How many students are in the class? ____

COLUMN GRAPHS

A column graph is a graph that shows the data in columns that can be vertical or horizontal.

Example 1: This is a vertical column graph.

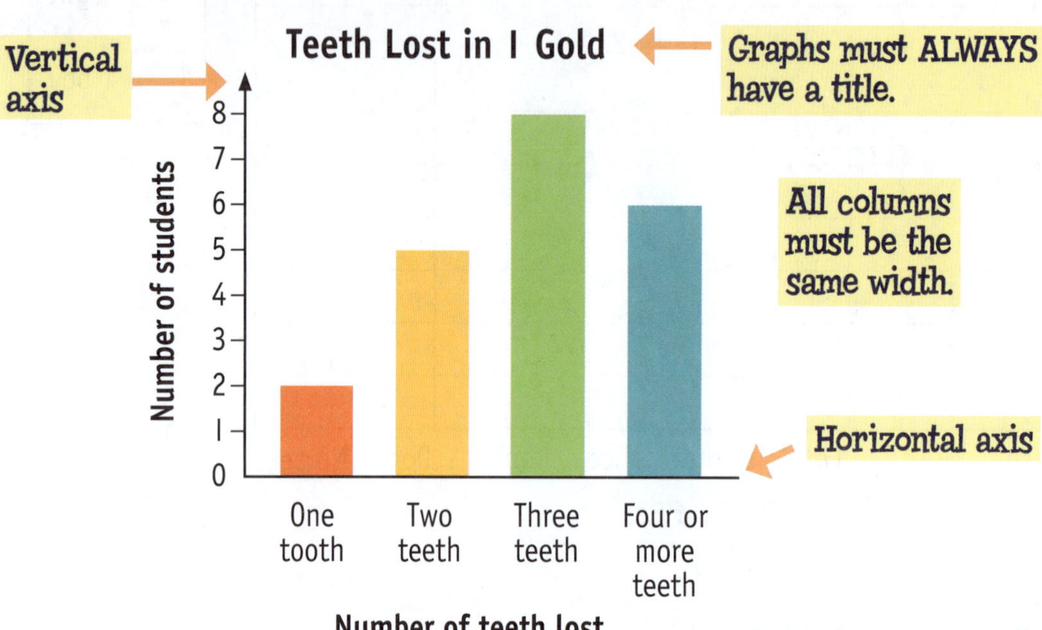

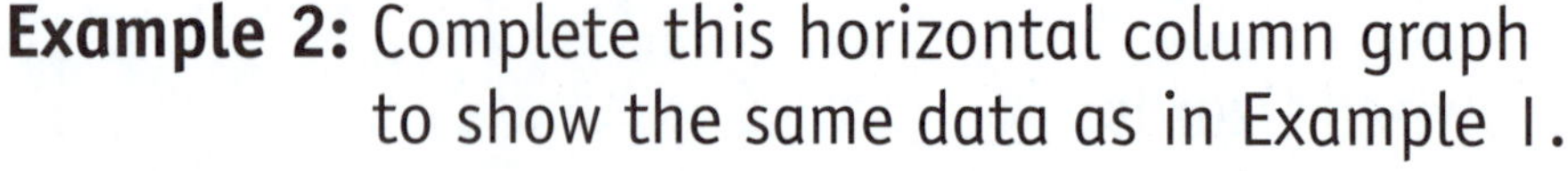
Example 2: Complete this horizontal column graph to show the same data as in Example 1.

Teeth Lost in 1 Gold

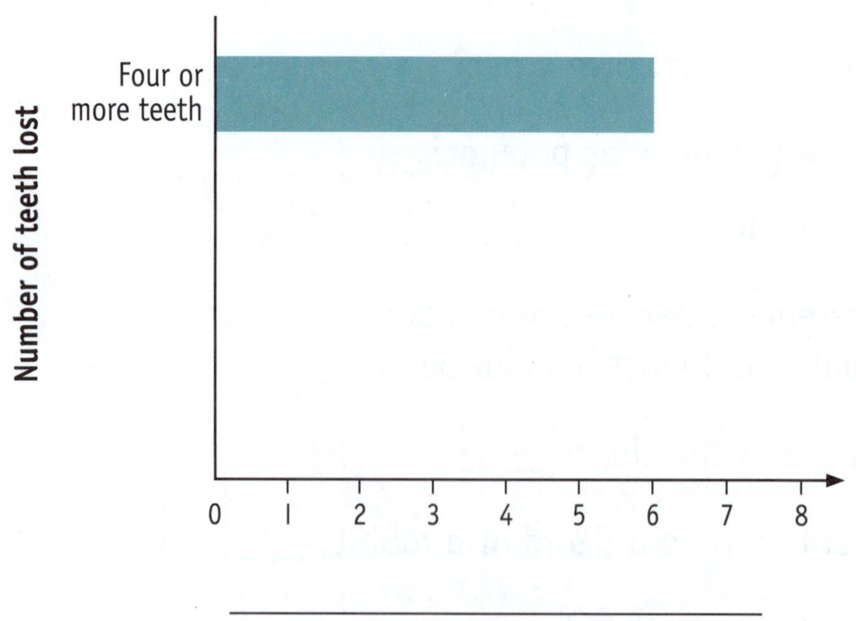

CATCH UP MATHS YEAR 4 BOOK A © PASCAL PRESS ISBN: 9781925726145

Example 3: Use the information in the column graph to complete the table.

Teeth Lost in 1 Gold

Number of Teeth Lost	Tally	Total
One tooth		
Two teeth		
Three teeth		
Four or more teeth		
	Total:	

Your turn

Answer the following questions about the column graphs.

- What does the yellow column represent?

  Two teeth lost

a What does the blue column represent?

____

b How many students in 1 Gold lost one tooth? ____

c How many students in 1 Gold lost more than three teeth?

____

d How many more students lost two teeth than one tooth?

____

e How many students lost three teeth? ____

f The least number of teeth lost is ____.

SELF CHECK Tick how you feel

Got it!	Need help...	I don't get it
☐	☐	☐

Check your answers

How many did you get correct? ☐

PRACTICE

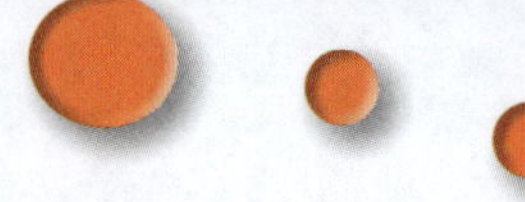

1 **Athena rolled a die and graphed her results in this column graph. Tally Athena's results in the table.**

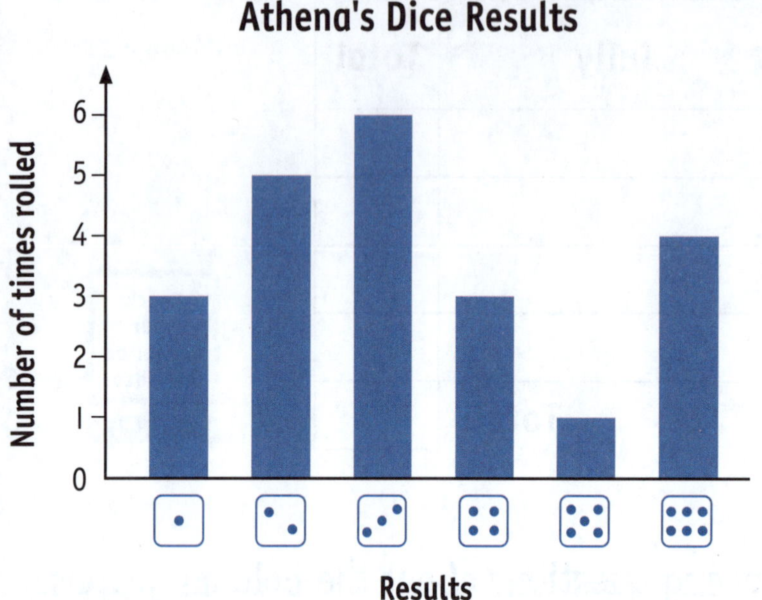

Athena's Dice Results

Roll	Tally	Total
⚀		
⚁		
⚂		
⚃		
⚄		
⚅		

Answer the questions about the data in the column graph above.

a Which number was rolled the most often? __

b How many more times was six rolled than five? __

c Which two numbers were rolled the same number of times?

__ and __

d Five was rolled the most/least often.

e Two was rolled __ times.

f The difference between the number of times three and five were rolled is __.

CATCH UP MATHS YEAR 4 BOOK A © PASCAL PRESS ISBN: 9781925726145

Year 4 surveyed the colour of the teachers' cars at their school and created the horizontal column graph.

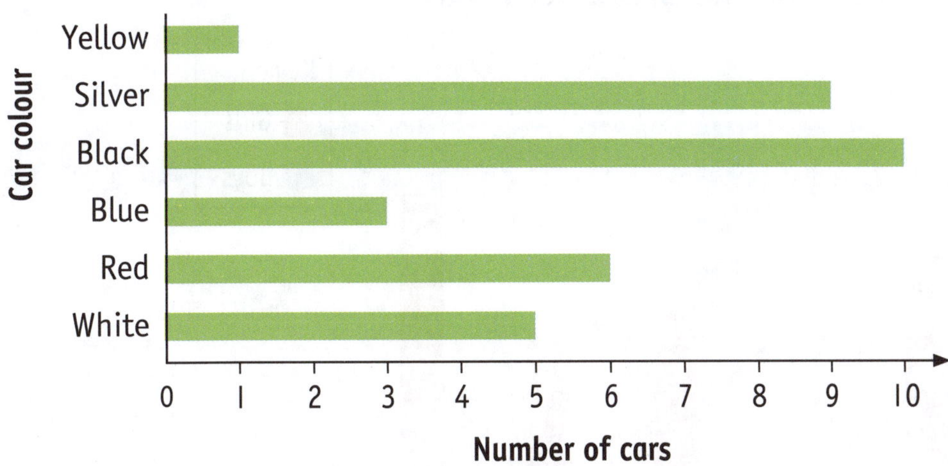

Answer the questions about the data in the column graph.

How many teachers have white cars? 5

a __ teachers have blue cars.

b The least popular colour of teachers' cars is ______________.

c The most popular colour of teachers' cars is ______________.

d Six teachers have ______________ cars.

e There are __ silver and black cars.

f The difference between the number of black and silver cars is __.

g The difference between the number of the most popular colour and the least popular colour is __.

This graph shows what food boys and girls in Year 4 liked to eat for breakfast. Answer the questions below about the data in the graph.

- The most popular food with the boys was toast.

a The most popular food with the girls was ____________.

b How many boys and girls in Year 4 liked pancakes for breakfast? ____

c What was the food least liked by boys and the least liked by girls?

____

d How many children altogether chose fruit? ____

e What is the difference in number between the most popular and least popular food chosen by boys? ____

f How many children are in Year 4? ____

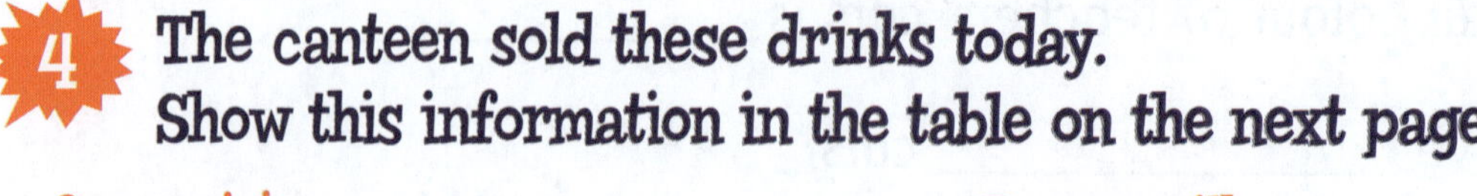

The canteen sold these drinks today. Show this information in the table on the next page.

CATCH UP MATHS YEAR 4 BOOK A © PASCAL PRESS ISBN: 9781925726145

Drinks Sold at the Canteen Today

Drink	Tally	Total
Orange juice		
Grape juice		
Banana milk		
Strawberry milk		
Chocolate milk		

5 **Construct a vertical column graph to represent the drinks sold at the canteen last week.**

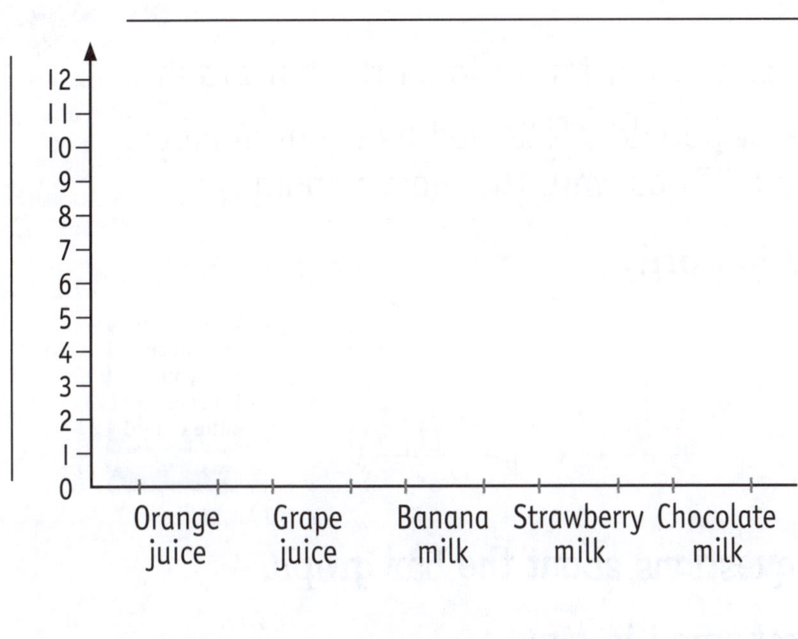

Answer the questions about the data in the column graph.

a Which drink was the most popular? ______________________

b Which drink had 4 sold? ______________________

c ___ cartons of chocolate milk were sold.

d The drink that was sold the least often was ______________________.

e The difference between the number of grape juices and strawberry milks sold is ___.

f How many drinks were sold altogether? ___

  ISBN: 9781925726145

BAR GRAPHS

A bar graph is a bar that has been divided into different lengths to show different quantities of data.

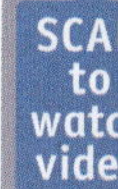

Example 1:

Animals rehomed in March

Dogs	Fish	Cats

The cat part of the bar graph is the longest and the fish part is the shortest.

It is easy to see that cats were the most rehomed pets in March and fish were the least.

Example 2: Use this information to label the bar graph:

Cats were the least popular, followed by equal numbers of horses and dogs. Birds were the most popular.

Animals rehomed in April

Answer the questions about the bar graph.

Animals rehomed in May

Birds	Fish	Guinea pigs	Cats	Dogs

<u>5</u> types of animals were rehomed in December.

a ____________ were the animals rehomed most often in May.

b ____________ were the animals rehomed least often in May.

c Half the animals rehomed in May were ____________.

SELF CHECK Tick how you feel

Got it!	Need help...	I don't get it
☐	☐	☐

Check your answers

How many did you get correct? ☐

CATCH UP MATHS YEAR 4 BOOK A © PASCAL PRESS ISBN: 9781925726145

PRACTICE

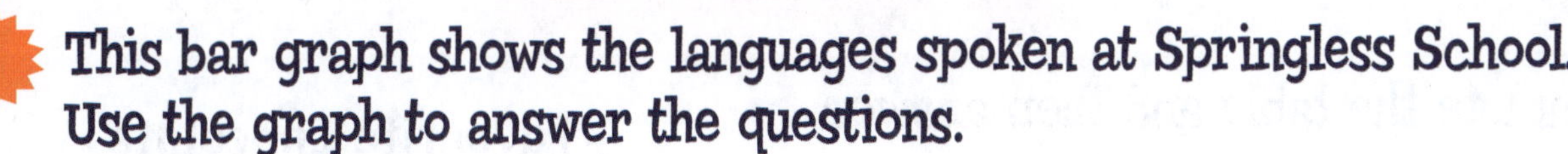

1 **This bar graph shows the languages spoken at Springless School. Use the graph to answer the questions.**

English	Italian	Maltese	Greek	Mandarin	Korean

a What title could be given to this graph?

____

b What language is spoken most at Springless School? ____________

c About ____________ of the students speak English at Springless School.

d The language spoken least at Springless School is ____________.

e If 100 students go to Springless School, about how many speak English? ___

f If $\frac{1}{4}$ of the students speak Italian, about how many students is this? ___

g How many languages are spoken at Springless School? ___

2 **Use the following information to help you draw lines on this bar graph to show how the children in 4T spend their free time.**

How children in 4T spend their free time

- There are five activities to choose from: Watching TV, Sport, iPad, Reading, Cooking.
- Half of the students chose iPad.
- Watching TV was the least popular choice.
- Sport and reading had about the same number of students.
- One-quarter of the students chose cooking.

DATA REVIEW

Complete the table and then answer the questions.

Favourite Chocolate

Flavour	Tally	Total
Plain	~~||||~~	
Peanut		7
Caramel	||||	
Coconut		12

a How many students were asked what their favourite chocolate was?

____

b What was the least popular chocolate?

____

c The most popular chocolate was ________________.

d ___ children chose peanut chocolate.

e The difference in number between the most popular and least popular chocolate was ___.

This table shows the favourite subjects of students in 4J.

4J's Favourite Subjects

Subject	Tally	Total
Maths	~~||||~~	
English	~~||||~~	
Science	|||	
PE	~~||||~~ |	
Art	~~||||~~ ||	
Music	|	

a Complete the table.

b Construct a picture graph showing the data. Use the key ■ = 2 students.

____

Maths English Science PE Art Music

Subjects

CATCH UP MATHS YEAR 4 BOOK A © PASCAL PRESS ISBN: 9781925726145

3 Antonia made this table from data she collected about jellybean colours.

a Fill in the missing information in the table.

b Write three questions about the data in Antonia's table.

Jellybean Colours

Colour	Tally	Total
Black	//	
Red	~~////~~ ////	
Yellow	////	
Green	~~////~~ ///	
Orange	////	
Blue	~~////~~	

____

____

____

____

____

____

c Construct vertical and horizontal column graphs showing this data.

Vertical column graph

____

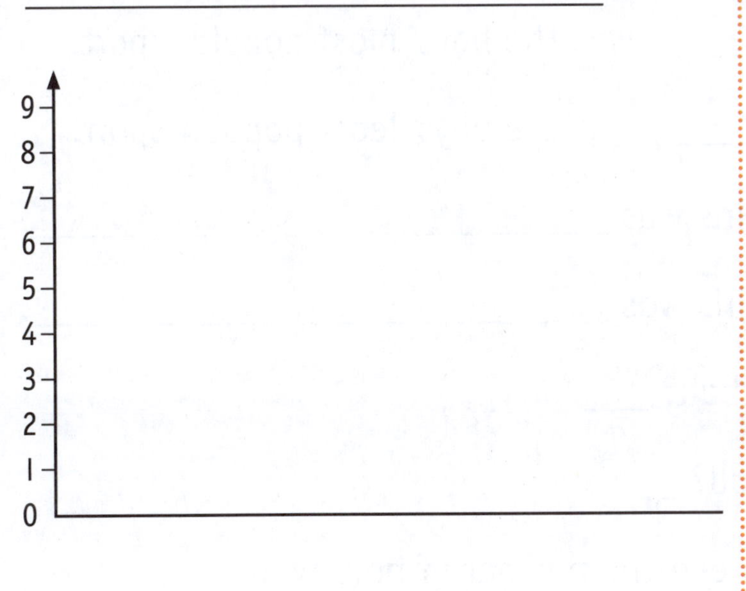

Horizontal column graph

____

REVIEW

4 Answer the following questions about the data shown in the vertical column graph.

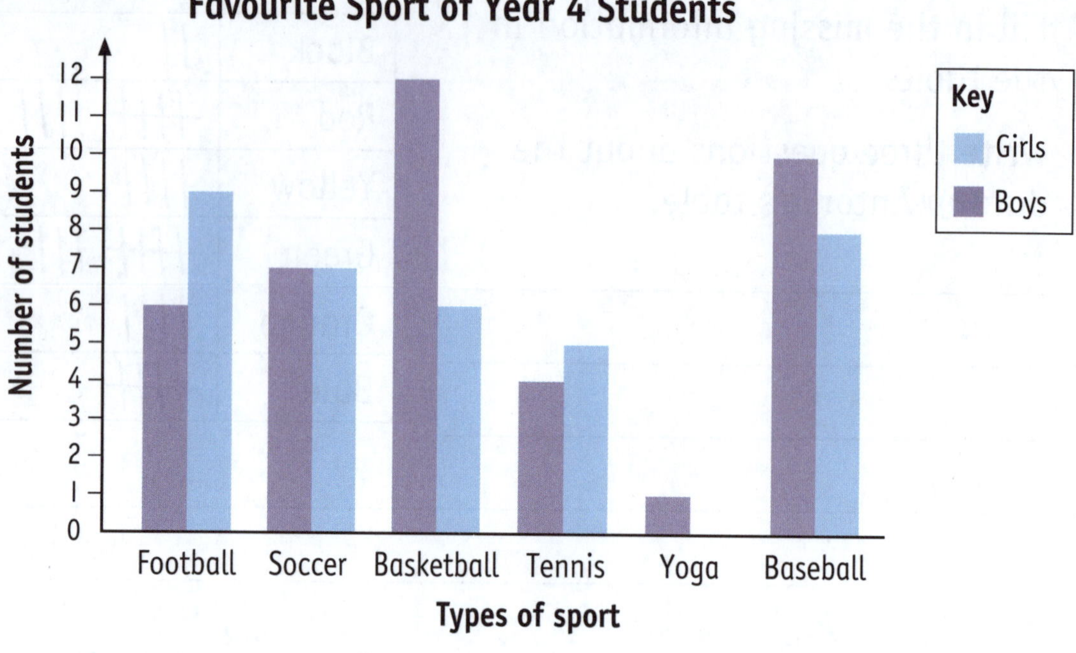

a How many students were surveyed? ____

b Which sport was equally popular for girls and boys?

____

c Which sport was not chosen by girls? ______________________

d ______________________ was the boys' most popular sport.

e ______________________ was the boys' least popular sport.

f The most popular sport for girls was ______________________.

g The least popular sport for girls was ______________________.

h How many girls chose basketball? ____

i How many boys chose baseball? ____

j What was the difference between the number of boys who chose football and the number of girls who chose tennis? ____

k The difference between the number of girls who chose basketball and the number of boys who chose yoga was ____.

CATCH UP MATHS YEAR 4 BOOK A © PASCAL PRESS ISBN: 9781925726145

5 Write three questions about the data shown in the bar graph.

Juice Sold at the Canteen Today

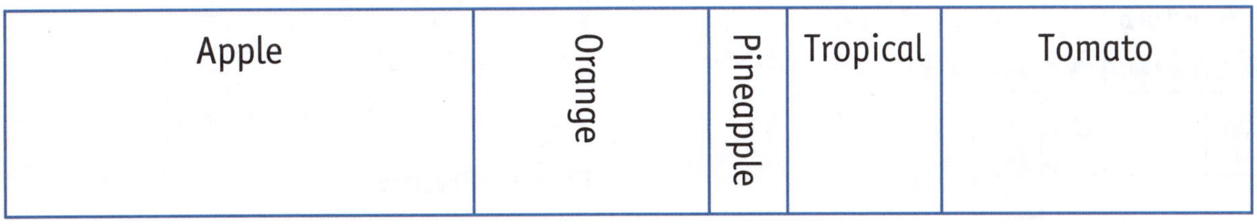

Apple	Orange	Pineapple	Tropical	Tomato

- ______________________________
  ____
- ______________________________
  ____
- ______________________________
  ____

6 This bar graph shows where 200 students went during the summer holidays.

Places Where Students Spent Their Summer Holidays

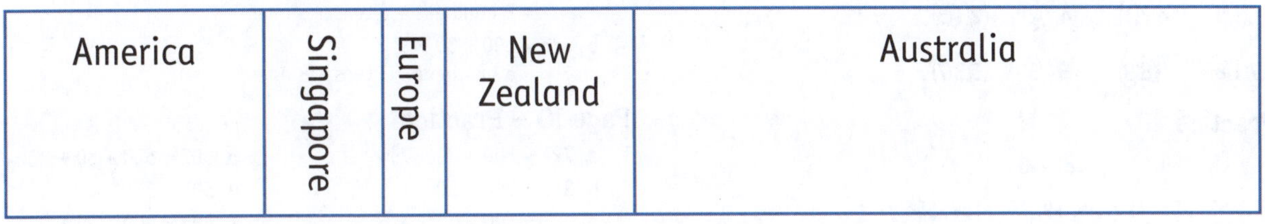

America	Singapore	Europe	New Zealand	Australia

a How many places did students spend their holidays? ___

b About how many students spent them in Australia? ___

c ______________________ was the least common place.

d If $\frac{1}{4}$ of the students went to New Zealand, how many students visited New Zealand? ___

e In which place did the most students spend their holidays?

____

ANSWERS

1 WHOLE NUMBERS

Three-digit Numbers

Page 1 – Your Turn

	Number	Hundreds	Tens	Ones
a	127	1	2	7
b	249	2	4	9
c	863	8	6	3
d	524	5	2	4
e	780	7	8	0

Page 2 – Practice

1 a two hundred and ninety-three
 b four hundred and fifty-one
 c seven hundred and sixty-four
 d five hundred and three
 e eight hundred and fifty
 f three hundred

2

	Number	Hundreds	Tens	Ones
a	410	4	1	0
b	324	3	2	4
c	568	5	6	8
d	879	8	7	9
e	903	9	0	3

3 Circled green:
359, 319, 362, 347, 300
Circled blue:
721, 924, 422, 923, 426
Circled red:
188, 218, 748, 838, 878

Place Value to 1000

Page 3 – Your Turn

1 a ones b hundreds c tens d ones e tens

2 936 415 671 472 215
324 716 989 543 537

Page 4 – Practice

1 a 213 b 758 c 960 d 505

2 Circled: 257, 452, 553, 159, 56, 59, 757, 157, 556

3 Crossed out: 937, 982, 993, 957, 909, 973, 987, 921

4 a ones
 b tens
 c ones
 d hundreds
 e tens
 f ones
 g hundreds
 h ones
 i tens
 j ones
 k hundreds

Value and Three-digit Numbers

Page 5 – Your Turn

a 243
b 714
c 139
d 287
e 324
f 903
g 874
h 492
i 736
j 649
k 999
l 710
m 118
n 247

Page 6 – Practice

1 a 371 b 623 c 258 d 184 e 989

2 26 326 246 6 16 746

3 78 872 79 777 175 474

4 832 849 863 847 888

5 a 40
 b 4
 c 400
 d 40
 e 40
 f 40
 g 400
 h 40
 i 400
 j 400
 k 4

Number Expanders and Three-digit Numbers

Page 7 – Your Turn

a

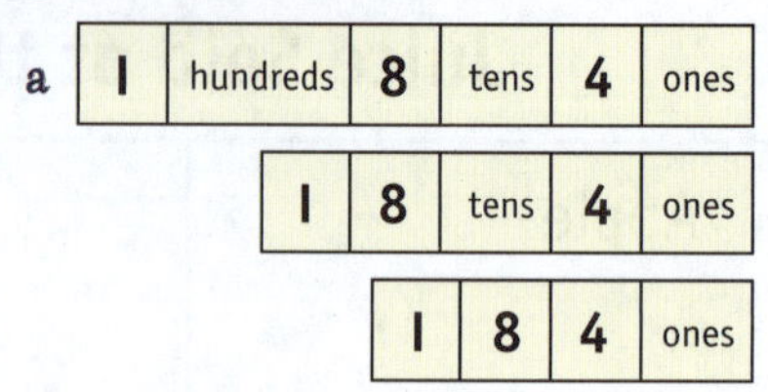

Page 8 – Practice

1 a

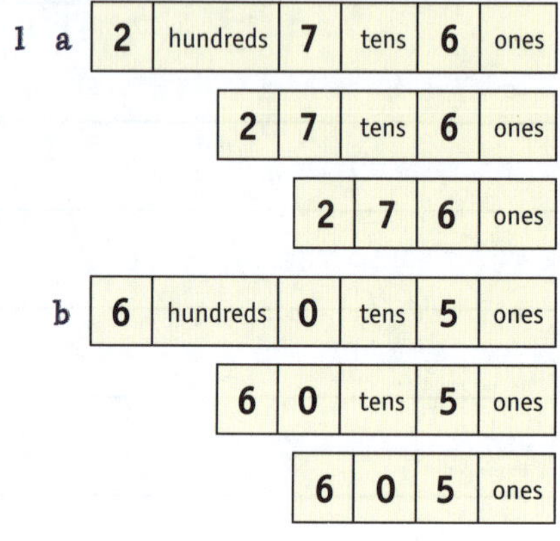

2 a 326 b 504 c 906

3 a 34 tens, 2 ones
 b 7 hundreds, 2 tens, 4 ones
 c 649 ones

Expanded Three-digit Numbers

Page 9 – Your Turn

a 265 = 200 + 60 + 5
 = 2 hundreds + 6 tens + 5 ones
b 638 = 600 + 30 + 8
 = 6 hundreds + 3 tens + 8 ones

Page 10 – Practice

1 a 727 = 700 + 20 + 7
 b 850 = 800 + 50
 c 607 = 600 + 7
 d 582 = 500 + 80 + 2
 e 500 = 500
 f 495 = 400 + 90 + 5

2 a 400 + 10
 b 300 + 50 + 9
 c 800 + 30 + 2
 d 200 + 6
 e 300
 f 700 + 80 + 3
 g 500 + 60 + 4

3 a 190
 b 430
 c 741
 d 252
 e 355
 f 802
 g 143
 h 984
 i 487
 j 376
 k 612
 l 538

Modelling Hundreds

Page 11 – Your Turn

a 524 b 231 c 982

Page 12 – Practice

1 a 1 hundred, 4 tens, 5 ones = 145
 b 3 hundreds, 1 ten, 9 ones = 319
 c 8 hundreds, 5 tens, 4 ones = 854
 d 4 hundreds, 4 tens, 6 ones = 446

2 a 340 b 302 c 987

Ordering Three-digit Numbers

Page 13 – Your Turn

1 a 147, 243, 763, 824
 b 124, 371, 431, 454

2 a 513, 335, 315, 153
 b 532, 235, 205, 105

CATCH UP MATHS YEAR 4 BOOK A © PASCAL PRESS ISBN: 9781925726145

Page 14 – Practice

1 a 872, 602, 215, 143, 136
  b 703, 411, 371, 204, 142
  c 711, 526, 341, 242, 101
  d 950, 624, 524, 139, 133

2 a 103, 588, 741, 954, 955
  b 111, 135, 207, 481, 740
  c 126, 299, 301, 531, 621
  d 145, 243, 321, 412, 702

3 a A b A c D d A e D f A g D h D i A

Counting by Hundreds

Page 15 – Your Turn

1 Circle: 243, 128, 920, 103, 349

2 a 400, 600 b 500, 400 c 325, 425 d 562, 462

Page 16 – Practice

1 a nine hundred, 900
  b six hundred and forty-two, 642
  c three hundred, 300

2 a 500, 600, 700
  b 447, 547, 647
  c 700, 800, 900
  d 600, 500, 400
  e 243, 143, 43
  f 471, 371, 271
  g 500, 700, 900
  h 445, 645, 745
  i 578, 678, 778
  j 892, 592, 492

Four-digit Numbers

Page 17 – Your Turn

	Number	Th	H	T	O	Words
a	4692	4	6	9	2	four thousand, six hundred and ninety-two
b	5300	5	3	0	0	five thousand, three hundred

Page 18 – Practice

1 a seven thousand, two hundred and nineteen
  b three thousand, four hundred and ninety
  c two thousand and five
  d eight thousand, nine hundred and forty-three

2 a 3, 4, 1, 2 b 4, 2, 3, 1 c 3, 4, 1, 2

3 a 4, 3, 2, 1 b 3, 1, 4, 2 c 3, 1, 4, 2

4 a 4922 b 8003 c 7469

Place Value and Four-digit Numbers

Page 19 – Your Turn

3921 4374 5500 9746 9525
2125 5251 6303 1809

Page 20 – Practice

1 a 1882
  b 3047
  c 6030
  d 9280

2

	Number	Th	H	T	O
a	3269	3	2	6	9
b	9038	9	0	3	8
c	4730	4	7	3	0
d	5476	5	4	7	6
e	6924	6	9	2	4

3

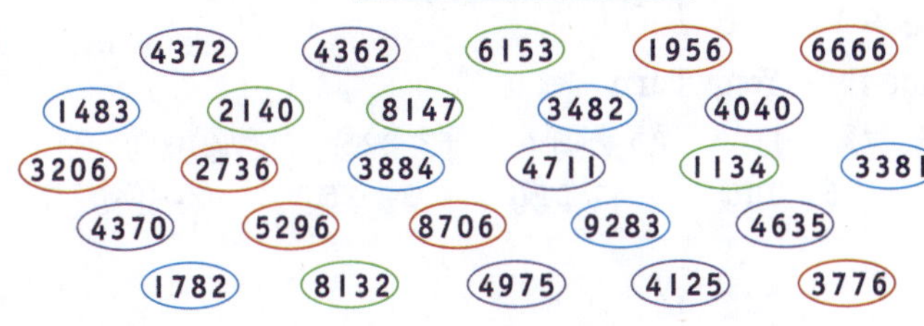

Value and Four-digit Numbers

Page 21 – Your Turn

1 8432, 8936

2 2294 1157 4458 3866 5490
  8745 8894 6128

Page 22 – Practice

1 Circled: 142, 640, 343

2 Circled: 524, 599, 507

3 a 400 b 4 c 40 d 4000 e 4 f 40 g 4 h 400 i 4000 j 40 k 400 l 4 m 4000 n 4 o 40 p 400 q 4000

4 a 1 b 300 c 9000 d 100 e 3 f 20 g 2000 h 0 i 600 j 3000 k 900 l 40 m 8000 n 7 o 700 p 5 q 20

Number Expanders and Four-digit Numbers

Page 23 – Your Turn

a

7	4	0	6
	74	0	6
		740	6
			7406

Page 24 – Practice

1 a 1 Th, 0 H, 3 T, 6 O
  b 4 Th, 5 H, 3 T, 2 O
  c 1 Th, 5 H, 8 T, 5 O
  d 2 Th, 2 H, 5 T, 1 O
  e 3 Th, 5 H, 4 T, 5 O

2 a 8532 b 2468 c 7564

Expanded Four-digit Numbers

Page 25 – Your Turn

a

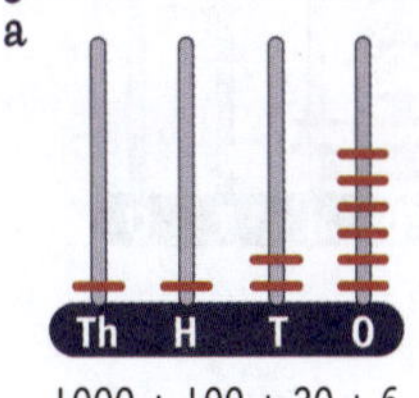

1000 + 100 + 20 + 6

b

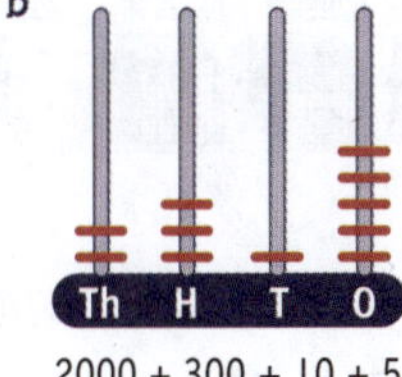

2000 + 300 + 10 + 5

Page 26 – Practice

1 a 3000 + 600 + 20 + 7
  b 4000 + 800 + 3
  c 5000 + 90 + 3
  d 2000 + 200 + 20 + 2
  e 7000 + 600 + 40

2 a 6312 b 7025 c 9900 d 4203 e 1527

3

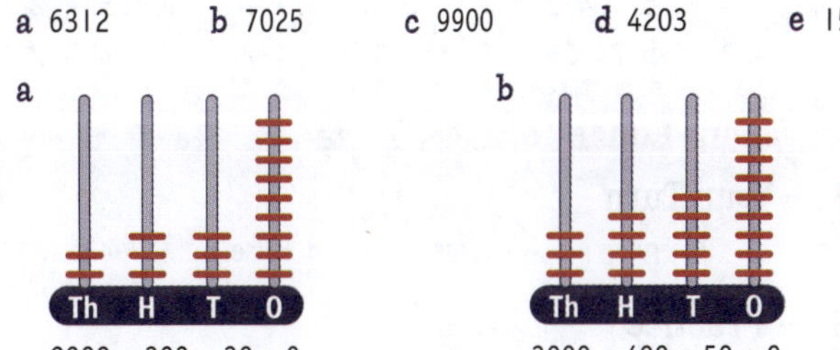

a

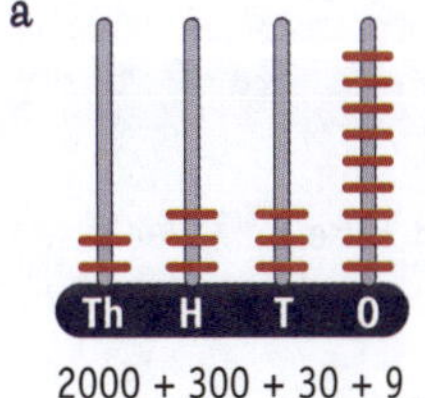

2000 + 300 + 30 + 9

b

3000 + 400 + 50 + 9

Modelling Four-digit Numbers

Page 27 – Your Turn

a 1437 b 3412 c 4513

1 WHOLE NUMBERS CONTINUED

Page 28 – Practice

1 a
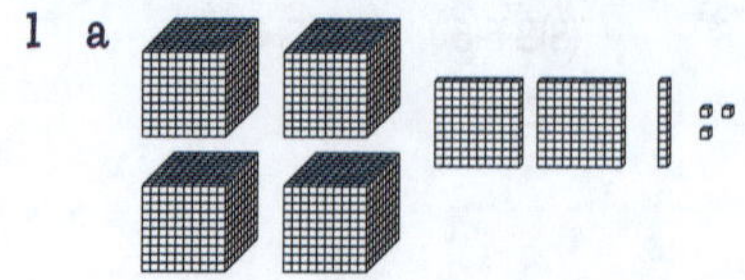
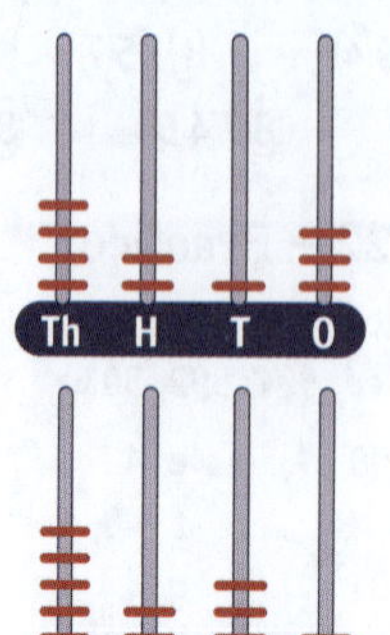

b
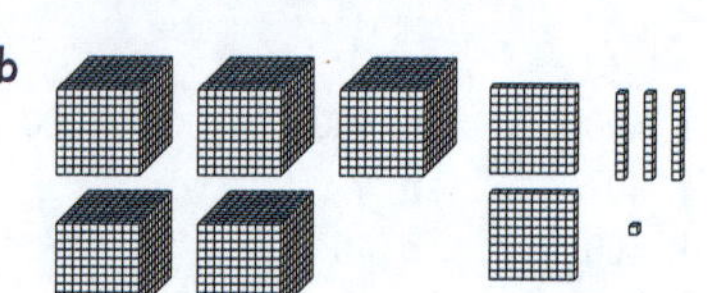

c
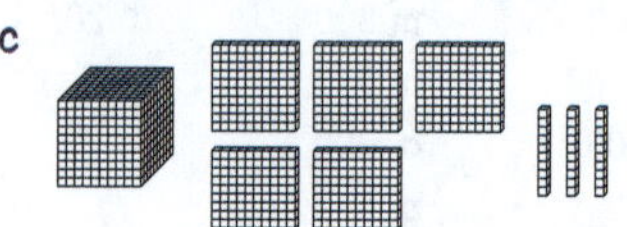
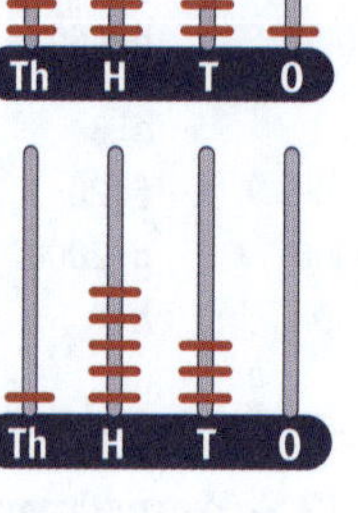

d
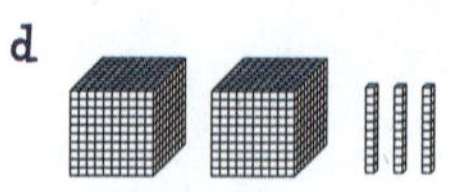
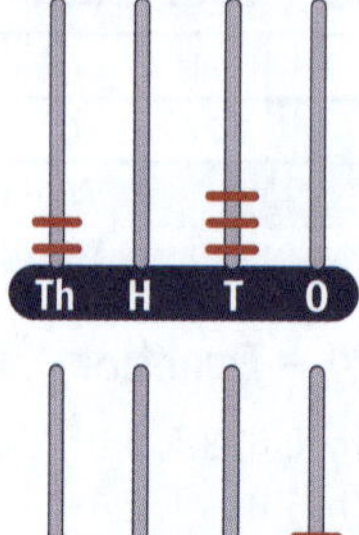

e
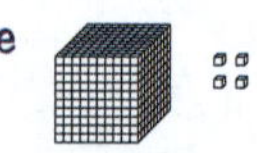

f
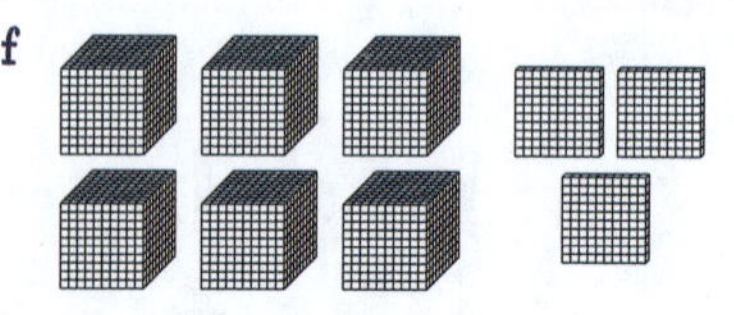
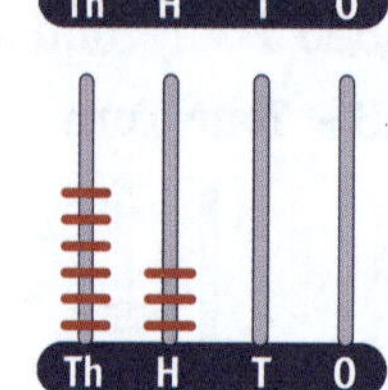

Ordering Four-digit Numbers

Page 29 – Your Turn

1 a 3788, 5253, 8873, 9691 b 2440, 6440, 6740, 9894
2 a 7791, 3719, 2020, 1936 b 7179, 2904, 1973, 1503

Page 30 – Practice

1 a 1, 5, 2, 3, 4 b 5, 4, 3, 2, 1 c 5, 1, 4, 3, 2 d 1, 5, 4, 2, 3
2 a 2, 5, 1, 4, 3 b 2, 1, 3, 4, 5 c 3, 1, 5, 4, 2 d 1, 4, 5, 2, 3

Greater Than, Equal To, Less Than

Page 31 – Your Turn

a True b True c True d False e False

Page 32 – Practice

1 a < b > c = d > e > f < g > h <
2 a is greater than b is greater than c is greater than d is less than e is equal to f is greater than g is less than h is equal to i is greater than

3 a 19, 16, 18
b 94, 95, 98
c 41, 17, 9, 82, 73
d 392, 497, 463
e 127, 127, 127
f 147, 214, 317, 234, 324
g 607, 607, 607

4 a > b = c > d < e < f > g < h < i > j < k <

5 Adult to check
6 Adult to check
7 Adult to check

8 a True b True c False d True e False f True g True h True i True

9 a 620, 262, 266, 594
b 714, 714, 714
c 224, 142, 242
d 928, 928, 928
e 5231, 1352, 1325, 1523
f 9729, 9772
g 3439, 3394
h 8352, 8352, 8352

10 a < b > c = d < e < f > g > h > i >

Rounding to 100 and 1000

Page 35 – Your Turn

1 a 700 b 2400
2 a 10 000 b 24 000

Page 36 – Practice

1 a 1700 b 300 c 1000 d 1000 e 2600 f 1600 g 4900 h 600
2 a 3000 b 5000 c 7000 d 6000 e 3000
3 a 4900, 5000 b 3100, 3000 c 8400, 8000 d 4700, 5000 e 6500, 6000

Five-digit Numbers

Page 37 – Your Turn

1 24 957, 42 056, 73 950, 26 435, 17 537
2 a seventy-three thousand, four hundred and twenty
b nineteen thousand, five hundred and ninety-six
c eighty-four thousand, two hundred and fifty-three

Page 38 – Practice

1 a forty-nine thousand, three hundred and fifty
b fifty-seven thousand, four hundred and twenty
c sixty-two thousand, nine hundred and forty-three
d eighty-one thousand, four hundred and sixty-two
e ninety-three thousand, two hundred and fifty-eight
f sixteen thousand, four hundred and twenty-two
2 a 53 847 fifty-three thousand, eight hundred and forty-seven
b 71 329 seventy one thousand, three hundred and twenty-nine
c 94 382 sixty-two thousand, four hundred and seventy-one
d 81 432 seventy-six thousand, two hundred and eighteen
e 76 218 eighty-one thousand, four hundred and thirty-two
f 62 471 ninety-four thousand, three hundred and eighty-two
g 31 240 thirty-one thousand, two hundred and forty

Place Value to 100 000

Page 39 – Your Turn

73 608 89 243 63 825 74 103
61 000 47 090 32 361 24 140

CATCH UP MATHS YEAR 4 BOOK A © PASCAL PRESS ISBN: 9781925726145

Page 40 – Practice

1 a 20 431 b 94 736 c 35 655 d 65 218

2

	Number	TT	Th	H	T	O
a	67 138	6	7	1	3	8
b	15 840	1	5	8	4	0
c	71 459	7	1	4	5	9
d	25 043	2	5	0	4	3
e	75 458	7	5	4	5	8

3

	TT	Th	H	T	O
a		5 000			
b					2
c				10	
d	80 000				
e					5
f				80	

The Value of Numbers to 100 000

Page 41 – Your Turn

28 491 10 398 24 937 82 358

90 009 12 683 73 469 44 444

35 293 47 247 64 293

Page 42 – Practice

1 Adult to check

2 a 2487 b 87 c 0 d 7

3 a 30 000 b 10 c 40 000 d 2000 e 0

f 8 g 90 000 h 70 i 400 j 3

k 1000 l 600 m 20 n 80 000 o 0

p 40 000 q 4000 r 50 s 7000

Number Expanders and Five-digit Numbers

Page 43 – Your Turn

7	TT	3	Th	4	H	6	T	3	O

7	3	Th	4	H	6	T	3	O

7	3	4	H	6	T	3	O

7	3	4	6	T	3	O

7	3	4	6	3	O

Page 44 – Practice

1 a 40 283 b 27 539 c 74 681 d 90 724 e 83 745

2

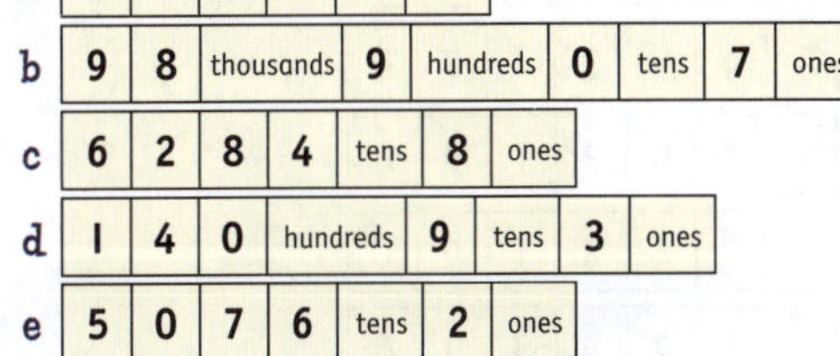

a 3 0 9 6 3 ones

b 9 8 thousands 9 hundreds 0 tens 7 ones

c 6 2 8 4 tens 8 ones

d 1 4 0 hundreds 9 tens 3 ones

e 5 0 7 6 tens 2 ones

3 a 52 692 b 20 360 c 28 253

Expanded Five-digit Numbers

Page 45 – Your Turn

a 57 = 50 + 7
= 5 tens + 7 ones

b 159 = 100 + 50 + 9
= 1 hundred + 5 tens + 9 ones

Page 46 – Practice

1 a 50 + 8
b 700 + 20 + 6
c 60 + 4
d 5
e 9000 + 300 + 70 + 3
f 9000 + 700 + 70 + 1
g 90 000 + 5000 + 200 + 50 + 6
h 30 000 + 300 + 50 + 7
i 100 + 20 + 8
j 70 000 + 2000 + 400 + 20 + 1

2 a 20 000 + 5000 + 200 + 90 + 5
b 60 000 + 8000 + 400 + 90 + 3
c 70 000 + 4000 + 2
d 10 000 + 400 + 90 + 3
e 10 000 + 8000 + 50 + 6
f 80 000 + 9000 + 700 + 3

Modelling Five-digit Numbers

Page 47 – Your Turn

a 72, 3, 4, 5 b 63 162 c 21 519 d 32, 4, 8, 1

Page 48 – Practice

1 a 94 620 b 67 809

2

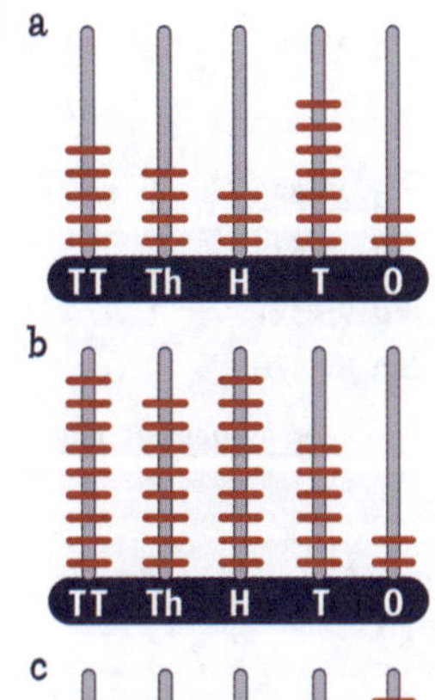

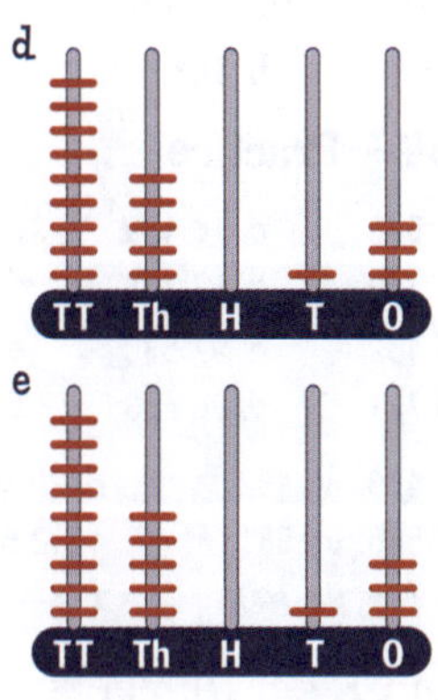

c

TT Th H T O

3

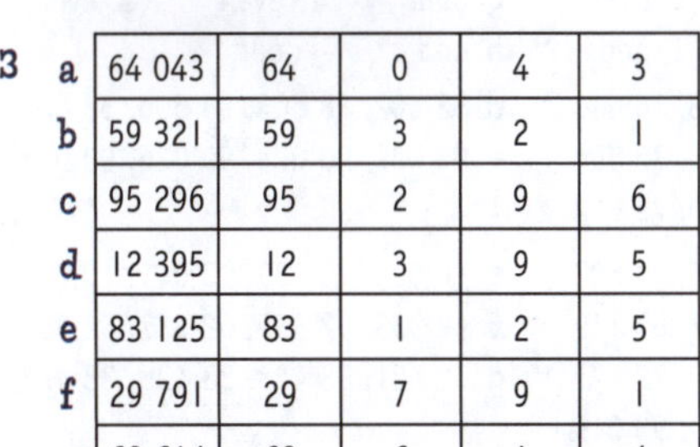

a	64 043	64	0	4	3
b	59 321	59	3	2	1
c	95 296	95	2	9	6
d	12 395	12	3	9	5
e	83 125	83	1	2	5
f	29 791	29	7	9	1
g	69 614	69	6	1	4

Ordering Five-digit Numbers

Page 49 – Your Turn

a descending b descending c ascending d ascending e ascending

Page 50 – Practice

1 a 13 578, 27 942, 62 435, 84 259
b 13 524, 21 531, 25 311, 31 524
c 57 462, 62 457, 72 624, 75 426
d 49 326, 58 385, 62 572, 85 358
e 27 345, 35 472, 42 573, 54 372

2 a 3, 2, 5, 4, 1 b 5, 4, 3, 1, 2 c 5, 1, 2, 3, 4 d 3, 4, 1, 5, 2

1 WHOLE NUMBERS CONTINUED

Greater Than, Equal To, Less Than: Numbers to 100 000

Page 51 – Your Turn

a No b Yes c Yes

Page 52 – Practice

1 a < c < e > g < i > k = m =
b = d < f < h < j < l <

2 a is less than c is less than
b is greater than d is equal to

3 Adult to check

Largest and Smallest Five-digit Numbers

Page 53 – Your Turn

a 27 b 103 c 577 d 2349 e 34 578

Page 54 – Practice

1 a 24 578 c 10 489 e 12 457 g 25 789 i 30 467
b 20 349 d 12 348 f 12 345 h 14 678

2 a 74 432 c 97 530 e 84 420 g 95 321 i 97 411
b 88 721 d 96 430 f 85 432 h 87 420

3 a 23 458, 85 432 d 20 468, 86 420 g 58 999, 99 985
b 13 566, 66 531 e 13 579, 97 531
c 40 789, 98 740 f 34 779, 97 743

Odd and Even Numbers to 100 000

Page 55 – Your Turn

Circled numbers: 34 686, 59 992, 52 430, 42 034, 95 018
Crossed-out numbers: 41 323, 24 985, 62 431, 57 597, 75 999

Page 56 – Practice

1 a odd c odd e even g odd i even k even
b even d even f odd h odd j odd

2 a 16 434, 16 436, 16 438, 16 440 d 58 602, 58 604, 58 606, 58 608
b 36 590, 36 592, 36 594, 36 596 e 90 022, 90 024, 90 026, 90 028
c 72 056, 72 058, 72 060, 72 062

3 a 59 601, 59 603, 59 605, 59 607 e 10 037, 10 039, 10 041, 10 043
b 61 113, 61 115, 61 117, 61 119 f 52 333, 52 335, 52 337, 52 339
c 72 465, 72 467, 72 469, 72 471 g 38 421, 38 423, 38 425, 38 427
d 90 009, 90 011, 90 013, 90 015

Rounding to 10 000

Page 57 – Your Turn

a 40 000 b 50 000 c 80 000

Page 58 – Practice

1 a 40 000 c 60 000 e 80 000 g 90 000
b 70 000 d 90 000 f 80 000 h 50 000

2 a 37 500, 37 500, 38 000, 40 000 d 26 610, 26 600, 27 000, 30 000
b 71 540, 71 500, 72 000, 70 000 e 30 430, 30 400, 30 000, 30 000
c 84 970, 85 000, 85 000, 80 000

Whole Numbers Review Page 59

1 a six hundred and seventy-five
b four hundred and twenty-nine
c three thousand and fifty-six
d two thousand, three hundred and fifty
e seven thousand, four hundred and thirty-eight
f eight thousand, five hundred and three
g twenty-six thousand, five hundred and ninety
h thirty-seven thousand, four hundred and twenty-nine

2

	Number	TT	Th	H	T	O
a	79	0	0	0	7	9
b	838	0	0	8	3	8
c	903	0	0	9	0	3
d	1430	0	1	4	3	0
e	2574	0	2	5	7	4
f	3827	0	3	8	2	7
g	12 507	1	2	5	0	7
h	35 639	3	5	6	3	9
i	40 256	4	0	2	5	6
j	57 007	5	7	0	0	7

3 Circled: 836, 842, 3863, 2862, 19 837, 5815, 29 836

4 Circled: 24 293, 4684, 4004, 94 362, 4937, 34 876

5 a hundreds f ten thousands k ones
b hundreds g thousands l ones
c thousands h hundreds m tens
d thousands i thousands n hundreds
e ten thousands j hundreds o tens

6 a 60 304 c 478 e 80 063 g 40 068
b 28 031 d 5070 f 53 002 h 70 590

7 a 20 f 200 k 2000 p 20 000
b 2 g 2 l 20 000 q 2000
c 20 h 20 000 m 2 r 200
d 2 i 2000 n 200
e 2000 j 200 o 2000

8
a | | 2 | thousands | 7 | hundreds | 5 | tens | 9 | ones |
b | 5 | 5 | thousands | 3 | hundreds | 2 | tens | 9 | ones |
c | | 3 | 2 | 4 | 8 | ones |
d | 4 | ten thousands | 6 | thousands | 8 | hundreds | 1 | tens | 9 | ones |
e | 7 | 2 | 3 | hundreds | 8 | tens | 5 | ones |
f | | 1 | thousands | 0 | hundreds | 3 | tens | 7 | ones |

9 a
1	TT	7	Th	4	H	3	T	9	0
1	7	Th	4	H	3	T	9	0
1	7	4	H	3	T	9	0
1	7	4	3	T	9	0
1	7	4	3	9	0

b
9	TT	8	Th	4	H	3	T	1	0
9	8	Th	4	H	3	T	1	0
9	8	4	H	3	T	1	0
9	8	4	3	T	1	0
9	8	4	3	1	0

CATCH UP MATHS YEAR 4 BOOK A © PASCAL PRESS ISBN: 9781925726145

10 **a** 600 + 30 + 2
b 200 + 7
c 900 + 10
d 1000 + 400 + 30 + 8
e 3000 + 500 + 80 + 9
f 7000 + 100 + 30 + 6
g 20 000 + 3000 + 400 + 50 + 8
h 60 000 + 2000 + 400 + 10
i 90 000 + 7
j 40 000 + 2000 + 70

11 **a**
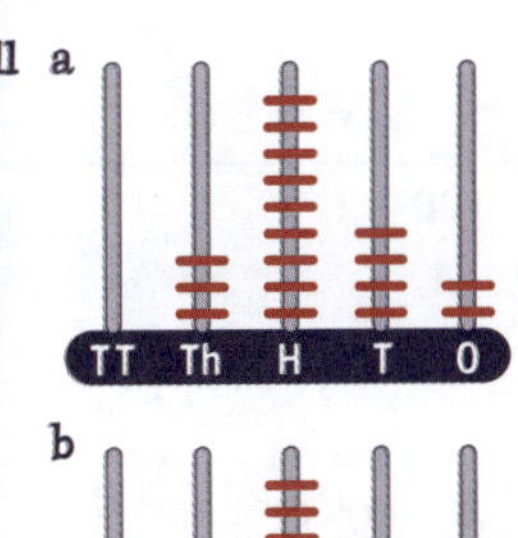

b
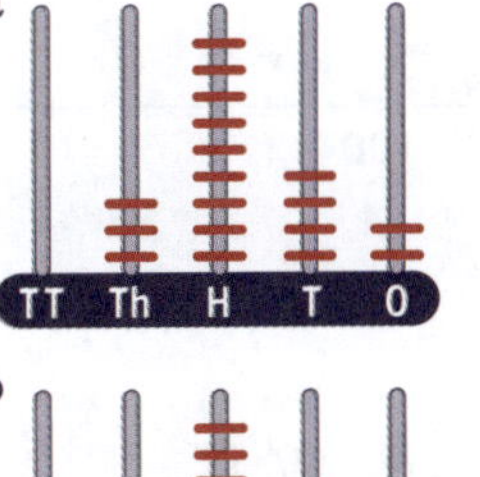

c
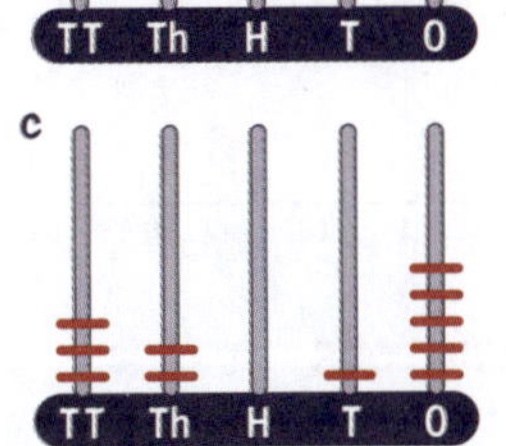

d
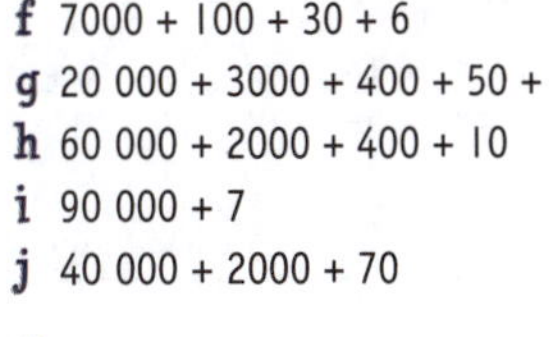

e
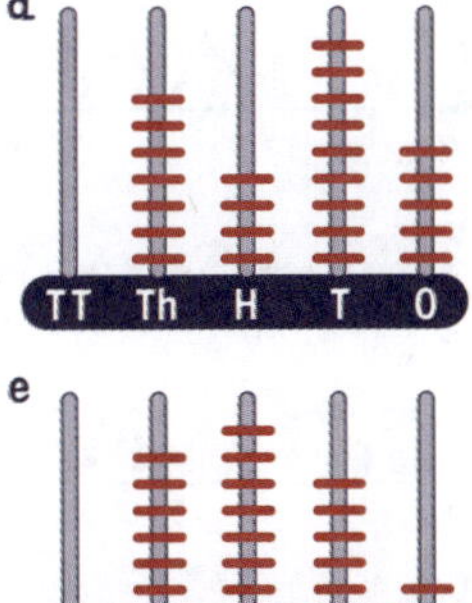

f
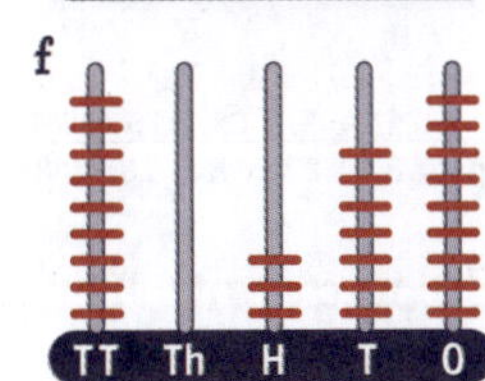

12 **a** 149, 325, 473, 632, 836
b 248, 632, 1432, 2593, 3025
c 103, 136, 316, 361, 631
d 10 425, 13 524, 27 626, 35 927, 41 341

13 **a** 62 538, 34 273, 8439, 1249, 763
b 84 327, 74 283, 42 438, 32 495, 14 327
c 84 823, 44 382, 38 428, 34 284, 28 434
d 63 294, 49 236, 36 492, 34 692, 24 936

14 **a** 1063, 1163, 1263, 1363
b 549, 649, 749, 849
c 632, 532, 432, 332
d 847, 747, 647, 547

15 **a** > **b** > **c** = **d** < **e** > **f** > **g** < **h** > **i** = **j** < **k** > **l** <

16

	Number	Nearest 10	Nearest 100	Nearest 1000	Nearest 10 000
a	13 427	13 430	13 400	13 000	10 000
b	49 238	49 240	49 200	50 000	50 000
c	56 502	56 500	56 500	57 000	60 000
d	60 348	60 350	60 300	60 000	60 000
e	99 037	99 040	99 000	99 000	100 000
f	74 295	74 300	74 300	74 000	70 000
g	86 697	86 700	86 700	87 000	90 000

17 **a** 246, 642
b 237, 732
c 368, 863
d 1378, 8731
e 4589, 9854
f 4056, 6540
g 20 356, 65 320
h 10 479, 97 410
i 45 678, 87 654

18 Odd: 243, 1343, 741, 8915, 71 245, 45 933, 21, 257, 7, 3
25 245, 93, 93 001
Even: 4, 16, 73 248, 51 424, 634, 58, 2462, 18, 838, 6410

19 one-digit (red): 9, 2, 8, 7
two-digit (blue): 27, 35, 82, 13, 98
three-digit (green): 492, 164, 613, 710, 103
four-digit (purple): 3524, 8211, 7363, 3711, 4940
five-digit (orange): 24 293, 41 321, 78 920, 67 341, 40 000

2 ADDITION

Addition Three or More Single-digit Numbers

Page 66 – Your Turn

a 6 + 4 + 5
= 6 + 4 + 5
= 10 + 5
= 15

b 1 + 8 + 9
= 1 + 9 + 8
= 10 + 8
= 18

c 6 + 3 + 7
= 3 + 7 + 6
= 10 + 6
= 16

d 5 + 5 + 4
= 5 + 5 + 4
= 10 + 4
= 14

e 1 + 9 + 1
= 1 + 9 + 1
= 10 + 1
= 11

Page 67 – Practice

1 **a** = 2 + 8 + 7
= 10 + 7
= 17

b = 4 + 6 + 8
= 10 + 8
= 18

c = 3 + 7 + 8
= 10 + 8
= 18

d = 9 + 1 + 7
= 10 + 7
= 17

e = 5 + 5 + 9
= 10 + 9
= 19

2 **a** = 5 + 5 + 6 + 2
= 10 + 8
= 18

b = 9 + 1 + 4 + 3
= 10 + 7
= 17

c = 5 + 5 + 6 + 4
= 10 + 10
= 20

d = 2 + 8 + 3 + 4
= 10 + 7
= 17

e = 4 + 6 + 1 + 4
= 10 + 5
= 15

Relating Addition to Subtraction

Page 68 – Your Turn

a 5 + 3 = 8
3 + 5 = 8
8 – 3 = 5
8 – 5 = 3

Page 69 – Practice

1 **a** 11 **b** 3 **c** 5 **d** 19 **e** 10

2 **a** 8, 24, 16
b 7, 7, 25
c 13, 6, 6
d 9, 9, 17
e 15, 6, 6
f 4 + 12 = 16
16 – 12 = 4
16 – 4 = 12
g 4 + 19 = 23
23 – 19 = 4
23 – 4 = 19
h 9 + 11 = 20
20 – 11 = 9
20 – 9 = 11

3 **a** 3 + 9 = 12
9 + 3 = 12
12 – 9 = 3
12 – 3 = 9
b 18 + 2 = 20
2 + 18 = 20
20 – 2 = 18
20 – 18 = 2
c 15 + 2 = 17
2 + 15 = 17
17 – 2 = 15
17 – 15 = 2
d 7 + 11 = 18
11 + 7 = 18
18 – 11 = 7
18 – 7 = 11
e 8 + 16 = 24
16 + 8 = 24
24 – 16 = 8
24 – 8 = 16
f 22 + 14 = 36
14 + 22 = 36
36 – 14 = 22
36 – 22 = 14
g 33 + 15 = 48
15 + 33 = 48
48 – 15 = 33
48 – 33 = 15
h 46 + 6 = 52
6 + 46 = 52
52 – 6 = 46
52 – 46 = 6

4 **a** 19 + 18 = 37
18 + 19 = 37
37 – 18 = 19
37 – 19 = 18
b 15 + 13 = 28
13 + 15 = 28
28 – 15 = 13
28 – 13 = 15
c 15 + 41 = 56
41 + 15 = 56
56 – 41 = 15
56 – 15 = 41
d 18 + 41 = 59
41 + 18 = 59
59 – 18 = 41
59 – 41 = 18
e 22 + 51 = 73
51 + 22 = 73
73 – 51 = 22
73 – 22 = 51
f 36 + 33 = 69
33 + 36 = 69
69 – 33 = 36
69 – 36 = 33
g 52 + 21 = 73
21 + 52 = 73
73 – 21 = 52
73 – 52 = 21
h 63 + 21 = 84
21 + 63 = 84
84 – 21 = 63
84 – 63 = 21

2 ADDITION CONTINUED

i 26 + 126 = 152
126 + 26 = 152
152 − 26 = 126
152 − 126 = 26

j 115 + 132 = 247
132 + 115 = 247
247 − 115 = 132
247 − 132 = 115

k 203 + 162 = 365
162 + 203 = 365
365 − 162 = 203
365 − 203 = 162

Addition without Trading

Page 72 – Your Turn

a 92 b 88 c 83 d 97 e 98

Page 73 – Practice

1 a 39 b 57 c 89 d 86 e 87 f 98 g 79

2 a 86 b 79 c 79 d 96 e 77 f 88 g 95 h 79 i 47 j 88 k 88 l 52 m 97 n 86 o 95 p 67 q 79 r 99 s 95

3 a 52 b 31 c 45 d 23 e 20 f 27 g 20

4 a 60 + 10, 70 b 80 + 20, 100 c 80 + 20, 100 d 70 + 20, 90 e 50 + 20, 70 f 70 + 20, 90 g 40 + 30, 70 h 60 + 30, 90 i 40 + 40, 80 j 40 + 20, 60 k 50 + 60, 110

5 a 978 b 979 c 685 d 987 e 979 f 694 g 559 h 988

6 a 280 b 233 c 231 d 121 e 572

7 a 260 + 20, 280 b 740 + 20, 760 c 590 + 10, 600

Addition with Trading

Page 76 – Your Turn

a 173 b 91 c 110

Page 77 – Practice

1 a 114 b 123 c 86 d 91 e 61 f 143 g 141

2 a 70 b 120 c 127 d 133 e 140 f 143 g 117 h 146 i 138 j 111 k 132 l 84 m 100 n 74 o 103 p 123 q 182 r 82 s 84

3 a 34 b 47 c 54 d 49 e 38 f 45 g 67 h 86 i 25 j 59 k 13 l 15 m 57 n 93 o 65 p 46 q 58 r 55 s 36

4 a 761 b 990 c 426 d 851 e 696 f 981 g 945 h 482 i 572 j 980 k 901

5 a 209 b 135 c 218 d 218 e 219 f 548 g 769 h 369 i 298 j 109 k 186 l 528 m 106 n 367

Jump Strategy to Solve Addition

Page 80 – Your Turn

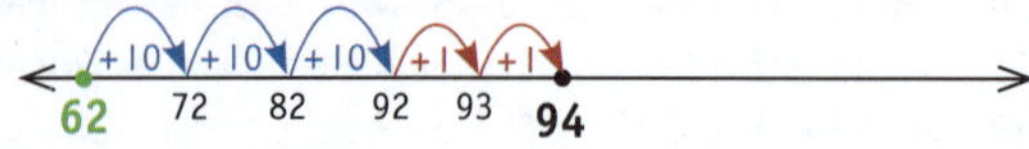

Page 81 – Practice

a 63

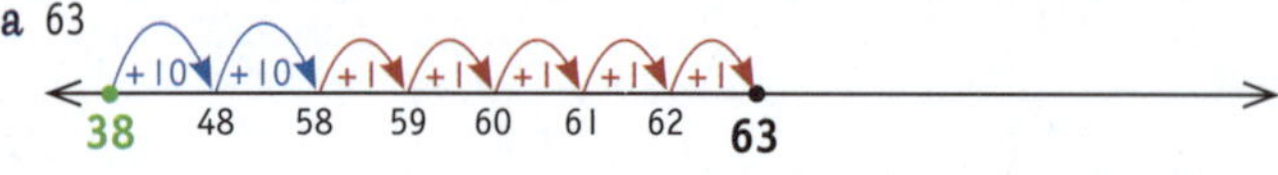

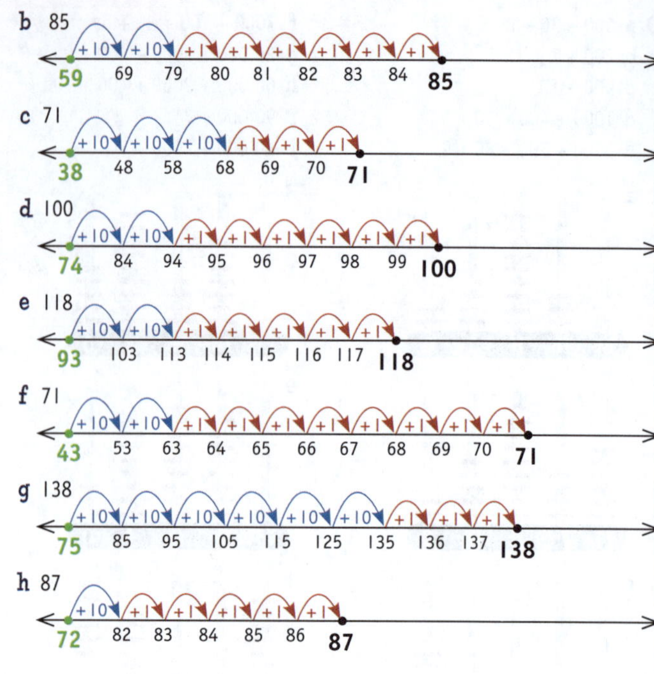

Jump Strategy with Tens and Hundreds

Page 82 – Your Turn

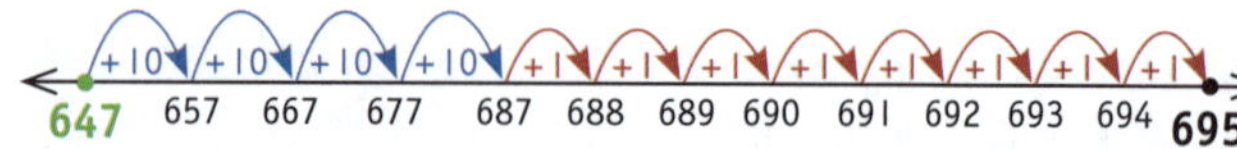

Page 83 – Practice

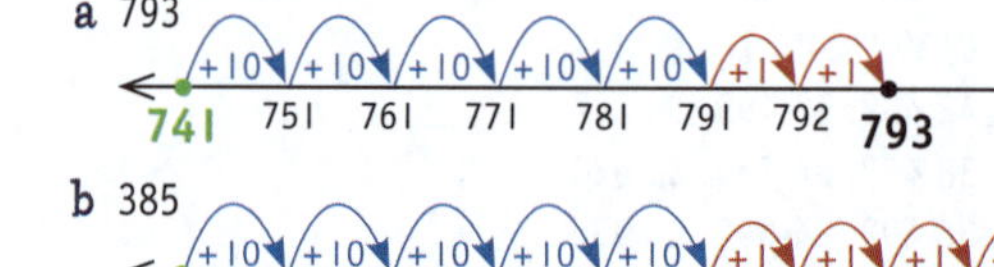

c 452
+10 +1 +1 +1
439 449 450 451 452

d 307
+10 +10 +10 +10 +10 +10 +10 +1 +1 +1
234 244 254 264 274 284 294 304 305 306 307

e 972
+10 +10 +10 +10 +10 +1 +1 +1 +1 +1 +1 +1 +1
914 924 934 944 954 964 965 966 967 968 969 970 971 972

f 477
+10 +10 +10 +10 +10 +10 +1 +1 +1 +1 +1
412 422 432 442 452 462 472 473 474 475 476 477

g 600
+10 +10 +10 +1 +1 +1 +1 +1 +1 +1 +1
562 572 582 592 593 594 595 596 597 598 599 600

h 676
+10 +10 +10 +10 +10 +1 +1 +1 +1 +1
621 631 641 651 661 671 672 673 674 675 676

Jump Strategy with Three- and Four-digit Numbers

Page 84 – Your Turn

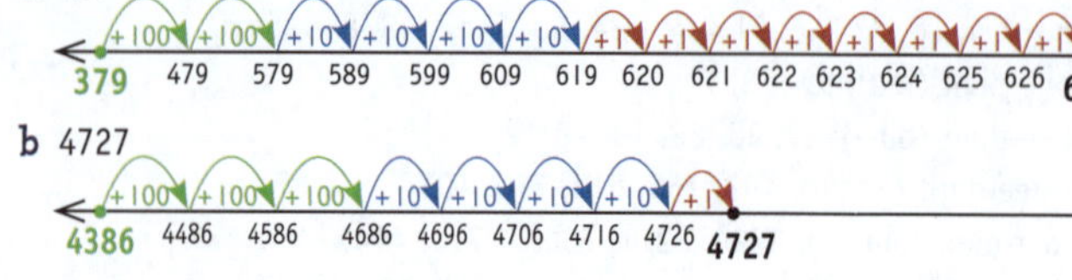

CATCH UP MATHS YEAR 4 BOOK A © PASCAL PRESS ISBN: 9781925726145

Page 85 – Practice

a 489

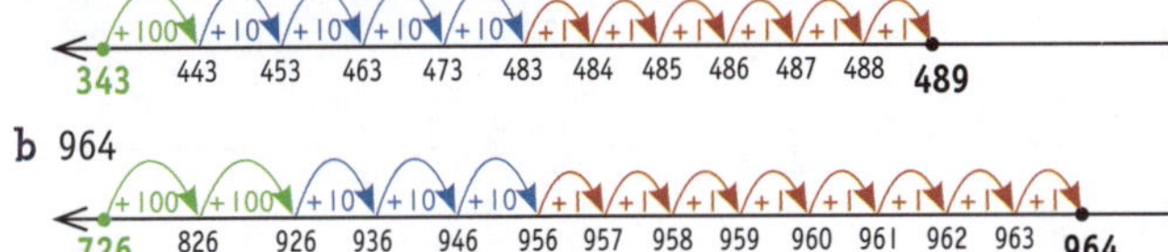

b 964

c 935

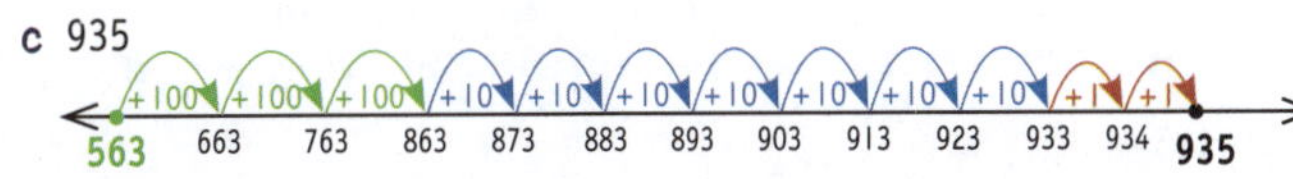

d 1274

+100 +100 +100 +100 +10 +10 +10 +10 +1 +1

832 932 1032 1132 1232 1242 1252 1262 1272 1273 1274

e 2492

+100 +10 +10 +1 +1 +1 +1 +1 +1 +1 +1 +1 +1

2364 2464 2474 2484 2485 2486 2487 2488 2489 2490 2491 2492

f 5885

+100 +100 +10 +10 +10 +10 +10 +1 +1 +1

5632 5732 5832 5842 5852 5862 5872 5882 5883 5884 5885

h 7866

+100 +100 +100 +10 +10 +10 +10 +1 +1 +1 +1 +1 +1

7520 7620 7720 7820 7830 7840 7850 7860 7861 7862 7863 7864 7865 7866

i 7199

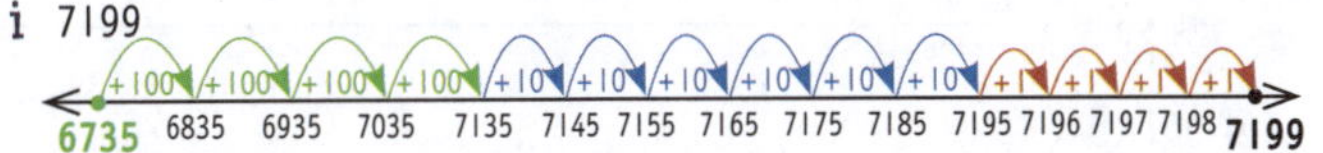

Split Strategy to Add Two-digit Numbers

Page 86 – Your Turn

a 52 + 33 = 50 + 30 = 80 and 2 + 3 = 5
= 80 + 5
= 85

Page 87 – Practice

1 a 58 + 11 = 50 + 10 = 60 and 8 + 1 = 9
= 60 + 9
= 69

b 63 + 24 = 60 + 20 = 80 and 3 + 4 = 7
= 80 + 7
= 87

c 81 + 14 = 80 + 10 = 90 and 1 + 4 = 5
= 90 + 5
= 95

d 75 + 23 = 70 + 20 = 90 and 5 + 3 = 8
= 90 + 8
= 98

e 81 + 15 = 80 + 10 = 90 and 1 + 5 = 6
= 90 + 6
= 96

f 43 + 25 = 40 + 20 = 60 and 3 + 5 = 8
= 60 + 8
= 68

g 52 + 36 = 50 + 30 = 80 and 2 + 6 = 8
= 80 + 8
= 88

Split Strategy to Add Three-digit Numbers

Page 88 – Your Turn

a 314 + 35 = 300 and 10 + 30 = 40 and 4 + 5 =9
= 300 + 40 + 9
= 349

b 715 + 23 = 700 and 10 + 20 = 30 and 5 + 3 = 8
= 700 + 30 + 8
= 738

Page 89 – Practice

1 a 432 + 43 = 400 and 30 + 40 = 70 and 2 + 3 = 5
= 400 + 70 + 5
= 475

b 472 + 17 = 400 and 70 + 10 = 80 and 2 + 7 = 9
= 400 + 80 + 9
= 489

c 541 + 35 = 500 and 40 + 30 = 70 and 1 + 5 = 6
= 500 + 70 + 6
= 576

d 632 + 47 = 600 and 30 + 40 = 70 and 2 + 7 = 9
= 600 + 70 + 9
= 679

e 706 + 52 = 700 and 0 + 50 = 50 and 6 + 2 = 8
= 700 + 50 + 8
= 758

f 902 + 83 = 900 and 0 + 80 = 80 and 2 + 3 = 5
= 900 + 80 + 5
= 985

Split Strategy with Three- and Four-digit Numbers

Page 90 – Your Turn

5672 + 213
= 5000 and 600 + 200 = 800 and 70 + 10 = 80 and 2 + 3 = 5
= 5000 + 800 + 80 + 5
= 5885

Page 91 – Practice

1 a 500 and 40 + 30 = 70 and 1 + 3 = 4
= 500 + 70 + 4
= 574

b 700 and 20 + 30 = 50 and 7 + 2 = 9
= 700 + 50 + 9
= 759

c 1000 and 300 + 400 = 700 and 50 + 20 = 70 and 2 + 3 = 5
= 1000 + 700 + 70 + 5
= 1775

d 2000 and 400 + 200 = 600 and 60 + 20 = 80 and 8 + 1= 9
= 2000 + 600 + 80 + 9
= 2689

e 8000 and 400 + 200 = 600 and 30 + 40 = 70 and 5 + 3 = 8
= 8000 + 600 + 70 + 8
= 8678

Rounding Money

Page 92 – Your Turn

a $10.20 b $18.50 c $21.15

Page 93 – Practice

1 a $7.25 b $9.10 c $4.15 d $3.30 e $1.20 f $4.45 g $8.65 h $9.90 i $4.70 j $3.80 k $2.55 l $6.40 m $1.05 n $5.55 o $7.90

2 a $24.70 b $36.55 c $78.65 d $81.85 e $53.00 f $135.70 g $411.25 h $536.75 i $2742.45 j $1352.50

3 a $5.15, $4.85 b $7.50, $2.50 c $8.65, $1.35 d $4.90, $5.10 e $3.30, $6.70 f $0.80, $9.20 g $1.25, $8.75 h $4.60, $5.40 i $6.50, $3.50 j $9.55, $0.45

4 a $0.75, $1.25 b $3.25, $1.75 c $15.50, $4.50 d $12.65, $7.35 e $7.30, $2.70 f $3.40, $1.60 g $1.25, $0.75 h $14.85, $5.15 i $3.75, $6.25 j $8.50, $11.50

2 ADDITION continued

5 a $2.47 + $2.47 + $2.47 + $2.47 = $9.88, $10.10
b $3.99 + $3.99 + $3.99 = $11.97, $8.05
c 77c + 77c + 77c + 77c + 77c + 77c = $4.62, $5.40

6 a $2.45, $7.55 b $7.35, $2.65 c $9.80, $10.20 d $2.45, $2.55 e $4.90, $15.10 f $7.35, $12.65 g $12.25, $7.75 h $2.45, $17.55

Addition Review Page 96

1 a 2 + 8 + 7 = 10 + 7 = 17
b 6 + 4 + 3 = 10 + 3 = 13
c 1 + 9 + 6 = 10 + 6 = 16
d 7 + 3 + 1 + 8 = 10 + 9 = 19
e 9 + 1 + 4 + 4 = 10 + 8 = 18
f 6 + 4 + 1 + 3 = 10 + 4 = 14

2 a 8, 8, 14 b 18, 9, 9 c 16, 7, 7
d 5 + 19 = 24, 24 − 5 = 19, 24 − 19 = 5
e 9 + 16 = 25, 25 − 9 = 16, 25 − 16 = 9
f 4 + 17 = 21, 21 − 17 = 4, 21 − 4 = 17

3 a 3 + 12 = 15, 12 + 3 = 15, 15 − 12 = 3, 15 − 3 = 12
b 32 + 20 = 52, 20 + 32 = 52, 52 − 20 = 32, 52 − 32 = 20
c 27 + 46 = 73, 46 + 27 = 73, 73 − 46 = 27, 73 − 27 = 46

4 a 23 + 17 = 40, 17 + 23 = 40, 40 − 17 = 23, 40 − 23 = 17
b 15 + 19 = 34, 19 + 15 = 34, 34 − 19 = 15, 34 − 15 = 19
c 123 + 254 = 377, 254 + 123 = 377, 377 − 254 = 123, 377 − 123 = 254
d 116 + 246 = 362, 246 + 116 = 362, 362 − 246 = 116, 362 − 116 = 246

5 a 99 b 94 c 99 d 99 e 759 f 898 g 858 h 999 i 897

6 a 107 b 161 c 141 d 490 e 780 f 1070

7 a 24 b 36 c 45 d 32 e 262 f 315 g 527 h 438 i 149

8 a 50, 30, 80 b 60, 30, 90 c 90, 40, 130 d 40, 40, 80 e 30, 80, 110 f 370, 80, 450 g 650, 20, 670 h 540, 30, 570 i 590, 20, 610 j 750, 50, 800

9 a 73 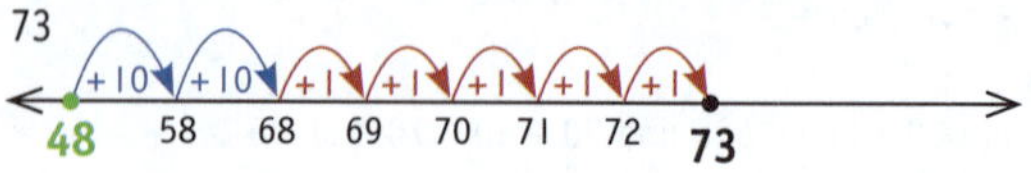
+10 +10 +1 +1 +1 +1 +1
48 58 68 69 70 71 72 73

b 80
+10 +10 +1
59 69 79 80

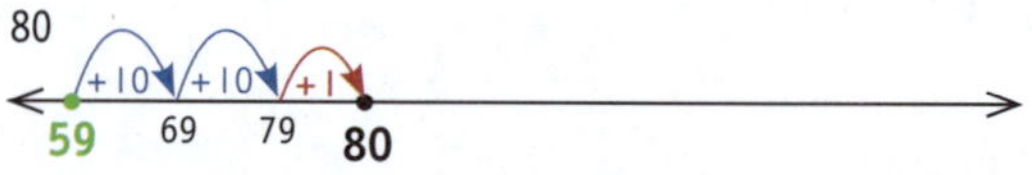

c 125 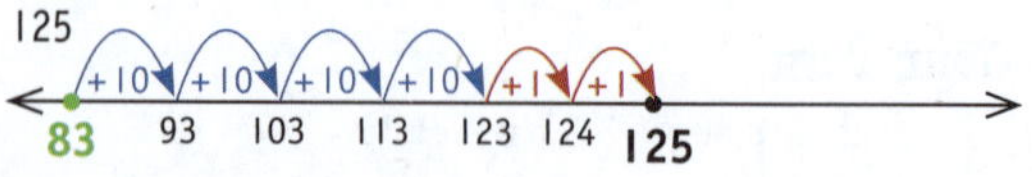
+10 +10 +10 +10 +1 +1
83 93 103 113 123 124 125

d 59
+10 +1 +1
47 57 58 59

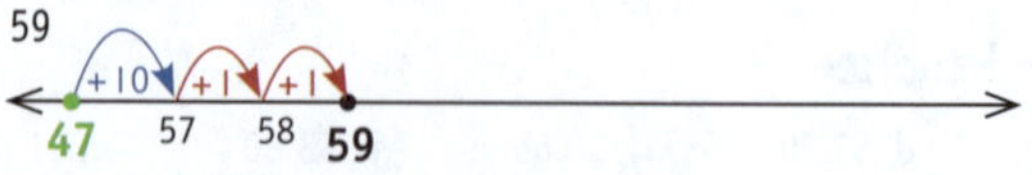

e 757 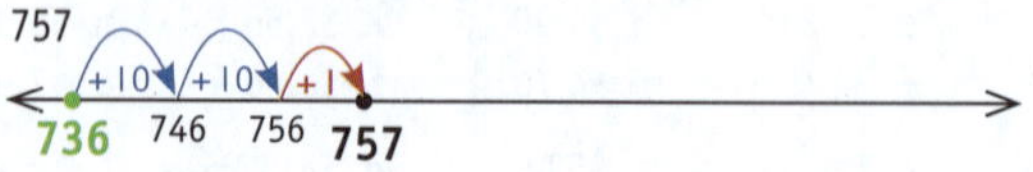
+10 +10 +1
736 746 756 757

f 581
+10 +10 +10 +1 +1 +1
548 558 568 578 579 580 581

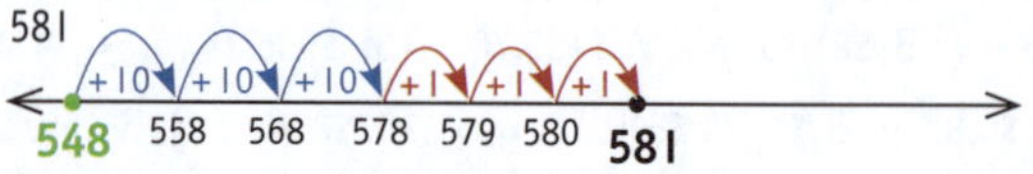

g 789 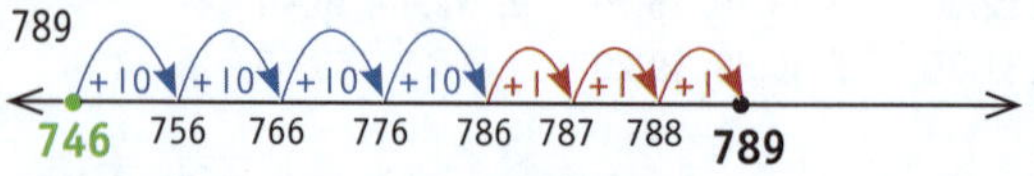
+10 +10 +10 +10 +1 +1 +1
746 756 766 776 786 787 788 789

h 1004
+100 +100 +100 +100 +10 +10 +1
583 683 783 883 983 993 1003 1004

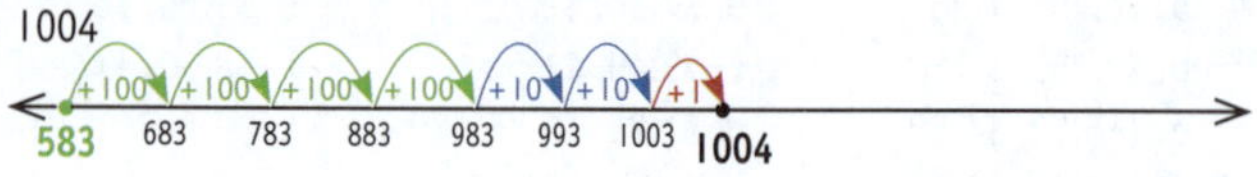

i 2097 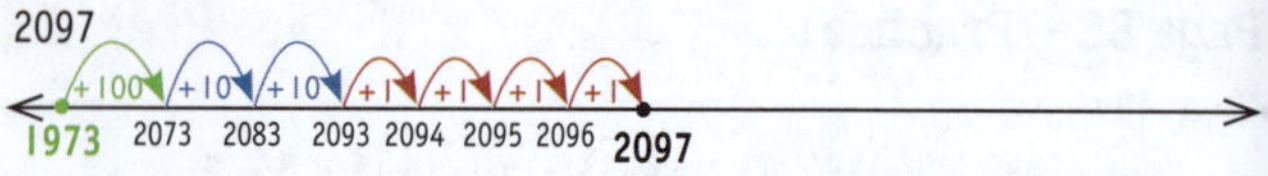
+100 +10 +10 +1 +1 +1 +1
1973 2073 2083 2093 2094 2095 2096 2097

j 3677
+100 +10 +10 +10 +10 +10 +1 +1 +1
3524 3624 3634 3644 3654 3664 3674 3675 3676 3677

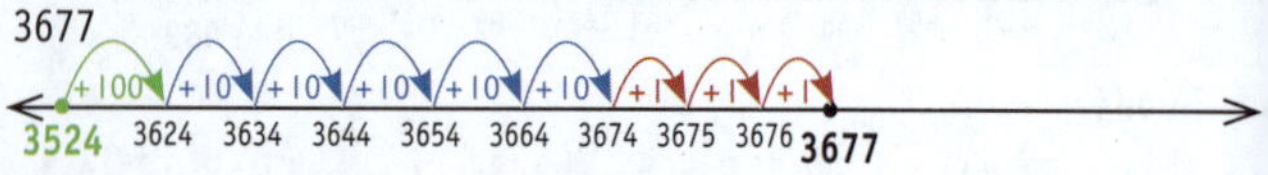

k 5508 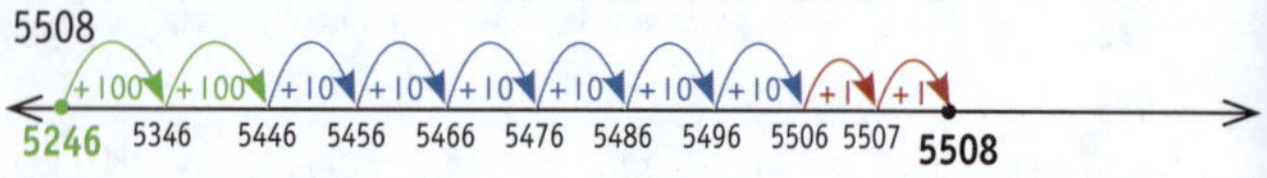
+100 +100 +10 +10 +10 +10 +10 +10 +1 +1
5246 5346 5446 5456 5466 5476 5486 5496 5506 5507 5508

l 183
+10 +10 +10 +1 +1
151 161 171 181 182 183

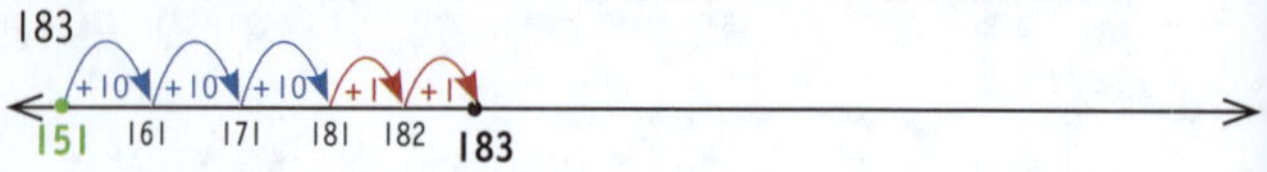

m 23 + 233 = 256 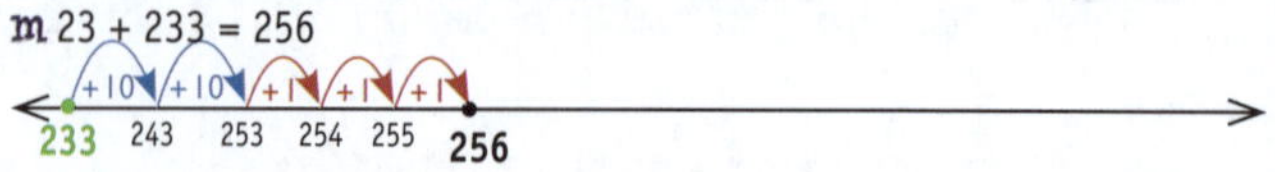
+10 +10 +1 +1 +1 +1
233 243 253 254 255 256

n 264
+10 +1 +1 +1 +1 +1 +1
248 258 259 260 261 262 263 264

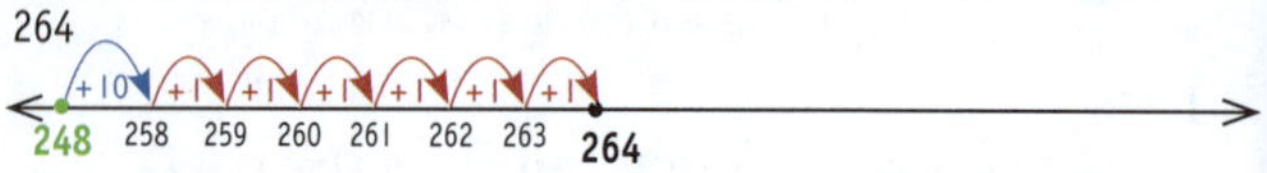

o 604 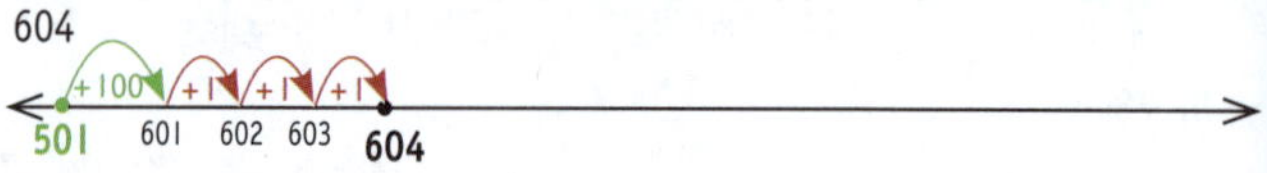
+100 +1 +1 +1
501 601 602 603 604

p 426
+100 +100 +10
216 316 416 426

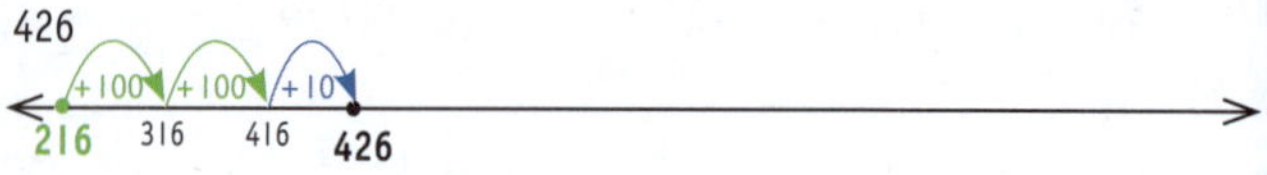

q 5077 
+100 +10 +10 +10 +10 +1 +1 +1 +1 +1
4932 5032 5042 5052 5062 5072 5073 5074 5075 5076 5077

r 6493
+100 +100 +100 +100 +1 +1 +1
6090 6190 6290 6390 6490 6491 6492 6493

10 b 25 + 63 = 20 + 60 = 80 and 5 + 3 = 8
= 80 + 8
= 88
c 42 + 34 = 40 + 30 = 70 and 2 + 4 = 6
= 70 + 6
= 76
d 713 + 44 = 700 and 10 + 40 = 50 and 3 + 4 = 7
= 700 + 50 + 7
= 757
e 824 + 32 = 800 and 20 + 30 = 50 and 4 + 2 = 6
= 800 + 50 + 6
= 856
f 642 + 16 = 600 and 40 + 10 = 50 and 2 + 6 = 8
= 600 + 50 + 8
= 658
g 4281 + 213 = 4000 and 200 + 200 = 400 and 80 + 10 = 90 and 1 + 3 = 4
= 4000 + 400 + 90 + 4
= 4494
h 2812 + 147 = 2000 and 800 + 100 = 900 and 10 + 40 = 50 and 2 + 7 = 9
= 2000 + 900 + 50 + 9
= 2959

11 a $4.15 b $2.50 c $6.50 d $43.75 e $89.90 f $142.00

12 a $0.65, $1.35 b $1.50, $0.50 c $3.65, $1.35 d $2.20, $2.80 e $0.80, $4.20 f $1.85, $8.15 g $5.90, $4.10 h $7.35, $2.65 i $12.15, $7.85 j $15.80, $4.20 k $17.40, $2.60

13 a $2.99 + $2.99 + $2.99 = $8.97, $1.05
b $5.98 + $5.98 + $5.98 = $17.94, $2.05
c 67c + 67c + 67c + 67c + 67c = $3.35, $6.65

3 SUBTRACTION

Jump Strategy to Solve Subtraction

Page 104 – Your Turn

a 51 b 41

Page 105 – Practice

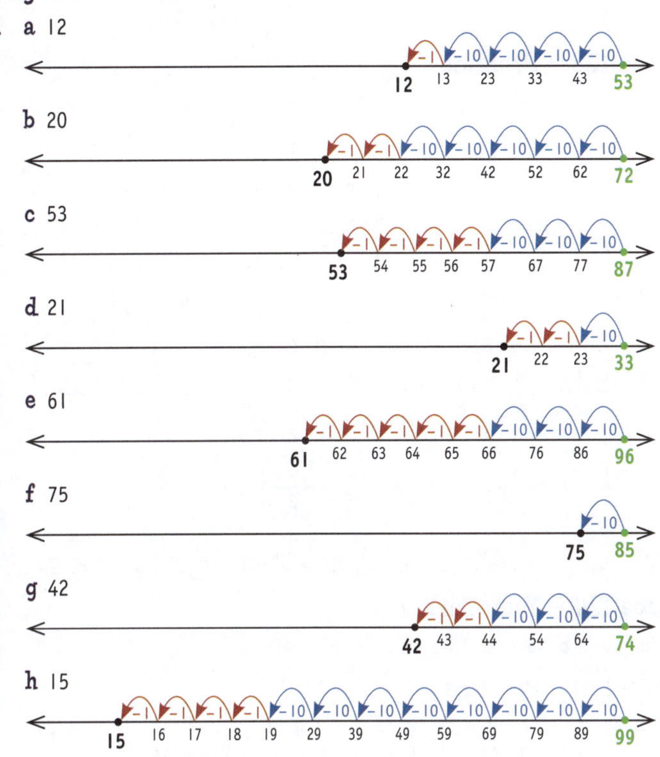

Jump Strategy with Three-digit Numbers

Page 106 – Your Turn

a 461 b 365

Page 107 – Practice

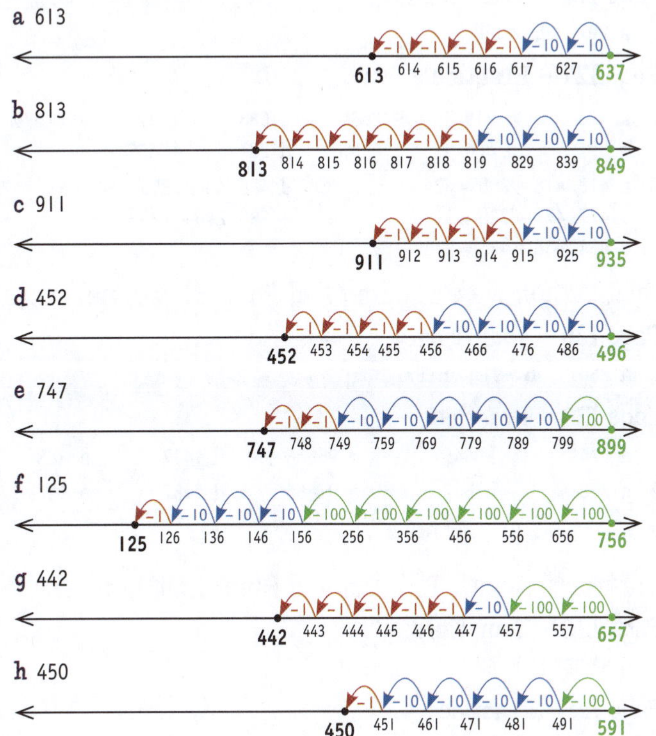

Jump Strategy with Larger Numbers

Page 108 – Your Turn

a 5451 b 5652

Page 109 – Practice

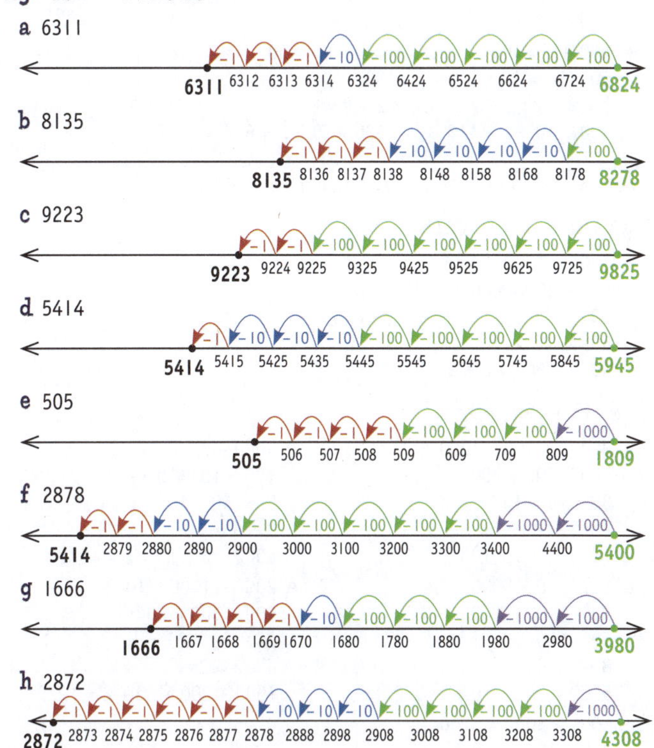

Subtraction Using the Split Strategy

Page 110 – Your Turn

a 89 – 32 = 57
80 – 30 = 50
9 – 2 = 7
50 + 7 = 57

Page 111 – Practice

1 a 20 – 10 = 10
7 – 3 = 4
10 + 4 = 14

b 50 – 20 = 30
9 – 0 = 9
30 + 9 = 39

c 70 – 30 = 40
8 – 2 = 6
40 + 6 = 46

d 60 – 30 = 30
6 – 2 = 4
30 + 4 = 34

e 90 – 20 = 70
2 – 0 = 2
70 + 2 = 72

f 80 – 10 = 70
3 – 2 = 1
70 + 1 = 71

g 40 – 10 = 30
9 – 8 = 1
30 + 1 = 31

h 60 – 20 = 40
7 – 6 = 1
40 + 1 = 41

i 50 – 40 = 10
9 – 2 = 7
10 + 7 = 17

Split Strategy with Three-digit Numbers

Page 112 – Your Turn

a 700 – 0 = 700
40 – 20 = 20
3 – 2 = 1
700 + 20 + 1 = 721

Page 113 – Practice

1 a 300 – 0 = 300
70 – 60 = 10
9 – 3 = 6
300 + 10 + 6 = 316

b 500 – 400 = 100
30 – 20 = 10
6 – 4 = 2
100 + 10 + 2 = 112

c 800 – 500 = 300
40 – 30 = 10
9 – 5 = 4
300 + 10 + 4 = 314

3 SUBTRACTION CONTINUED

d 700 – 0 = 700
50 – 40 = 10
7 – 4 = 3
700 + 10 + 3 = 713

e 400 – 0 = 400
60 – 50 = 10
5 – 3 = 2
400 + 10 + 2 = 412

f 100 – 100 = 0
70 – 60 = 10
8 – 2 = 6
10 + 6 = 16

g 600 – 0 = 600
80 – 10 = 70
2 – 0 = 2
600 + 70 + 2 = 672

Split Strategy with Four-digit Numbers

Page 114 – Your Turn

a 5000 – 1000 = 4000
800 – 200 = 600
30 – 30 = 0
6 – 4 = 2
4000 + 600 + 0 + 2 = 4602

Page 115 – Practice

1 a 3000 – 2000 = 1000
400 – 200 = 200
80 – 70 = 10
7 – 5 = 2
1000 + 200 + 10 + 2 = 1212

b 5000 – 0 = 5000
300 – 100 = 200
70 – 60 = 10
8 – 7 = 1
5000 + 200 + 10 + 1 = 5211

c 6000 – 3000 = 3000
200 – 100 = 100
60 – 40 = 20
5 – 3 = 2
3000 + 100 + 20 + 2 = 3122

d 4000 – 0 = 4000
100 – 100 = 0
50 – 40 = 10
8 – 6 = 2
4000 + 0 + 10 + 2 = 4012

e 2000 – 1000 = 1000
500 – 300 = 200
70 – 50 = 20
4 – 2 = 2
1000 + 200 + 20 + 2 = 1222

Subtraction without Trading

Page 116 – Your Turn

a 27 b 22 c 23

Page 117 – Practice

1 a 32 b 16 c 21 d 22 e 32 f 19 g 11

2 a 35 b 23 c 34 d 13 e 33 f 22 g 51

3 b Incorrect c Incorrect d Correct e Correct f Correct g Correct h Incorrect
Mario got 5 out of 8.

Subtraction without Trading and Three-digit Numbers

Page 118 – Your Turn

a 415 b 856

Page 119 – Practice

1 a 402 b 400 c 134 d 921 e 612 f 116 g 100 h 701 i 401 j 181 k 144

2 a 416 b 521 c 845 d 16 e 511 f 0 g 463

Subtraction without Trading and Four-digit Numbers

Page 120 – Your Turn

a 6136

Page 121 – Practice

1 a 4462 b 9411 c 2441 d 2811 e 5103 f 7200 g 2065 h 7111 i 6120 j 4300 k 3104

2 a 4224 b 7551 c 9075 d 5117 e 1111

Subtraction without Trading and Five-digit Numbers

Page 122 – Your Turn

a 17 111

Page 123 – Practice

1 a 11 136 b 28 201 c 21 141 d 10 112 e 206 f 81 111 g 61 016 h 37 001 i 11 411 j 11 009 k 62 422

2 a

	7	4	3	8	2
–	2	3	1	7	1
	5	1	2	1	1

b

	8	9	4	7	3
–	7	1	2	5	2
	1	8	2	2	1

c

	5	9	3	4	6
–	1	3	2	3	5
	4	6	1	1	1

d

	9	9	8	7	9
–	5	6	2	4	5
	4	3	6	3	4

e

	6	5	4	8	7
–	1	4	3	2	3
	5	1	1	6	4

f

	7	8	5	9	3
–	4	2	3	7	1
	3	6	2	2	2

g

	6	5	4	3	2
–	5	4	3	2	1
	1	1	1	1	1

h

	1	9	4	3	2
–	1	3	0	1	0
	0	6	4	2	2

Subtraction with Trading and Two-digit Numbers

Page 124 – Your Turn

a 45 b 19 c 18

Page 125 – Practice

1 a 6 b 6 c 19 d 19 e 33 f 19 g 55 h 9 i 18 j 38 k 29

2 a 38, 38 + 33 = 71
b 39, 39 + 43 = 82
c 29, 29 + 64 = 93
d 16, 16 + 25 = 41
e 26, 26 + 34 = 60

Subtraction with Trading and Three-digit Numbers

Page 126 – Your Turn

a 297 b 286

Page 127 – Practice

1 a 416 b 391 c 219 d 684 e 263 f 486 g 688 h 287 i 171 j 507 k 679

2 a 418, 418 + 206 = 624
b 477, 477 + 235 = 712
c 488, 488 + 425 = 913
d 488, 488 + 356 = 844
e 286, 286 + 123 = 409

Subtraction with Trading and Four-digit Numbers

Page 128 – Your Turn

1 a 2507 b 191 c 527

Page 129 – Practice

1 a 5749 b 4457 c 9366 d 1477 e 888

2 a 7083 b 842 c 9188 d 7457 e 1928

3 a Cross (4349) b Tick

Subtraction with Trading and Five-digit Numbers

Page 130 – Your Turn

a 53 334 b 42 517 c 43 457

Page 131 – Practice

1 a 57 591 b 63 424 c 38 836

2 a 87 758 c 59 657 e 66 905
b 37 836 d 9222 f 40 286

3 a 21 915, 23 847 + 21 915 = 45 762
b 54 875, 1949 + 54 875 = 56 824

What's the Difference?

Page 132 – Your Turn

a 1516 b 610 c 2916

Page 133 – Practice

1 a 624 – 110 = 514 c 1526 – 12 =1514 e 98 – 13 = 85
b 8374 – 2242 = 6132 d 153 – 21 = 132

2 a 5738 b $1346 c $4229 d $7383

3 a $464 b 6202 c $3618

Estimating Subtraction Answers

Page 134 – Your Turn

1 a 610, 420, 190 2 a 500, 200, 300

Page 135 – Practice

1 a 1350, 380, 970 e 29 390, 26 420, 2970
b 5950, 3270, 2680 f 540, 170, 370
c 700, 390, 310 g 63 530, 62 350, 1180
d 3270, 2130, 1140

2 a 7400, 4200, 3200 e 24 400, 19 400, 5000
b 51 800, 34 600, 17 200 f 900, 200, 700
c 600, 300, 300 g 4400, 2500, 1900
d 4400, 1200, 3200 h 6100, 6000, 100

Subtraction Review Page 137

1 a 59
b 27
c 279
d 547
e 152
f 7917
g 6938
h 7422

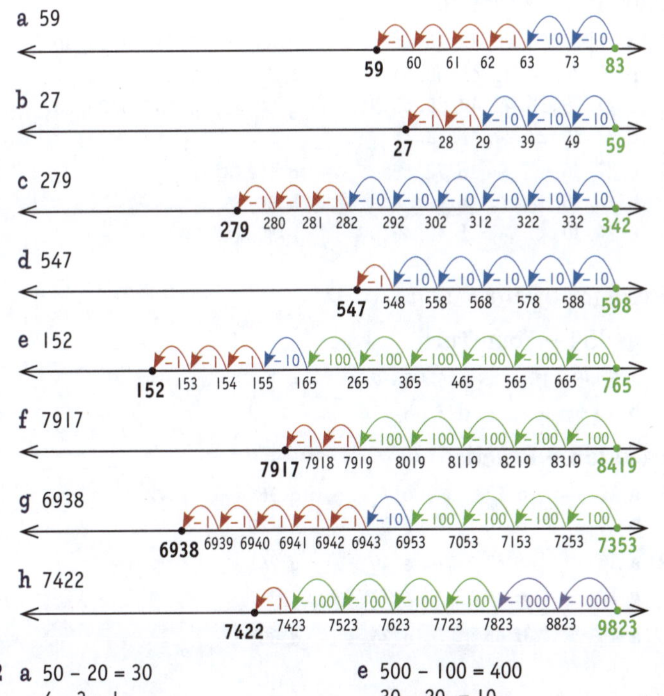

2 a 50 – 20 = 30
4 – 3 = 1
30 + 1 = 31
b 80 – 40 = 40
5 – 2 = 3
40 + 3 = 43
c 500 – 0 = 500
60 – 30 = 30
4 – 2 = 2
500 + 30 + 2 = 532
d 700 – 0 = 700
50 – 40 = 10
4 – 1 = 3
700 + 10 + 3 = 713
e 500 – 100 = 400
30 – 20 = 10
7 – 3 = 4
400 + 10 + 4 = 414
f 3000 – 0 = 3000
400 – 100 = 300
90 – 70 = 20
5 – 3 = 2
3000 + 300 + 20 + 2 = 3322
g 7000 – 5000 = 2000
400 – 200 = 200
60 – 30 = 30
7 – 1 = 6
2000 + 200 + 30 + 6 = 2236
h 8000 – 5000 = 3000
300 – 200 = 100
40 – 40 = 0
2 – 1 = 1
3000 + 100 + 0 + 1 = 3101

3 a 32 c 62 e 37 g 14 i 111 k 289
b 62 d 52 f 19 h 17 j 311 l 589

4 a 2465 c 2322 e 4779 g 27 955 i 88 742 k 57 607
b 5481 d 6218 f 6611 h 60 540 j 68 687 l 10 480

5 a 26, 26 + 32 = 58 c 424, 424 + 123 = 547
b 49, 49 + 25 = 74 d 361: 361 + 253 = 614

6 a 4214: 4214 + 143 = 4357 d 58 565: 58 568 + 1245 = 59 810
b 6429: 6429 + 2305 = 8734 e 82 547: 82 547 + 6989 = 89 536
c 26 613: 26 613 + 3243 = 29 856

7 a 11 136 c 21 141 e 206 g 61 016 i 68 702
b 28 201 d 10 112 f 81 111 h 37 001

8 a wrong (39) e wrong (650) i wrong (61 731)
b correct f wrong (3781) j correct
c wrong (109) g correct Score: 4 out of 10
d correct h wrong (14 881)

9 a 16 c 25 e 489 g 5106 i 6621 k 20 404
b 18 d 95 f 223 h 5931 j 33 899 l 11 140

10 a 530, 140, 390 d 7440, 4320, 3120
b 320, 210, 110 e 8270, 1650, 6620
c 3520, 2990, 530

11 a 600, 200, 400 e 24 400, 15 700, 8700
b 500, 300, 200 f 53 300, 45 300, 8000
c 1800, 400, 1400 g 54 300, 51 800, 2500
d 2900, 300, 2600 h 73 200, 41 300, 31 900

4 MULTIPLICATION

Groups and Rows

Page 144 – Your Turn

1 a 5 b 6 2 a 2 b 3

Page 145 – Practice

1 a b c d e

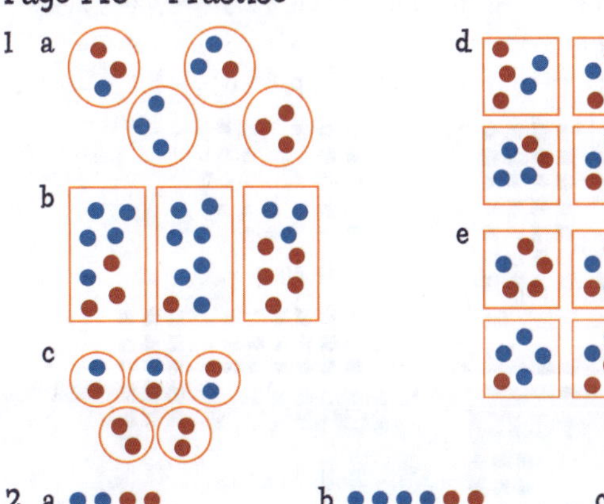

2 a b c

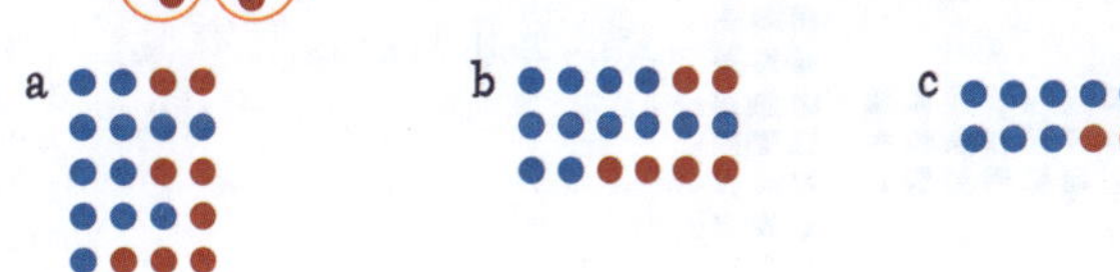

3 a 3 rows of 5 c 4 rows of 1 e 3 groups of 4 g 3 rows of 5
b 4 rows of 6 d 2 groups of 3 f 4 groups of 6 h 2 rows of 9

4 MULTIPLICATION CONTINUED

Repeated Addition to Solve Multiplication

Page 146 – Your Turn

1 a 3 groups of 4 = 12
4 + 4 + 4 = 12
3 × 4 = 12

b 3 rows of 4 = 12
4 + 4 + 4 = 12
3 × 4 = 12

c 1 groups of 4 = 4
4 = 4
1 × 4 = 4

Page 147 – Practice

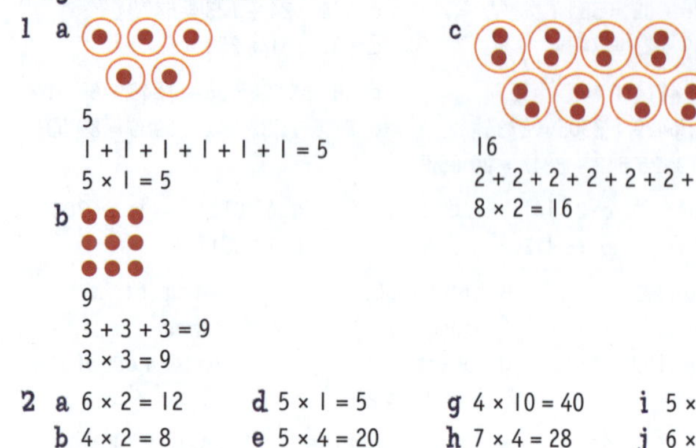

1 a 5
1 + 1 + 1 + 1 + 1 = 5
5 × 1 = 5

b 9
3 + 3 + 3 = 9
3 × 3 = 9

c 16
2 + 2 + 2 + 2 + 2 + 2 + 2 + 2 = 16
8 × 2 = 16

2 a 6 × 2 = 12
b 4 × 2 = 8
c 3 × 9 = 27
d 5 × 1 = 5
e 5 × 4 = 20
f 5 × 5 = 25
g 4 × 10 = 40
h 7 × 4 = 28
i 5 × 7 = 35
j 6 × 6 = 36

Commutative Property

Page 148 – Your Turn

1 a b c d e

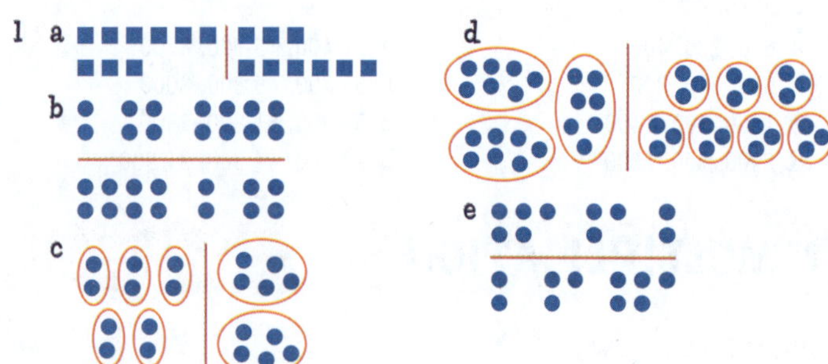

Page 149 – Practice

1 a 3 × 9
b 5 + 4
c 1 × 7
d 2 + 8
e 3 + 9
f 9 × 8
g 10 × 9
h 11 × 1
i 6 × 2
j 4 + 6
k 8 × 3
l 3 + 9 + 1
m 2 + 4 + 7
n 4 × 3
o 1 × 5
p 8 + 8
q 4 × 9
r 9 × 5
s 7 × 6

2 a

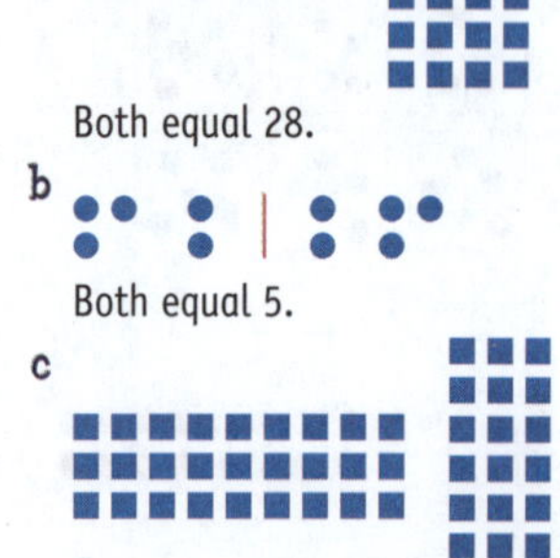

Both equal 28.

b

Both equal 5.

c

Both equal 27.

d

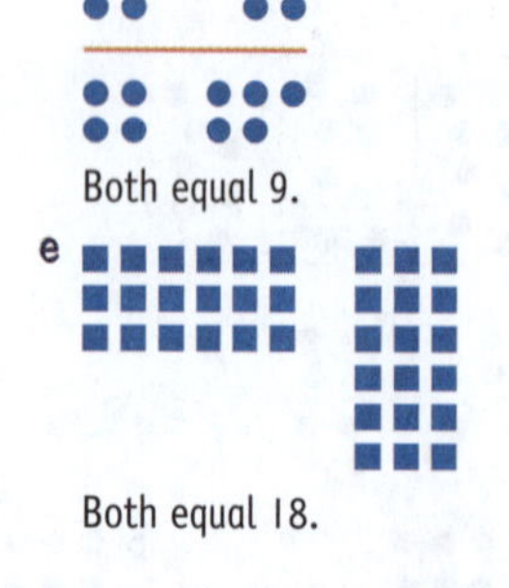

Both equal 9.

e

Both equal 18.

Inverse Operations of Multiplication and Division

Page 150 – Your Turn

1 a 4, 4 b 4, 4 c 4, 4 d 4, 4 e 3, 3

Page 151 – Practice

1 a 25 ÷ 5 = 5
b 21 ÷ 3 = 7
c 40 ÷ 4 = 10
d 36 ÷ 6 = 6
e 9 ÷ 1 = 9
f 48 ÷ 4 = 12
g 24 ÷ 8 = 3

2

a	6 × 2 = 12	12 ÷ 2 = 6	2 × 6 = 12	12 ÷ 6 = 2
b	2 × 5 = 10	10 ÷ 5 = 2	5 × 2 = 10	10 ÷ 2 = 5
c	8 × 5 = 40	40 ÷ 5 = 8	5 × 8 = 40	40 ÷ 8 = 5
d	1 × 7 = 7	7 ÷ 7 = 1	7 × 1 = 7	7 ÷ 1 = 7
e	2 × 12 = 24	24 ÷ 12 = 2	12 × 2 = 24	24 ÷ 2 = 12
f	4 × 5 = 20	20 ÷ 5 = 4	5 × 4 = 20	20 ÷ 4 = 5
g	7 × 4 = 28	28 ÷ 4 = 7	4 × 7 = 28	28 ÷ 7 = 4
h	8 × 4 = 32	32 ÷ 4 = 8	4 × 8 = 32	32 ÷ 8 = 4
i	3 × 11 = 33	33 ÷ 11 = 3	11 × 3 = 33	33 ÷ 3 = 11
j	6 × 7 = 42	42 ÷ 7 = 6	7 × 6 = 42	42 ÷ 6 = 7
k	9 × 10 = 90	90 ÷ 10 = 9	10 × 9 = 90	90 ÷ 9 = 10
l	2 × 9 = 18	18 ÷ 2 = 9	9 × 2 = 18	18 ÷ 9 = 2

Skip Counting

Page 152 – Your Turn

1 a 12, 18, 24, 30 b 8, 12, 16, 20
2 a 22, 20, 18, 16 b 20, 16, 12, 8

Page 153 – Practice

1 a 16, 20, 24, 28, 32, 36, 40
b 18, 24, 30, 36, 42, 48, 54
c 21, 28, 35, 42, 49, 56, 63
d 33, 30, 27, 24, 21, 18, 15
e 88, 80, 72, 64, 56, 48, 40
f 81, 72, 63, 54, 45, 36, 27
g 50, 45, 40, 35, 30, 25, 20

2 a 27, 36
b 21, 35, 42
c 24, 30
d 72, 56
e 3,4
f 40, 48
g 15, 18
h 50, 45
i 30, 24

Product of Numbers

Page 154 – Your Turn

a 3 × 6 = 18
b 1 × 9 = 9
c 5 × 5 = 25
d 2 × 8 = 16
e 7 × 7 = 49

Page 155 – Practice

1 a 42 b 48 c 27 d 16 e 7 f 0 g 36 h 120 i 72
2 a 35 b 18 c 63 d 30 e 40 f 7 g 24
3 a 45 b 56 c 24 d 80

Factors and Multiples

Page 156 – Your Turn

a F b T c T d F e T

Page 157 – Practice

1 a 1, 2, 3, 4, 6, 8, 12, 24
b 1, 2, 5, 10
c 1, 2, 4, 5, 10, 20
d 1, 2, 3, 6, 9, 18
e 1, 5

2 a 4, 8, 12, 16, 20
b 7, 14, 21, 28, 35
c 6, 12, 18, 24, 30
d 5, 10, 15, 20, 25
e 10, 20, 30, 40, 50

3 a 5 b 6 c 9 d 9 e 7 f 4 g 5 h 9 i 8 j 8 k 7

4 a 14, 35, 49, 56, 21, 63
b 18, 30, 54, 48, 42, 24
c 9, 81, 90, 108, 54, 72

Doubling and Halving

Page 158 – Your Turn

1 a 10 b 1 c 50
2 a 12 b 20 c 100

Page 159 – Practice

1 a 4 b 12, 12 c 32 d 15, 15 e 72, 72 f 65, 65 g 400 h 60, 60 i 90 j 125, 125 k 175, 175

2

	Half	Number	Double
a	10	20	40
b	19	38	76
c	45	90	180
d	80	160	320
e	135	270	540
f	240	480	960
g	250	500	1000
h	25	50	100
i	55	110	220

3 a

In	Out
20	40
30	60
40	80
150	300
125	250
400	800

b

In	Out
16	8
28	14
29	58
212	106
44	22
190	380

Equivalent Number Relationships

Page 160 – Your Turn

a 2 × 9 b 2 × 14 c 3 × 12 d 3 × 2 e 4 × 6

Page 161 – Practice

1 Adult to check

2 a 9 × 2 b 1 × 12 c 4 × 11 d 3 × 7 e 5 × 3 f 6 × 2 g 12 × 11

3 a 8 b 3 c 24 d 4 e 5 f 3 g 5 h 3 i 2 j 4 k 7

Multiplying Three Single-digit Numbers

Page 162 – Your Turn

a = 5 × 6 × 4
= 30 × 4
= 120

b = 5 × 2 × 6
= 10 × 6
= 60

Page 163 – Practice

1 a = 5 × 2 × 7
= 10 × 7
= 70

b = 5 × 4 × 8
= 20 × 8
= 160

c = 4 × 5 × 4
= 20 × 4
= 80

d = 6 × 5 × 7
= 30 × 7
= 210

e = 5 × 4 × 9
= 20 × 9
= 180

f = 2 × 5 × 8
= 10 × 8
= 80

g = 4 × 5 × 3
= 20 × 3
= 60

h = 4 × 5 × 6
= 20 × 6
= 120

i = 2 × 5 × 8
= 10 × 8
= 80

j = 2 × 5 × 2
= 10 × 2
= 20

k = 5 × 2 × 5
= 10 × 5
= 50

l = 8 × 10 × 3
= 80 × 3
= 240

m = 4 × 5 × 7
= 20 × 7
= 1400

n = 6 × 5 × 2
= 30 × 2
= 60

Multiplying Two-digit by Single-digit Numbers

Page 164 – Your Turn

a 40 × 7 = 280
280 + 7 + 7
= 294

= 4 tens × 7 + 7 twos
= 280 + 14
= 294

	40	2
7	280	14

= 280 + 14
= 294

Page 165 – Practice

1 a 52 × 5
50 × 5 = 250
50 + 5 + 5
= 260

b 19 × 6
10 × 6 = 60
60 + 6 + 6 + 6 + 6 + 6 + 6 + 6 + 6 + 6
= 114

c 28 × 3
20 × 3 = 60
60 + 3 + 3 + 3 + 3 + 3 + 3 + 3 + 3
= 84

d 97 × 4
90 × 4 = 360
360 + 4 + 4 + 4 + 4 + 4 + 4 + 4
= 388

e 88 × 8
80 × 8 = 640
640 + 8 + 8 + 8 + 8 + 8 + 8 + 8 + 8
= 704

f 34 × 7
30 × 7 = 210
210 + 7 + 7 + 7 + 7
= 238

g 45 × 8
40 × 8 = 320
320 + 8 + 8 + 8 + 8 + 8
= 360

2 a 58 × 4
= 4 × 5 tens + 4 eights
= 200 + 32
= 232

b 463 × 3
= 3 × 4 tens + 3 sixes
= 120 + 18
= 138

c 39 × 7
= 7 × 3 tens + 7 nines
= 210 + 63
= 273

d 71 × 8
= 8 × 7 tens + 8 ones
= 560 + 8
= 568

e 62 × 5
= 5 × 6 tens + 5 twos
= 300 + 10
= 310

3 a 63 × 7

	60	3
7	420	21

= 420 + 21
= 441

b 92 × 6

	90	2
6	540	12

= 540 + 12
= 552

c 32 × 8

	30	2
8	240	16

= 240 + 16
= 256

d 19 × 4 = 76

	10	9
4	40	36

= 40 + 36
= 76

e 86 × 3

	80	6
3	240	18

= 240 + 18
= 258

Multiplication Review Page 167

1 a 4, 3 b 5, 2 c 3, 4

2 a 6 rows of 2
2 + 2 + 2 + 2 + 2 + 2 = 12
2 × 6 = 12

b 4 groups of 2
2 + 2 + 2 + 2 = 8
2 × 4 = 8

c 3 rows of 5
5 + 5 + 5 = 15
3 × 5 = 15

d 6 groups of 3
3 + 3 + 3 + 3 + 3 + 3 = 18
3 × 6 = 18

3 a 4 × 8 = 32 b 5 × 5 = 25 c 4 × 6 = 24 d 6 × 2 = 12 e 4 × 1 = 4 f 6 × 9 = 54

4 MULTIPLICATION CONTINUED

4 a 9 × 5 b 4 × 7 c 4 + 3 + 2 d 2 + 5 e 8 × 6 f 8 + 7 + 5

5 a 36, 36, 4 b 3, 3 c 30, 30, 5 d 7, 7 e 7, 7 f 10, 10 g 4, 32 h 11, 11

6 a 9, 18, 27, 36, 45 b 5, 10, 15, 20, 25 c 8, 16, 24, 32, 40 d 3, 6, 9, 12, 15 e 4, 8, 12, 16, 20 f 6, 12, 18, 24, 30 g 10, 20, 30, 40, 50 h 12, 24, 36, 48, 60

7 a 20, 18, 16, 14, 12 b 40, 36, 32, 28, 24 c 42, 36, 30, 24, 18 d 70, 63, 56, 49, 42 e 36, 33, 30, 27, 24 f 50, 45, 40, 35, 30 g 72, 63, 54, 45, 36 h 88, 77, 66, 55, 44

8 a 27 b 72 c 49 d 11 e 0 f 110

9 a 1, 2, 3, 4, 6, 12 b 1, 2, 4, 5, 10, 20 c 1, 2, 3, 4, 6, 9, 12, 18, 36 d 1, 2, 4, 5, 8, 10, 20, 40

10 a T b T c T d F e F f T g T h F

11 a 9, 18, 27, 36, 45 b 3, 6, 9, 12, 15 c 10, 20, 30, 40, 50 d 2, 4, 6, 8, 10 e 1, 2, 3, 4, 5 f 5, 10, 15, 20, 25

12 a

×	1	6	3	8	11	4
5	5	30	15	40	55	20
4	4	24	12	32	44	16
7	7	42	21	56	77	28
2	2	12	6	16	22	8
9	9	54	27	72	99	36

b

×	5	2	12	10	0	7
8	40	16	96	80	0	56
3	15	6	36	30	0	21
10	50	20	120	100	0	70
1	5	2	12	10	0	7
6	30	12	72	60	0	42

13 a 24 b 36 c 24, 24 d 30, 30 e 150 f 145, 145 g 75, 75 h 340

14

	Half	Number	Double
a	5	10	20
b	20	40	80
c	8	16	32
d	25	50	100
e	40	80	160
f	10	20	40
g	4	8	16
h	6	12	24

	Half	Number	Double
i	8	16	32
j	9	18	36
k	22	44	88
l	24	48	96
m	32	64	128
n	68	136	272
o	175	350	700
p	300	600	1200

15 Adult to check

16 a 6 × 5 × 2
= 30 × 2
= 60

b 4 × 5 × 8
= 20 × 8
= 160

c 5 × 2 × 9
= 10 × 9
= 90

17 a 16 × 3
= 10 × 3 = 30
= 30 + 3 + 3 + 3 + 3 + 3 + 3
= 48
16 × 3
= 3 × 1 tens + 3 sixes
= 30 + 18
= 48
16 × 3

	10	6
3	30	18

= 30 + 18
= 48

b 42 × 6
= 40 × 6 = 240
= 240 + 6 + 6
= 252
42 × 6
= 6 × 4 tens + 2 sixes
= 240 + 12
= 252
42 × 6

	40	2
6	240	12

= 240 + 12
= 252

5 DIVISION

Grouping

Page 172 – Your Turn

a 2 b 3

Page 173 – Practice

1 a 3, 4 b 2, 6 c 4, 2 d 3, 6
2 a 5, 5 b 4, 4
3 a 36, 6 b 14, 2

Equal Rows

Page 174 – Your Turn

a

b 

Page 175 – Practice

1 a 5, 5 b 12, 12 c 2, 2 d 1, 1 e 3, 3
2 a 4 rows of 3 b 3 rows of 4 c 8 rows of 5 d 2 rows of 9

Repeated Subtraction to Solve Division

Page 176 – Your Turn

a 9 ÷ 3 = 3
Start at 9
9 – 3 = 6
6 – 3 = 3
3 – 3 = 0
3

b 14 ÷ 7 = 2
Start at 14
14 – 7 = 7
7 – 7 = 0
2

c 12 ÷ 4 = 3
Start at 12
12 – 4 = 8
8 – 4 = 4
4 – 4 = 0
3

Page 177 – Practice

1 a 45 ÷ 5 = 9
45 – 5 = 40
40 – 5 = 35
35 – 5 = 30
30 – 5 = 25
25 – 5 = 20
20 – 5 = 15
15 – 5 = 10
10 – 5 = 5
5 – 5 = 0
9

b 24 ÷ 8 = 3
24 – 8 = 16
16 – 8 = 8
8 – 8 = 0
3

c 30 ÷ 6 = 5
30 – 6 = 24
24 – 6 = 18
18 – 6 = 12
12 – 6 = 6
6 – 6 = 0
5

d 35 ÷ 7 = 5
35 – 7 = 28
28 – 7 = 21
21 – 7 = 14
14 – 7 = 7
7 – 7 = 0
5

e 12 ÷ 2 = 6
12 – 2 = 10
10 – 2 = 8
8 – 2 = 6
6 – 2 = 4
4 – 2 = 2
2 – 2 = 0
6

Inverse Operations of Multiplication and Division

Page 178 – Your Turn

a 28 ÷ 7 = 4 b 18 ÷ 9 = 2 c 42 ÷ 6 = 7 d 45 ÷ 9 = 5 e 24 ÷ 8 = 3

Page 179 – Practice

1 a 6 × 8 = 48
8 × 6 = 48
48 ÷ 8 = 6
48 ÷ 6 = 8

b 6 × 4 = 24
4 × 6 = 24
24 ÷ 4 = 6
24 ÷ 6 = 4

c 5 × 6 = 30
6 × 5 = 30
30 ÷ 6 = 5
30 ÷ 5 = 6

d 8 × 9 = 72
9 × 8 = 72
72 ÷ 9 = 8
72 ÷ 8 = 9

e 7 × 8 = 56
8 × 7 = 56
56 ÷ 8 = 7
56 ÷ 7 = 8

2 a ÷ = b ÷ = c ÷ = d ÷ = e ÷ = f ÷ = g ÷ = h ÷ =

3 a 14 b 18 c 12 d 110 e 60 f 70 g 64 h 9

CATCH UP MATHS YEAR 4 BOOK A © PASCAL PRESS ISBN: 9781925726145

Formal Division

Page 180 – Your Turn

a 6, 6 b 6, 6 c 9, 9 d 12, 12

Page 181 – Practice

1 a 4 b 11 c 8 d 10 e 9 f 9 g 7 h 8 i 5 j 7 k 6

2 a 28, 28 b 8, 8 c 6, 6 d 5, 5 e 5, 5 f 22, 22 g 60, 60

3 a False b True c True d False e False f True g False h True i False j True k False

Division with Remainders

Page 182 – Your Turn

a 2, 2
12 ÷ 5 = 2 remainder 2

Page 183 – Practice

1 a 15 ÷ 2 = 7 remainder 1
b 18 ÷ 4 = 4 remainder 2
c 48 ÷ 8 = 6 remainder 0
d 25 ÷ 3 = 8 remainder 1
e 21 ÷ 7 = 3 remainder 0
f 29 ÷ 4 = 7 remainder 1
g 83 ÷ 9 = 9 remainder 2

2 a 4 r 2 b 9 r 2 c 4 r 3 d 5 r 2 e 9 r 2 f 10 r 1 g 12 r 2

3 a remainder 3 b remainder 2 c remainder 1 d remainder 4 e remainder 5

Division Review Page 184

1 a 2, 5 b 3, 4 c 4, 3

2 a 5, 5 b 6, 6 c 7, 7

3 a
5 rows of 4 = 20
20 ÷ 5 = 4

b
3 rows of 4 = 12
12 ÷ 3 = 4

c 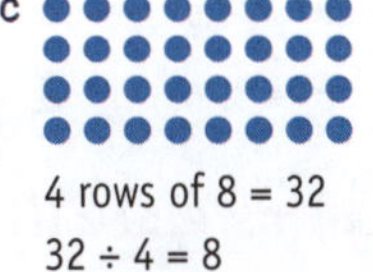
4 rows of 8 = 32
32 ÷ 4 = 8

4 a 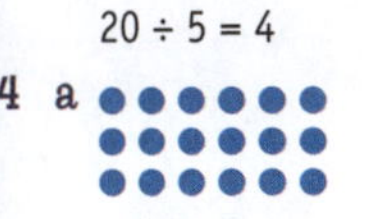
b 
c 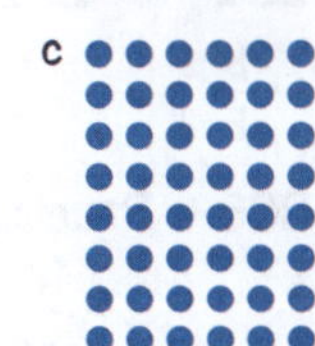

5 a 32 ÷ 4 = 8
32 – 4 = 28
28 – 4 = 24
24 – 4 = 20
20 – 4 = 16
16 – 4 = 12
12 – 4 = 8
8 – 4 = 4
4 – 4 = 0

b 48 ÷ 12 = 4
48 – 12 = 36
36 – 12 = 24
24 – 12 = 12
12 – 12 = 0

c 24 ÷ 6 = 4
24 – 6 = 18
18 – 6 = 12
12 – 6 = 6
6 – 6 = 0

d 18 ÷ 9 = 2
18 – 9 = 9
9 – 9 = 0

e 6 ÷ 6 = 1
6 – 6 = 0

6 a 30 ÷ 5 = 6 or 30 ÷ 6 = 5
b 36 ÷ 9 = 4 or 36 ÷ 4 = 9
c 24 ÷ 12 = 2 or 24 ÷ 2 = 12
d 33 ÷ 11 = 3 or 33 ÷ 3 = 11
e 72 ÷ 9 = 8 or 72 ÷ 8 = 9
f 32 ÷ 4 = 8 or 32 ÷ 8 = 4
g 48 ÷ 4 = 12 or 48 ÷ 12 = 4
h 20 ÷ 2 = 10 or 20 ÷ 10 = 2
i 36 ÷ 6 = 6
j 16 ÷ 4 = 4
k 27 ÷ 3 = 9 or 27 ÷ 9 = 3

7 a 4 × 8 = 32
8 × 4 = 32
32 ÷ 8 = 4
32 ÷ 4 = 8

b 10 × 4 = 40
4 × 10 = 40
40 ÷ 4 = 10
40 ÷ 10 = 4

c 12 × 2 = 24
2 × 12 = 24
24 ÷ 2 = 12
24 ÷ 12 = 2

d 7 × 7 = 49
7 × 7 = 49
49 ÷ 7 = 7
49 ÷ 7 = 7

8 a 3 × 10 b 8 × 9 c 10 × 8 d 5 × 8 e 4 × 6 f 7 × 3 g 8 × 8 h 5 × 4 i 6 × 2 j 12 × 3

9 a 12 b 8 c 18 d 27 e 28 f 77 g 50 h 6 i 96 j 14

10 a ÷ = b ÷ = c ÷ = d ÷ = e ÷ = f ÷ = g ÷ = h ÷ = i ÷ = j ÷ =

11 a 12 b 4 c 8 d 7 e 8 f 12 g 10 h 6 i 8

12 a 4, 4 b 5, 5 c 7, 7 d 40, 40 e 7, 7 f 12, 12 g 8, 8 h 3, 3 i 99, 99 j 6, 6

13 a 23 ÷ 4 = 5 r 3
b 37 ÷ 12 = 3 r 1
c 43 ÷ 5 = 8 r 3
d 92 ÷ 10 = 9 r 2
e 14 ÷ 7 = 2 r 0
f 28 ÷ 9 = 3 r 1
g 63 ÷ 6 = 10 r 3
h 48 ÷ 10 = 4 r 8
i 25 ÷ 5 = 5 r 0
j 59 ÷ 12 = 4 r 11

14 a 5 r 3 b 5 r 3 c 7 r 3 d 8 r 6 e 5 r 2 f 11 r 1 g 10 r 6 h 10 r 3 i 8 r 6 j 11 r 1 k 8 r 5 l 6 r 1 m 10 r 1 n 6 r 2

6 FRACTIONS

Numerators and Denominators

Page 190 – Your Turn

Colour all the top numbers red, colour all the horizontal lines green and colour all the bottom numbers blue.

Page 191 – Practice

1 a 3 b 1 c 7 d 2 e 4 f 4 g 3

2 a 8 b 4 c 5 d 8 e 2 f 4 g 5

3 a three-eighths b one-fifth c seven-eighths d three-fifths e three-quarters f two-quarters g one-eighth h five-eighths i one-third

4 a $\frac{3}{4}$ b $\frac{4}{5}$ c $\frac{3}{8}$ d $\frac{3}{5}$ e $\frac{3}{3}$

5 a three-quarters b four-fifths c three-eighths d three-fifths e three-fifths

Fractions—Halves

Page 192 – Your Turn

Circle: **d, e, f, g**

Page 193 – Practice

1 Tick: **a, f, g**

2 a 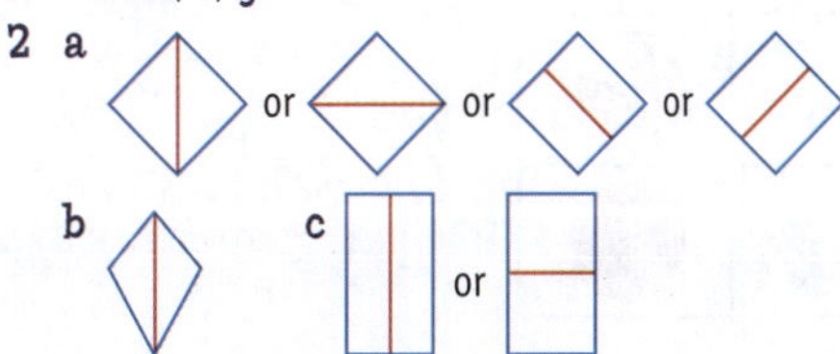

b

c or

3 Cross out **b, f, h, j**

Fractions—Quarters and Eighths

Page 194 – Your Turn

Colour blue (quarters): **g** Colour red (eighths): **a, c, e** Cross out: **b, d, f**

6 FRACTIONS CONTINUED

Page 195 – Practice

1

	Fraction name	Fraction	Picture
a	seven-eighths	$\frac{7}{8}$	
b	three-quarters	$\frac{3}{4}$	
c	five-eighths	$\frac{5}{8}$	
d	two-quarters	$\frac{2}{4}$	
e	six-eighths	$\frac{6}{8}$	
f	four-quarters	$\frac{4}{4}$	

2 (Sample answers provided; other answers are possible. All parts must be equal in size.)

a  b  c 

3 (Sample answers provided; other answers are possible. All parts must be equal in size.)

a 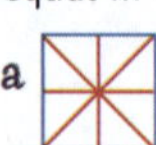 b 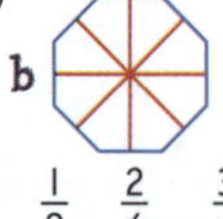 c 

4

$\frac{3}{8}$	$\frac{4}{4}$	$\frac{7}{8}$	$\frac{1}{8}$	$\frac{2}{4}$	$\frac{3}{4}$
2	6	5	1	3	4

Fractions—Thirds and Fifths

Page 196 – Your Turn

Colour purple (thirds): **b, f, g** Colour green (fifths): **a, e** Cross out: **c, d**

Page 197 – Practice

1

	Fraction name	Fraction	Picture
a	two-fifths	$\frac{2}{5}$	
b	two-thirds	$\frac{2}{3}$	
c	three-thirds	$\frac{3}{3}$	
d	four-fifths	$\frac{4}{5}$	
e	five-fifths	$\frac{5}{5}$	

2

$\frac{2}{3}$	$\frac{1}{5}$	$\frac{3}{5}$	$\frac{1}{3}$	$\frac{3}{3}$	$\frac{4}{5}$
4	**1**	**3**	**2**	**6**	**5**

3 **a** $\frac{2}{5}$ **b** $\frac{4}{5}$ **c** $\frac{5}{5}$ **d** $\frac{3}{3}$ **e** $\frac{1}{3}$

Halves of a Collection

Page 198 – Your Turn

a 12, 6 **b** 20, 10

Page 199 – Practice

1 **a** 2 **b** 1 **c** 5 **d** 10 **e** 3 **f** 12 **g** 6

2 **a** 12, 6 **b** 16, 8

3 **a** 8, 16 **b** 2, 4 **c** 10, 20

Quarters and Eighths of a Collection

Page 200 – Your Turn

1 **a** 3 2 **a** 5

Page 201 – Practice

1 **a** Colour 2, $\frac{1}{8} = 2$ **b** Colour 5, $\frac{1}{8} = 5$ **c** Colour 4, $\frac{1}{8} = 4$ **d** Colour 7, $\frac{1}{8} = 7$

2 **a** Colour 3, $\frac{1}{4} = 3$ **b** Colour 6, $\frac{1}{4} = 6$ **c** Colour 5, $\frac{1}{4} = 5$ **d** Colour 1, $\frac{1}{4} = 1$ **e** Colour 7, $\frac{1}{4} = 7$

3 **a** 4 **b** 7 **c** 12 **d** 5 **e** 8 **f** 3 **g** 6

4 **a** 11 **b** 1 **c** 9 **d** 12 **e** 6 **f** 2 **g** 7

Simplifying Fractions

Page 202 – Your Turn

a 5 **b** 3 **c** $\frac{2}{7}$

Page 203 – Practice

1 **a** 3, 24, $\frac{1}{8}$ **b** 5, 20, $\frac{1}{4}$ **c** 4, 40, $\frac{1}{10}$

2 **a** ÷ 5, $\frac{3}{4}$ **b** ÷ 3, $\frac{3}{4}$ **c** ÷ 50, $\frac{1}{2}$ **d** ÷ 7, $\frac{3}{4}$ **e** ÷ 6, $\frac{3}{5}$ **f** ÷ 8, $\frac{3}{5}$ **g** ÷ 5, $\frac{8}{9}$

Thirds and Fifths of a Collection

Page 204 – Your Turn

1 **a** 3 **b** 2

2 **a** 1 **b** 3

Page 205 – Practice

1 **a** 12, 4 **b** 24, 8 **c** 18, 6

2 **a** 20, 4 **b** 30, 6 **c** 5, 1

3 **a** 1 **b** 3 **c** 5 **d** 7 **e** 9 **f** 11 **g** 10

4 **a** 2 **b** 5 **c** 10 **d** 20 **e** 6 **f** 7 **g** 8

5 **a** 11 **b** 1 **c** 2 **d** 12

Comparing Fractions

Page 206 – Your Turn

a 4, 4 **b** 3, 3 **c** 5, 5

Page 207 – Practice

1 **a** 3, 2, 5, 4, 6, 1 **b** 5, 6, 4, 2, 1, 3

2 **a** = **b** > **c** < **d** > **e** < **f** < **g** > **h** = **i** = **j** > **k** =

3 Adult to check

4 Adult to check

5 Adult to check

CATCH UP MATHS YEAR 4 BOOK A © PASCAL PRESS ISBN: 9781925726145

6 a
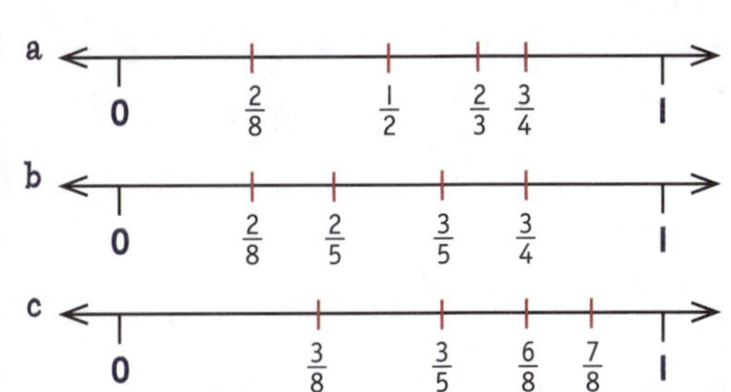

b 0, $\frac{2}{8}$, $\frac{2}{5}$, $\frac{3}{5}$, $\frac{3}{4}$, 1

c 0, $\frac{3}{8}$, $\frac{3}{5}$, $\frac{6}{8}$, $\frac{7}{8}$, 1

7 a 4, 4 b 3, 2 c 2, 1 d 1, 4 e 5, 0

8 a $\frac{3}{8}, \frac{2}{4}, \frac{3}{5}, 1$ c $\frac{2}{4}, \frac{6}{8}, \frac{7}{8}, \frac{5}{5}$ e $\frac{1}{5}, \frac{1}{4}, \frac{3}{5}, \frac{7}{8}$
b $\frac{2}{8}, \frac{2}{5}, \frac{3}{5}, \frac{3}{4}$ d $\frac{2}{8}, \frac{2}{5}, \frac{2}{4}, \frac{2}{3}$

9 a $\frac{1}{4}$ c $\frac{1}{4}$ e $\frac{2}{5}$ g $\frac{1}{4}$ i $\frac{1}{4}$ k $\frac{3}{8}$
b $\frac{3}{4}$ d $\frac{6}{10}$ f $\frac{1}{5}$ h $\frac{3}{5}$ j $\frac{2}{5}$

10 a $\frac{2}{3}$ c $\frac{3}{4}$ e $\frac{7}{8}$ g $\frac{3}{4}$ i $\frac{1}{3}$ k $\frac{7}{8}$
b $\frac{3}{4}$ d $\frac{2}{3}$ f $\frac{9}{10}$ h $\frac{7}{8}$ j $\frac{4}{5}$

11 Adult to check
12 Adult to check

Equivalent Fractions

Page 210 – Your Turn

a 2 b 2 c 1

Page 211 – Practice

1 a $\frac{3}{4}$ is equivalent to $\frac{6}{8}$
d $\frac{4}{8}$ is equivalent to $\frac{2}{4}$
b $\frac{1}{2}$ is equivalent to $\frac{2}{4}$
e $\frac{1}{4}$ is equivalent to $\frac{2}{8}$
c $\frac{6}{10}$ is equivalent to $\frac{3}{5}$

2 a 2 d 4 g 4 j 2 m 9 p 24 s 4
b 6 e 1 h 4 k 5 n 3 q 5
c 1 f 4 i 2 l 2 o 9 r 9

Fractions Review Page 212

1 a 1 b 2 c 3 d 2 e 1 f 3
2 a 3 b 8 c 8 d 4 e 5 f 2
3 a three-fifths c four-fifths e three-quarters
b two-eighths d one-half f seven-eighths
4 (Sample answers provided; other answers are possible for **b** and **d**. All parts must be equal in size.)

a 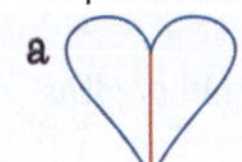  b 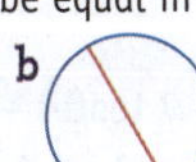 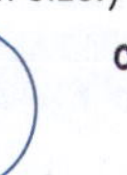 c 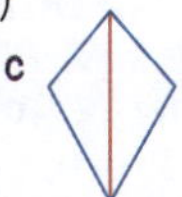 d 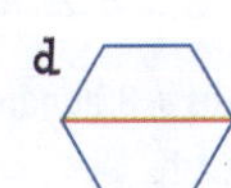 e 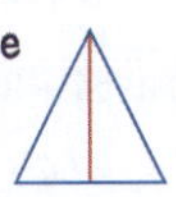

5 (Sample answers provided; other answers are possible for **a**, **b**, **c** and **e**. All parts must be equal in size.)

a 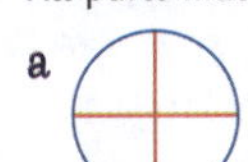 b 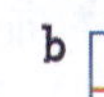 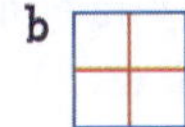 c 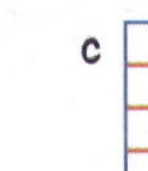 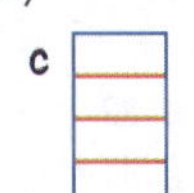 d 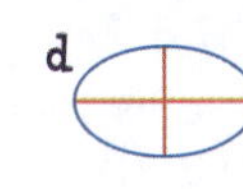 e 

6 a, d, e
7 a purple b green c purple d green
8 b, c, d

9 a 12, 6 b 16, 8 c 4, 2 d 22, 11
10 a 10, 20 b 14, 28 c 3, 6
11 a 24, 6 b 32, 8 c 40, 10
12 a 4 b 2 c 5
13 a 15, 5 b 9, 3 c 18, 6
14 a 15, 3 b 5, 1 c 20, 4
15 a 1 b 11 c 4 d 10
16 a 2 b 5 c 10 d 7
17 a > c > e < g =
b > d = f > h =
18 a 6, 2 b 4, 1 c 3, 0

19 a $\frac{1}{8}, \frac{1}{4}, \frac{3}{8}, \frac{1}{2}, \frac{3}{5}$ c $\frac{1}{8}, \frac{1}{5}, \frac{1}{4}, \frac{1}{3}, \frac{1}{2}$ e $\frac{1}{3}, \frac{3}{8}, \frac{2}{5}, \frac{4}{5}, \frac{4}{4}$
b $\frac{2}{8}, \frac{2}{5}, \frac{1}{2}, \frac{3}{4}, 1$ d $\frac{1}{8}, \frac{1}{5}, \frac{2}{4}, \frac{3}{5}, \frac{6}{8}$

20 a $\frac{1}{5}$ c $\frac{1}{2}$ e $\frac{1}{4}$ g $\frac{4}{8}$
b $\frac{1}{4}$ d $\frac{3}{8}$ f $\frac{1}{2}$ h $\frac{3}{8}$

21 a $\frac{2}{3}$ c $\frac{3}{5}$ e 1 whole g $\frac{2}{3}$
b $\frac{7}{8}$ d $\frac{2}{3}$ f $\frac{3}{5}$ h $\frac{7}{8}$

22 Adult to check
23 Adult to check

24 a 6 c 4 e 2 g 4 i 2 k 8
b 2 d 4 f 4 h 6 j 2 l 2

7 DECIMALS

Decimals to Hundredths

Page 216 – Your Turn

a 18, 0.18 b 97, 0.97 c 152, 1.52

Page 217 – Practice

1 a b c d e

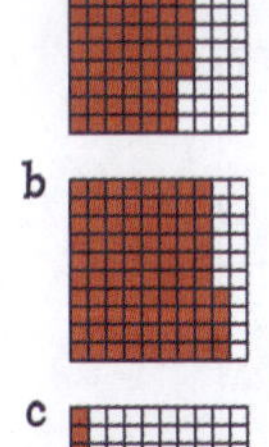

2 a 1.04 b 1.99 c 0.44 d 2.43 e 2.15
3 a 91 b 72 c 36 d 41 e 50
4 a 71 d 166 g 120 j 174 m 70
b 127 e 2 h 25 k 36
c 46 f 89 i 152 l 101

Relating Tenths to Hundredths

Page 219 – Your Turn

a $\frac{1}{10}$, 0.10 b 60, 0.60 c 40, 0.40

Page 220 – Practice

1 a 0.70 b 1.00

7 DECIMALS CONTINUED

2 a  1.00 b 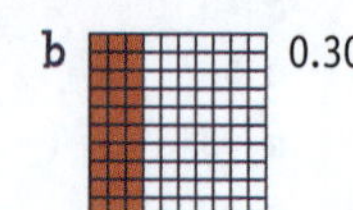 0.30

3

	Words	Fraction	Decimal
a	two-tenths	$\frac{2}{10}$	0.20
b	five-tenths	$\frac{5}{10}$	0.50
c	six-tenths	$\frac{6}{10}$	0.60
d	ten-tenths	$\frac{10}{10}$	1.00

4 a $\frac{7}{10} = \frac{70}{100}$
7 tenths = 70 hundredths

b $\frac{8}{10} = \frac{80}{100}$
8 tenths = 80 hundredths

c $\frac{9}{10} = \frac{90}{100}$
9 tenths = 90 hundredths

Writing Decimals

Page 221 – Your Turn

	Fraction	Decimal Fraction	Out of 100
a	$\frac{28}{100}$	0.28	28 out of 100
b	$\frac{10}{100}$	0.10	10 out of 100
c	$\frac{93}{100}$	0.93	93 out of 100
d	$\frac{312}{100}$	3.12	312 out of 100

Page 222 – Practice

1 a True b True c False d True e True f False g True h True i False

2

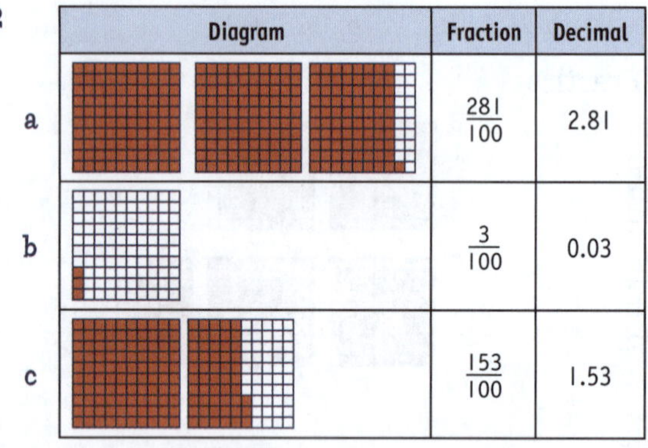

	Diagram	Fraction	Decimal
a		$\frac{281}{100}$	2.81
b		$\frac{3}{100}$	0.03
c		$\frac{153}{100}$	1.53

3 a 0.73 b 0.14 c 2.77 d 4.03 e 7.1 f 0.08 g 0.1 h 9.83

Decimal Fractions in Words

Page 223 – Your Turn

	Fraction	Fraction in words	Decimal fraction	Decimal in words
a	$\frac{58}{100}$	fifty-eight hundredths	0.58	zero point five eight
b	$\frac{197}{100}$	one hundred and ninety-seven hundredths	1.97	one point nine seven

Page 224 – Practice

1 a zero point nine five
b zero point two zero
c zero point seven one
d four point three three
e six point one two

2 a 2 b 48 c 93 d 70 e 102

3 a 0.82 b 0.07 c 0.30 d 1.79 e 2.64

Place Value

Page 225 – Your Turn

1 a 243.⑥2 b 120.③2 c 40.⑧4

2 a 341.2⑥ b 203.2① c 83.0④

3 a 1④.36 b 32④.23 c 43⑥.24

4 a ⑦0.34 b 1②3.58 c 5④2.63

Page 226 – Practice

1

	Decimal	Tens	Ones	Point	Tenths	Hundredths
a	63.81	6	3	.	8	1
b	57.90	5	7	.	9	0
c	40.75	4	0	.	7	5
d	38.69	3	8	.	6	9
e	94.21	9	4	.	2	1
f	83.06	8	3	.	0	6

2 a 13.09 b 47.36 c 72.09 d 83.95 e 57.60 f 92.72 g 64.58

3 a 5.03, 5.05, 5.15, 5.50, 5.85
b 7.38, 8.30, 8.37, 8.70, 8.73
c 6.19, 6.91, 9.16, 9.61, 19.6
d 1.26, 2.61, 6.12, 6.21, 12.62
e 5.57, 5.75, 7.55, 7.57, 7.75
f 5.03, 10.35, 10.53, 13.10, 13.35
g 0.34, 1.07, 1.34, 4.56, 7.89
h 13.46, 14.36, 14.69, 14.96, 16.34

Partitioning Decimals

Page 227 – Your Turn

	Mixed number	Decimal	Wholes	Tenths	Hundredths	Wholes	Hundredths
a	$7\frac{26}{100}$	7.26	7	$\frac{2}{10}$	$\frac{6}{100}$	7	$\frac{26}{100}$
b	$8\frac{14}{100}$	8.14	8	$\frac{1}{10}$	$\frac{4}{100}$	8	$\frac{14}{100}$
c	$6\frac{25}{100}$	6.25	6	$\frac{2}{10}$	$\frac{5}{100}$	6	$\frac{25}{100}$
d	$9\frac{70}{100}$	9.70	9	$\frac{7}{10}$	$\frac{0}{100}$	9	$\frac{70}{100}$
e	$2\frac{16}{100}$	2.16	2	$\frac{1}{10}$	$\frac{6}{100}$	2	$\frac{16}{100}$

Page 228 – Practice

1 a $3\frac{76}{100}$ b 4, $\frac{1}{10}$, $\frac{5}{100}$ c $7\frac{48}{100}$ d 6, $\frac{3}{10}$, $\frac{8}{100}$ e $5\frac{99}{100}$ f 9, $\frac{2}{10}$, $\frac{3}{100}$

2 a $2\frac{4}{100}$ b 5, $\frac{12}{100}$ c $6\frac{82}{100}$ d 2, $\frac{4}{100}$ e $8\frac{16}{100}$

3 a 54 + 9 tenths + 8 hundredths
$= 54 + \frac{9}{10} + \frac{8}{100}$

b 82 + 5 tenths + 3 hundredths
$= 82 + \frac{5}{10} + \frac{3}{100}$

c 61 + 7 tenths + 5 hundredths
$= 61 + \frac{7}{10} + \frac{5}{100}$

d 284 + 1 tenth + 8 hundredths
$= 284 + \frac{1}{10} + \frac{8}{100}$

4 a 57 b 36 c 98 d 64 e 22 f 51 g 35

5 a \$39.42 b \$54.68 c \$1.36 d \$2.49 e \$5.78 f \$2.93 g 16.65

CATCH UP MATHS YEAR 4 BOOK A © PASCAL PRESS ISBN: 9781925726145

Decimals Review Page 230

1 a $\frac{4}{10}$, 0.40 c $\frac{2}{10}$, 0.20 e $\frac{32}{100}$, 0.32 g $\frac{90}{100}$, 0.90
b $\frac{60}{100}$, 0.60 d $\frac{70}{100}$, 0.70 f $\frac{1}{10}$, 0.10 h $\frac{3}{10}$, 0.30

2 a $3\frac{72}{100} = 3.72$ b $2\frac{99}{100}$, 2.99

3

	Words	Fraction	Decimal
a	three-tenths	$\frac{3}{10}$	0.30
b	eight-tenths	$\frac{8}{10}$	0.80
c	seven-tenths	$\frac{7}{10}$	0.70
d	ten-tenths	$\frac{10}{10}$	1.00
e	one-tenth	$\frac{1}{10}$	0.10
f	five-tenths	$\frac{5}{10}$	0.50
g	six-tenths	$\frac{6}{10}$	0.60
h	nine-tenths	$\frac{9}{10}$	0.90
i	sixty-seven hundredths	$\frac{67}{100}$	0.67
j	twenty-two hundredths	$\frac{22}{100}$	0.22
k	eighty-nine hundredths	$\frac{89}{100}$	0.89
l	one hundred and twenty-one hundredths	$\frac{121}{100}$	1.21
m	three hundred and sixty-one hundredths	$\frac{361}{100}$	3.61
n	five hundred and thirty-seven hundredths	$\frac{537}{100}$	5.37
o	eight hundred and nine hundredths	$\frac{809}{100}$	8.09
p	seven hundred and sixteen hundredths	$\frac{716}{100}$	7.16

4 a 76 b 19 c 95 d 87
5 a 65 b 179 c 53 d 121 e 161 f 120

6

	Coloured squares	Fraction	Decimal
a		$\frac{123}{100}$	1.23
b		$\frac{56}{100}$	0.56
c		$\frac{229}{100}$	2.29
d		$\frac{143}{100}$	1.43
e		$\frac{15}{100}$	0.15
f		$\frac{287}{100}$	2.87

7 a 1.27 c 3.56 e 4.95 g 2.30
b 0.73 d 1.09 f 0.09 h 0.17
8 a False b False c True d False e False f True

9

	Fraction	Fraction in words	Decimal fraction	Decimal in words
a	$\frac{91}{100}$	ninety-one hundredths	0.91	zero point nine one
b	$\frac{37}{100}$	thirty-seven hundredths	0.37	zero point three seven
c	$\frac{77}{100}$	seventy-seven hundredths	0.77	zero point seven seven
d	$\frac{183}{100}$	one hundred and eighty-three hundredths	1.83	one point-eight three
e	$\frac{339}{100}$	three hundred and thirty-nine hundredths	3.39	three point three nine
f	$\frac{478}{100}$	four hundred and seventy-eight hundredths	4.78	four point seven eight
g	$\frac{646}{100}$	six hundred and forty-six hundredths	6.46	six point four six
h	$\frac{862}{100}$	eight hundred and sixty-two hundredths	8.62	eight point six two

10 a zero point four two
b one point four zero
c one point six three
d one point zero six
e four point eight two
f three point nine five
g nine point nine nine

11 a 61 b 92 c 78 d 102 e 381 f 203
12 a 0.72 b 1.53 c 0.94 d 3.92 e 2.77 f 1.48
13 a 3.⑧2 b 0.④6 c 38.⑨5 d 132.⑧6 e 493.⑧9
14 a 27.4① b 137.2⑤ c 0.9③ d 1.9② e 437.2⓪
15 a 9②.37 b ⓪.95 c 2④.29 d 14⑦.38 e 8⑦.40
16 a ①0.36 b ⑨2.37 c 1④3.73 d 1⓪7.24 e ⑦1.31
17 a 3.47, 3.74, 4.37, 7.34, 7.43
b 1.03, 1.06, 1.24, 1.30, 1.68
c 2.86, 6.28, 6.82, 8.26, 8.62

18

	Mixed number	Wholes	Tenths	Hundredths	Wholes	Hundredths
a	$8\frac{16}{100}$	8	$\frac{1}{10}$	$\frac{6}{100}$	8	$\frac{16}{100}$
b	$2\frac{63}{100}$	2	$\frac{6}{10}$	$\frac{3}{100}$	2	$\frac{63}{100}$
c	$1\frac{33}{100}$	1	$\frac{3}{10}$	$\frac{3}{100}$	1	$\frac{33}{100}$
d	$7\frac{46}{100}$	7	$\frac{4}{10}$	$\frac{6}{100}$	7	$\frac{46}{100}$
e	$5\frac{47}{100}$	5	$\frac{4}{10}$	$\frac{7}{100}$	5	$\frac{47}{100}$
f	$9\frac{78}{100}$	9	$\frac{7}{10}$	$\frac{8}{100}$	9	$\frac{78}{100}$
g	$3\frac{56}{100}$	3	$\frac{5}{10}$	$\frac{6}{100}$	3	$\frac{56}{100}$
h	$8\frac{86}{100}$	8	$\frac{8}{10}$	$\frac{6}{100}$	8	$\frac{86}{100}$

7 DECIMALS CONTINUED

	Mixed number	Wholes	Tenths	Hundredths	Wholes	Hundredths
i	$4\frac{24}{100}$	4	$\frac{2}{10}$	$\frac{4}{100}$	4	$\frac{24}{100}$
j	$5\frac{6}{100}$	5	$\frac{0}{10}$	$\frac{6}{100}$	5	$\frac{6}{100}$
k	$10\frac{71}{100}$	10	$\frac{7}{10}$	$\frac{1}{100}$	10	$\frac{71}{100}$
l	$12\frac{36}{100}$	12	$\frac{3}{10}$	$\frac{6}{100}$	12	$\frac{36}{100}$

19 a 5 b 7 c 1

20 a 2 b 5 c 2

21 a 34 b 41 c 27

22 a 49 + 3 tenths + 8 hundredths = 49 + $\frac{3}{10}$ + $\frac{8}{100}$
b 56 + 59 hundredths = 56 + $\frac{59}{100}$
c 81 + 2 tenths + 7 hundredths = 81 + $\frac{2}{10}$ + $\frac{7}{100}$

23 a 3 + $\frac{52}{100}$ c 14 + $\frac{50}{100}$ e 20 + $\frac{80}{100}$
b 1 + $\frac{57}{100}$ d 82 + $\frac{45}{100}$ f 143 + $\frac{26}{100}$

23 a 10.75 b 2.53 c 1.40 d 28.54 e 6.07 f 102.05

8 PATTERNS AND ALGEBRA

Patterns

Page 237 – Your Turn

1 a △□△□△ b ○□☆○□

2 a 20, 24, 28, 32, 36. Rule: + 4 b 70, 60, 50, 40, 30. Rule: – 10

Page 238 – Practice

1 a 15, 10, 5. Rule: – 5
b 36, 32, 28. Rule: – 4
c 54, 45, 36. Rule: – 9
d 8, 10, 12. Rule: + 2
e 70, 80, 90. Rule: + 10
f 37, 47, 57. Rule: + 10
g 61, 51, 41. Rule: – 10

2 a ● ▲ ● ● ▲
b ▲ ■ ▲ ▲ ▲
c D P A D P
d ● ▲ ★ ● ▲
e, f
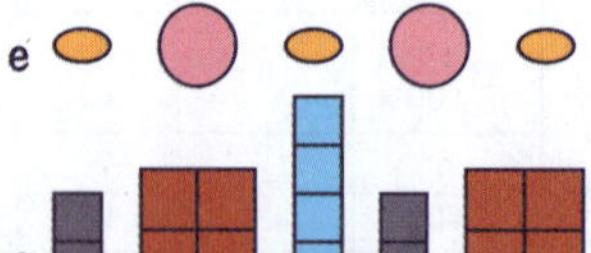
g ▲ ♥ ● ▲ ♥

3 a 42, 52, 62
b 80, 70, 40
c 18, 21, 24
d 19, 29, 39
e 95, 80, 70
f 77, 73, 67
g 70, 77, 91
h 24, 18, 6
i 19, 39, 59
j 8, 32, 40
k 120, 110, 100
l 17, 22, 37
m 19, 24, 29

4 a b c
d
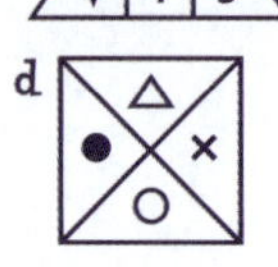
e

f
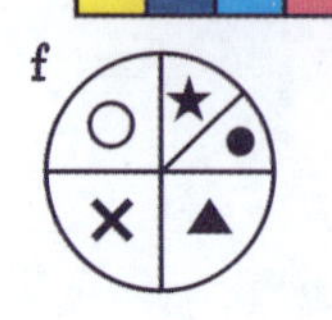

5 a Subtract 3
b Add 6
c Subtract 2
d Add 3
e Subtract 4
f Subtract 5

6 a

b X O Y P X O Y P X O Y P X O Y P X
c ▲ ● ◗ ▲ ● ◗ ▲ ● ◗ ▲ ● ◗ ▲ ● ◗ ▲
d ♥ × ○ ♠ ♥ × ○ ♠ ♥ × ○ ♠ ♥ × ○ ♠ ♥
e ⊞ ◫ ⊟ ⊠ ⊞ ◫ ⊟ ⊠ ⊞ ◫ ⊟ ⊠ ⊞ ◫ ⊟
f ⦶ ⊖ ⊗ ⊕ ⦶ ⊖ ⊗ ⊕ ⦶ ⊖ ⊗ ⊕ ⦶ ⊖ ⊗
g B A ■ ● B A ■ ● B A ■ ● B A ■ ● B A ■
h 4 5 7 9 4 5 7 9 4 5 7 9 4 5 7 9 4 5 7 9

Pattern Tables

Page 241 – Your Turn

a

●	3	5	7	9
★	9	15	21	27

b

■	4	5	6	7
★	10	11	12	13

c

▲	2	4	6	8
●	13	11	9	7

Page 242 – Practice

1 a 24, 23, 22, 21, 20 b 48, 56, 64, 72, 80 c 24, 22, 20, 18, 16

2 a

▲	3	4	5	6	7
× 2	6	8	10	12	14
× 4	12	16	20	24	28
× 6	18	24	30	36	42
× 7	21	28	35	42	49
× 8	24	32	40	48	56

b

●	1	2	3	4	5
× 3	3	6	9	12	15
× 5	5	10	15	20	25
× 9	9	18	27	36	45
× 10	10	20	30	40	50
× 12	12	24	36	48	60

3 a 54, 102, 198 b 106, 218, 442

4 a 18, 36, 54, 108, 135, 90, 99 b 8, 10, 12, 18, 21, 16, 17

Equivalent Number Sentences

Page 243 – Your Turn

a 30 b 3 c 9

Page 244 – Practice

1 a 2 b 58 c 2 d 45 e 12 f 11 g 5 h 5 i 4 j 2 k 3

2 a 26 b 12 c 8 d 11 e 30 f 2 g 10 h 15 i 1

Terms in Patterns

Page 245 – Your Turn

a 9 b 15 c 12 d 5 e 21 f 63 g 80

Page 246 – Practice

1 a 12 b 28 c 32

2 a 20, 25, 30
b 9, 11, 13
c 16, 20, 40
d 24, 16, 8
e 8, 10, 20

CATCH UP MATHS YEAR 4 BOOK A © PASCAL PRESS ISBN: 9781925726145

3 a 1, 3, 6, 10, 15, 21, 28, 36, 45, 55 b 78
4 a 36, 49, 64, 81, 100 b 144

Patterns and Algebra Review Page 247

1 a ☆, B, A, ☆
b □, ♡, ○, □
c △, ○, ○, △
d △, ○, ⬯, △
e 40, 48, 56, 64
f 60, 50, 40, 30
g 18, 12, 6, 0
h 63, 72, 81, 90

2 a 15, 20, 25
b 28, 38, 48
c 21, 26, 36
d 25, 34, 52
e 110, 90, 60
f 75, 63, 59
g 126, 118, 110
h 86, 76, 46

3 a
b
c
d
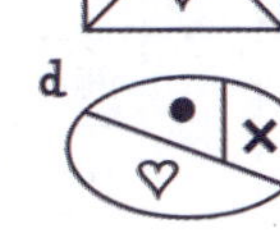
e
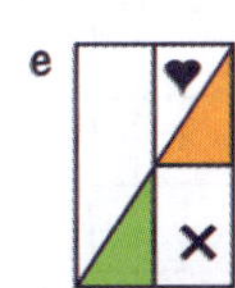
f
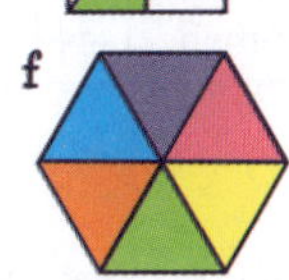

4 a Add 3 b Subtract 4
5 a 8, 12, 16, 20, 24, 28, 32 b 10, 9, 8, 7, 6, 5, 4
6

Rule	1	4	8	7	3	9	10	6	11	12	5
+ 6	7	10	14	13	9	15	16	12	17	18	11
× 5	5	20	40	35	15	45	50	30	55	60	25
× 2, + 3	5	11	19	17	9	21	23	15	25	27	13
× 3, + 1	4	13	25	22	10	28	31	19	34	37	16

7 a 29, 61, 125, 253
b 63, 313, 1563, 7813
c 30, 84, 246, 732

8 a 7 b 6 c 12 d 14 e 5 f 14
9 a 12 b 8 c 5 d 12
10 a 8 b 10 c 7 d 19
11 a 40 b 48 c 80

9 CHANCE

Chance

Page 250 – Your Turn

a Circled blue for 'won't happen'
b Circled green for 'might happen'
c Circled red for 'will happen'

Page 251 – Practice

1 a won't happen
b won't happen
c might happen
d will happen
2 Adult to check

Certain and Uncertain Events

Page 252 – Your Turn

a Because the month after February is definitely March and no other month
b Because you can eat something else, go out for dinner or the pasta may bur

Page 253 – Practice

1 Coloured green for certain: **b e f g**. Coloured blue for uncertain: **a c d h**
2 Adult to check
3 a Certain b Uncertain c Certain d Uncertain

Likely and Unlikely Events

Page 254 – Your Turn

a Likely b Unlikely c Likely

Page 255 – Practice

1 Likely:
d The sun will be shining during the day.
e In summer the weather is hot.
f I will wear a jumper in winter.

Unlikely:
a I will eat waffles for breakfast every day.
b There is a giraffe outside my bedroom window.
c I will go to school on Saturday.
g I will get a cat for my birthday.

2 Adult to check

Probability

Page 256 – Your Turn

a No b Yes

Page 257 – Practice

1 a It is a boy.
b I draw a green car.
c The shoe shop only sells high heels.
d We didn't have a spelling test this week.
e There are only red motorbikes for sale.
f I turn right.
g I look down at the ground.
h The coin lands on heads.
i The sun sets.

Related Events

Page 258 – Your Turn

a I like eating popcorn.

Page 259 – Practice

1 a Roisin's sister Ann won a medal in the Olympics.
b My favourite dinner is curry.
c Christian loves mountain biking.
d Elise's mum is a pro-surfer.
e Danny likes painting.
f Chris decorates cakes.

Outcomes

Page 260 – Your Turn

OOO, GGO, GOO, GOG

Page 261 – Practice

1 Chocolate and Strawberry, Chocolate and Mint, Chocolate and Bubblegum, Vanilla and Strawberry, Vanilla and Mint, Vanilla and Bubblegum, Strawberry and Mint, Strawberry and Bubblegum, Mint and Bubblegum
2 Orange shirt + blue shorts
Orange shirt + green shorts
Orange shirt + purple shorts
Pink shirt + blue shorts
Pink shirt + green shorts
Pink shirt + purple shorts
6 possible outcomes
3 18
4 Heads + Heads, Heads + Tails, Tails + Tails
5
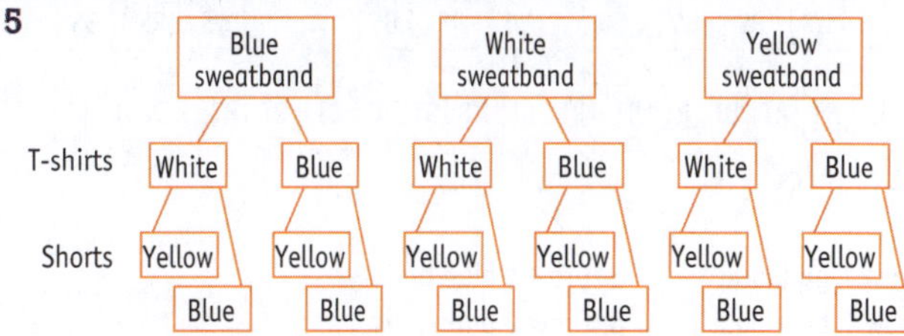

Even Chance

Page 264 – Your Turn

a, b

9 CHANCE CONTINUED

Page 265 – Practice

1 a Draw 2 pink
b Draw 5 orange
c Draw 5 purple
d Draw 2 aqua
e Draw 5 rainbow
f Draw 3 pink
g Draw 3 purple

2 Circled: **a b d e**

3 a 1, 2, 2 b 1, 2, 2

Unequal Chance

Page 266 – Your Turn

3, 5, unequal

Page 267 – Practice

1 a 3 b blue, green c yellow
2 a 8 b 3, 8, 3 c 5, 8, 5 d unequal
3 a 10 b 6, 10, 6 c 4, 10, 4 d unequal

Chance Review Page 268

1 Might happen:
a It will rain on a cloudy day.
d My dad will make a cake for dessert.
e You will go swimming in summer.

Will not happen:
b My mum will turn into a horse.
f My pet chicken will learn to sing.

Will happen:
c The day after Thursday is Friday.

2 a certain b uncertain c uncertain d certain

3 Likely:
a You eat pizza on Friday night.
d You eat breakfast.
g You go to sleep at night.
i You have a shower.
j You swim in summer.
l You eat something today.
n You use an umbrella when it rains.

Unlikely:
b You dance with your teacher at assembly.
c You don't go to school.
e You ride a horse to school.
f You fly an aeroplane after school.
h You turn blue if you eat too many blueberries.
k You watch TV after midnight.
m There is a cow in your classroom.

4 a It is a cat.
b It is dry.
c I draw a yellow flower.
d The light switch is off.
e It is red or amber.

5 a False b False c true d False e True

6 purple + pink, purple + green, purple + red, yellow + pink, yellow + green, yellow + red, blue + pink, blue + green, blue + red.

7 red + pink, red + green, red + black, red + blue, pink + green, pink + black, pink + blue, green + black, green + blue, black + blue

8

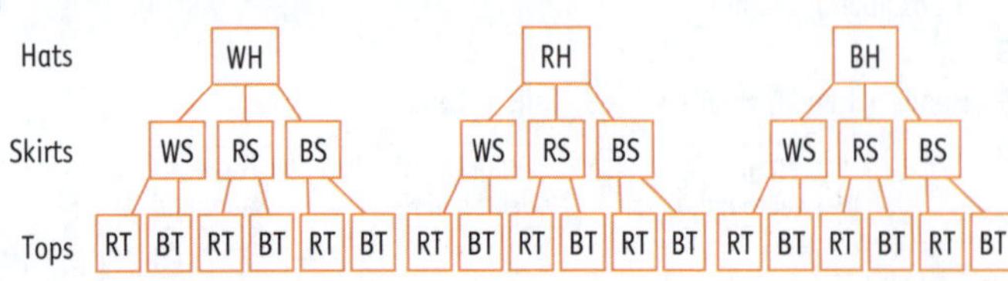

9 Adult to check

10 d

11 a Draw 1 pink ball
b Draw 2 blue balls
c Draw 4 blue balls
d Draw 2 pink balls

12 a d e

13 a 4, 4, equal b 4, 8, unequal

10 DATA

Data and Tables

Page 273 – Your Turn

a Friday b Monday c 10 d 95

Page 274 – Practice

1 a **Lolly Colour in Packet**

Colour	Tally	Total
Red	//	2
Blue	////	4
Yellow	~~////~~ /	6
Green	~~////~~	5
		17

b **4S's Favourite Colour**

Colour	Tally	Total
Black	~~////~~ ~~////~~	10
Pink	~~////~~	5
Orange	///	3
Blue	~~////~~ ///	8
Yellow	/	1
		27

2 a cricket b 4 c 6

3 a 3 b 1

Picture Graphs

Page 276 – Your Turn

Number of Books Read in Book Week

Class	Total
4C	35
4D	50
4T	55
4J	40
4M	50
	230

Page 277 – Practice

1 a 33
b January, May, November: two birthdays each month
c 5
d March
e September, October and December
f 4
g 6
h 6

2 **Bags of Rubbish Collected on Clean-up Australia Day**

Allan	2 bags
Dani	3 bags
Christian	4 bags
Tarka	3 bags
Antonia	2 bags

3 a 15 b 2020 c 20 d 2018 e 70 f 2018

4 a 5
b 3
c cat
d 4
e birds and mice
f 7
g 20
h 6
i 38

CATCH UP MATHS YEAR 4 BOOK A © PASCAL PRESS ISBN: 9781925726145

Column Graphs

Page 281 – Your Turn

a Four or more teeth
b 2
c 6
d 3
e 8
f one

Page 282 – Practice

1 **Athena's Dice Results**

Roll	Tally	Total
⚀	|||	3
⚁	卌	5
⚂	卌		6
⚃	|||	3
⚄			1
⚅	||||	4
		22

a 3
b 3
c 1 and 4
d least
e 5
f 5

2 a 3
b yellow
c black
d red
e 19
f 1
g 9

3 a smoothie
b 10
c boys fruit, girls cereal
d 5
e 9
f 42

4 **Drinks Sold at the Canteen Today**

Drink	Tally	Total
Orange juice	卌		6
Grape juice	卌 卌	10
Banana milk	卌 卌 ||	12
Strawberry milk	||||	4
Chocolate milk	卌 卌	10
		42

5

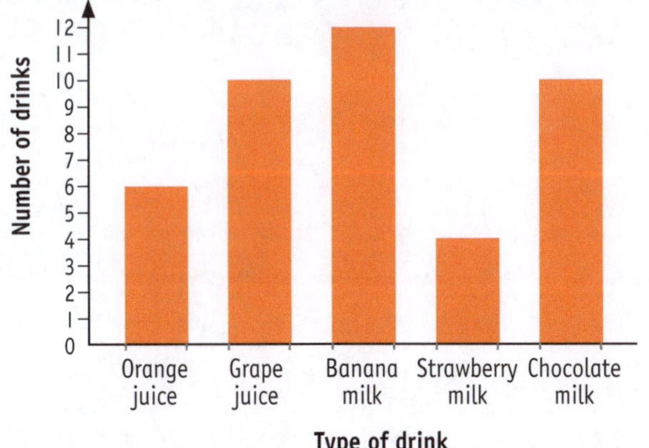

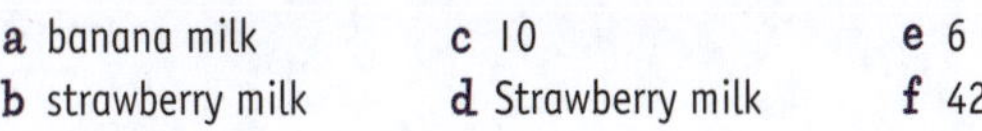

a banana milk
b strawberry milk
c 10
d Strawberry milk
e 6
f 42

Bar Graphs

Page 286 – Your Turn

a Dogs
b Birds
c dogs

Page 287 – Practice

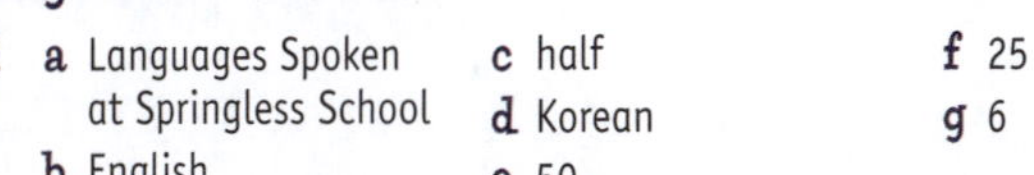

1 a Languages Spoken at Springless School
b English
c half
d Korean
e 50
f 25
g 6

2 **How children in 4T spend their free time**

iPad	Reading	Play sport	Watch TV	Cooking

Data Review Page 288

1 **Favourite Chocolate**

Flavour	Tally	Total
Plain	卌	5
Peanut	卌 ||	7
Caramel	||||	4
Coconut	卌 卌 ||	12
		28

a 28
b plain
c coconut
d 7
e 8

2 a Maths 5, English 5, Science 3, PE 6, Art 7, Music 1
b

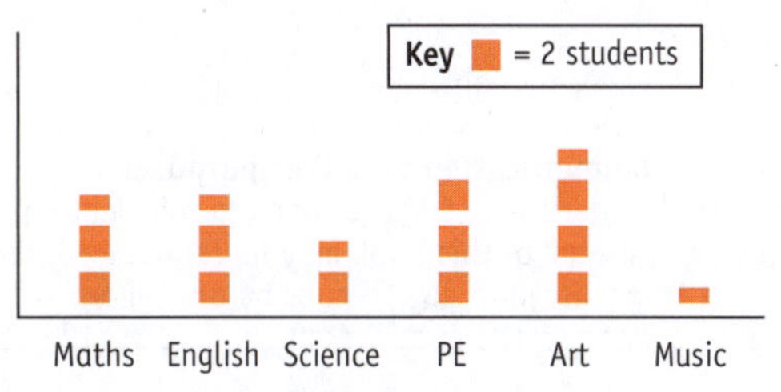

3 a black 2, red 9, yellow 4, green 8, orange 4, blue 5
b Adult to check
c

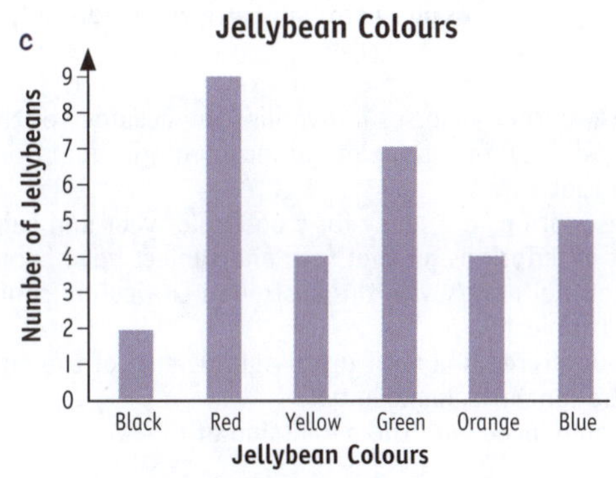

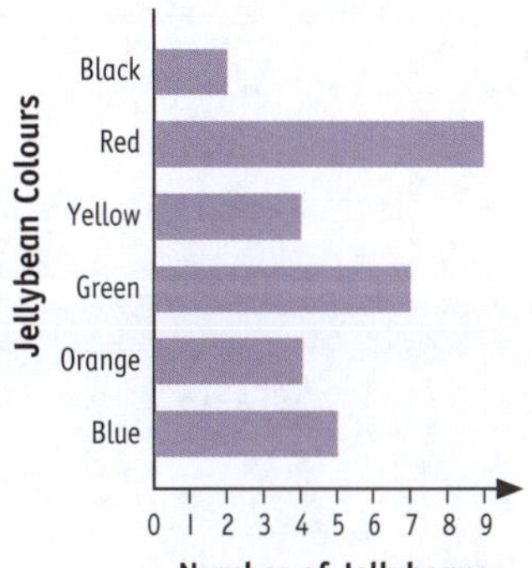

4 a 75
b Soccer
c Yoga
d Basketball
e Yoga
f Football
g Yoga
h 6
i 10
j 1
k 5

5 Adult to check

6 a 5
b 100
c Europe
d 50
e Australia

Catch-Up Maths Book 4A

ISBN: 9781925726145

Published by Pascal Press
PO Box 250
Glebe NSW 2037
www.pascalpress.com.au
contact@pascalpress.com.au

Design: Janice Bowles
Author: Deborah Frendo-Toman
Publisher: Lynn Dickinson
Typesetter: Ruth Schultz
Editor: Rosemary Peers, Vaishali Batra

Printed by Wai Man Book Binding (China) Ltd.